BRITISH AIRCRAFT
ARMAMENT

By the same author

British Aircraft Armament
Volume 1: RAF gun turrets from 1914 to the present day

Patrick Stephens Limited, an imprint of Haynes Publishing, has published authoritative, quality books for enthusiasts for more than a quarter of a century. During that time the company has established a reputation as one of the world's leading publishers of books on aviation, maritime, military, model-making, motor cycling, motoring, motor racing, railway and railway modelling subjects. Readers or authors with suggestions for books they would like to see published are invited to write to: The Editorial Director, Patrick Stephens Limited, Sparkford, Nr. Yeovil, Somerset BA22 7JJ.

BRITISH AIRCRAFT ARMAMENT

VOLUME 2: RAF GUNS AND GUNSIGHTS FROM 1914 TO THE PRESENT DAY

R Wallace Clarke

Patrick Stephens Limited

© R. Wallace Clarke 1994

All rights reserved. No part of this publication may be reproduced, stored in a retrieval system or transmitted, in any form or by any means, electronic, mechanical, photocopying, recording or otherwise without prior permission in writing from Patrick Stephens Limited.

First published in 1994

British Library Cataloguing in Publication Data
A catalogue record for this book is available from the British Library

ISBN 1 85260 402 6

Patrick Stephens Limited is an imprint of Haynes Publishing, Sparkford, Nr Yeovil, Somerset BA22 7JJ.

Typeset by MS Filmsetting Limited, Frome, Somerset
Printed and bound in Great Britain
by Butler & Tanner Ltd,
London and Frome

Contents

Dedication   6
Acknowledgements   7

Part 1: RAF Guns   9
  Introduction   11
  The Guns   13
  Summary of guns used by the British Air Services   105
  Appendices   107

Part 2: Gunsighting Systems   113
  Introduction   115
  The gunsights   117
  Summary of the Development History of British Aircraft Gunsights   217

Bibliography   220
Index   221

This is the song of the gun,
the muttering, stuttering gun,
the maddening, gladdening gun
that chuckles with evil glee
as its muzzle spits lead in the sun,
till a last long dive has begun,
with its end in eternity.

Gordon Alchin, RFC   1917

Muzzle of the Vickers Mk I.*

Acknowledgements

I am grateful for the kind assistance and material supplied by the following experts in their field: Mr Bill Gunston, whose advice on the subjects covered and mastery of the written word were invaluable, Mr Jack Bruce, Desmond Molins, S/Ldr. H. F. King, Mr C. F. Weide, Mr Bill Ainley, Mr Michael Goodall, Mr Terry Heffernan, Mr Roy Bonser, Mr Leonard Hutchinson, Mr Bert Woodend MOD, Mr Peter Labbet, Col. Ross Whistler USAF Ret, Mr Peter Ward (Rolls-Royce), Mr Michael Gibson, Mr Bert Aggas, Mr Don Randal USAF Ret, Mr Kenneth Ward, and Mr Harry Ramsden.

I would also like to express my appreciation to the following companies and government establishments, without whose help it would not have been possible to obtain details from ex employees and archive material:

Holland & Holland Ltd, Barr and Stroud Ltd, Glasgow; Ferranti Edinburgh; Dowty Defence and Air Systems Ltd, RAFM Hendon; Vickers Ltd, Crayford; British Aerospace; Molins Ltd, Peterborough; RNZAF Museum, Wigram; Smiths Industries Aerospace and Defence Systems; Imperial War Museum; Colt Firearms, USA; Bundesarchiv, Frieburg; GEC Avionics; A&AEE Farnborough; Mr Ken Ellis of *Flypast* magazine; Mr Richard Riding of *Aeroplane* magazine; MOD Fort Halstead; Palmer Tyre Co and Dunlop Ltd.

I must also thank my wife Mary, who has endured prolonged bursts of typewriter fire, and stoppages in our recreational arrangements.

Ron Clarke
Kettering, February 1994

Part One

RAF GUNS

Introduction

A military aircraft would be of little use without the means to engage the enemy, or to defend itself against hostile aircraft. The following chapters give the history and details of the guns used by the British military air services. Very few of these weapons were designed by British ordnance technicians but, apart from a few American designs and standard NATO guns, all have been manufactured in Britain. There has been a rigidly enforced rule to this effect, for obvious reasons: no country can afford to be dependent on another for the means to carry out its military action.

The United Kingdom is not renowned for having a large armament industry, but in times of national emergency British companies have, usually after a slow start, been remarkably good at producing the necessary armaments. This is especially true of air weaponry. From 1914, using mainly female workers on a continual shift system, the Vickers factory at Crayford and the BSA works at Birmingham began to turn out the thousands of guns required to equip the land and air forces in the First World War. Thus, after some initial disastrous hold-ups, Vickers and Lewis guns began to arm squadron aircraft.

In the Second World War the German Luftwaffe received a far greater variety of guns than the RAF: British (and American) warplanes relied mainly on just three types of guns. Each side had its own advantages. The Germans were the first to introduce large-calibre weapons firing advanced ammunition, but these complex guns required specialist technicians to maintain them. The RAF guns, all developed from 1914–18 designs, made possible long and rapid production runs and high reliability. The RAF also used a few large-calibre weapons, but these were phased out in favour of rocket projectiles.

After 1945 the Cold War ensured that the development of air weapons would continue. Although guided missiles were seen as being the main fighter armament of the future, virtually all the world's air forces adopted copies of a gun designed in the late war period, the German Mauser MG213c. Featuring a revolving breech mechanism and electric firing, these guns were in widespread use until the appearance in 1953 of a revolutionary American gun based on the Gatling principle of the 19th century, the General Electric M16 Vulcan. The 20 mm version of this weapon was used on RAF Phantoms in a detachable pod. Then came the ADEN 25 mm. Doubtless after this book is published there will be others, yet 40 years ago armament experts expected the gun to be replaced by guided weapons. Guns would go the way of swords, consigned to museums. This theory had one flaw: once a guided missile is launched, it is dependent on its guidance system to reach its target. As the makers of burglar alarms have discovered, even the most sophisticated systems can be overcome, provided enough money is available. Radars can be jammed, infra-red seekers can be diverted, but once a gun button is pressed, no amount of black boxes can divert the projectile from its path ...

The Guns

When war broke out in 1914 the Royal Naval Air Service and Royal Flying Corps were desperately short of suitable weapons with which to arm the few aeroplanes available for their use. The RNAS was particularly concerned, having been given the task of intercepting the expected attacks from Zeppelins. A few Vickers FB.5s, armed with Vickers machine guns, were delivered to the RNAS at Eastchurch, but, apart from these and a few other experimental installations, aviators took to the air armed with an assortment of pistols, carbines and shotguns.

In the years before the war, experimental firing from the air against ground targets had been carried out with automatic guns of various types. It had been found that the vibration and movement of the host aircraft made aiming difficult. However, the much greater problems of air to air firing had not yet been encountered. The guns used on these trials were usually Vickers Maxims, secured by an assortment of pivoted mountings. They weighed about 25 kg (55 lb) and were fed by canvas ammunition belts which writhed and flapped in the slipstream. It soon became clear that the Vickers was not an ideal air weapon, and other guns such as the Colt-Browning, Hotchkiss, and Madsen were tried, but the gun which seemed to meet all the requirements for air firing was the Lewis, which was made under licence by the BSA Co. of Birmingham.

Production at the new BSA factory was just beginning, and the first guns were needed by the Army, in France. The only other gun available was the Vickers, but output from the Crayford factory in August 1914 amounted to a mere 12 guns, and the Army was also in desperate need of these for specialist corps in France. The War Office was opposed to the procurement of foreign-made weapons. So, lacking in political clout, the Air

Account submitted by Holland & Holland for seven specially made 'Aero' guns. They had been ordered by the Admiralty, and were basically 12-bore Paradox guns. (Holland & Holland)

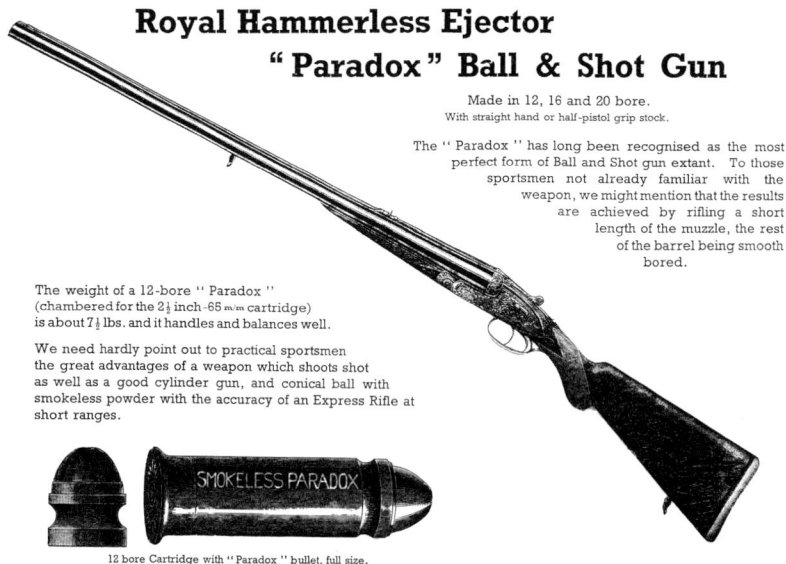

Above *Holland & Holland Paradox shotgun the service Aero version was more basic.* (Holland & Holland)

Below *The Grahame White Type VI Warplane of 1913 fitted with a Colt Model 1895. This and other contemporary aircraft were fitted with experimental gun mountings at this time, the Colt Browning being one of the more successful.*

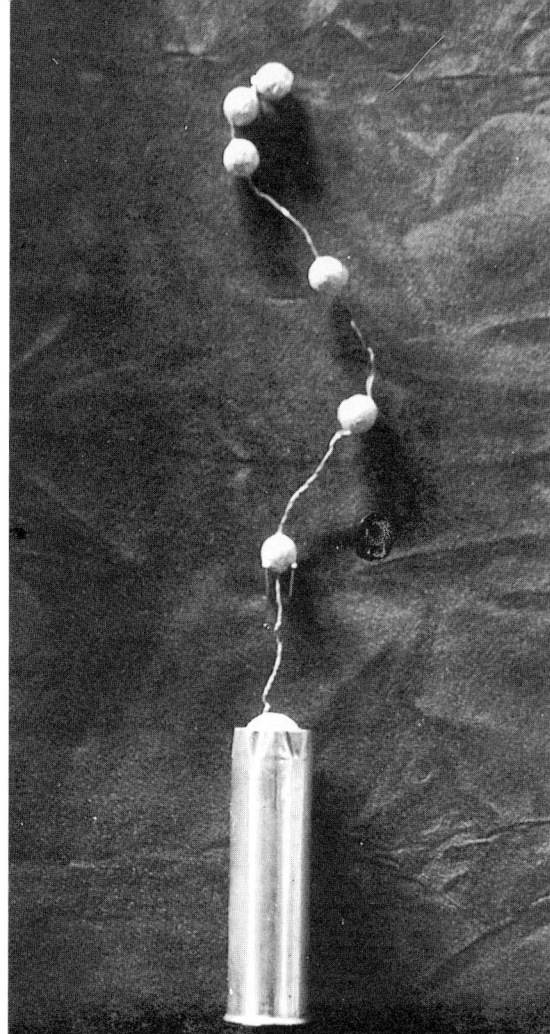

Chain shot cartridges used without success by Naval airmen.

Services had to buy single-shot weapons. One supplier was Holland & Holland of Harrow Rd, London, who dispatched their first guns on 23 November 1914. These were Paradox and Aero patterns, firing 12-bore (.707 in) cartridges containing single lead balls through barrels choked and rifled at the muzzle to give improved range and accuracy. Later incendiary, explosive, and even chain shot cartridges were issued. It was hoped that incendiary cartridges would be effective against the Zeppelins, but the big airships could outclimb any of the home defence aircraft. Martini Henry 0.45 in carbines, and 0.303 Lee Enfields and pistols, were among the assortment of weapons used in the first months of the war. From 1915 the Lewis became available, and was soon in operational use by the RNAS and RFC.

As the war progressed it was found that the most effective destroyers of enemy aircraft were machines armed with fixed guns firing through the arc of the propeller. The weapons most suited to synchronising mechanisms of both Allied and German air arms were the various versions of the Maxim gun. The Vickers gun became the standard British fixed pilot's gun, remaining in service until the advent of the 0.303 Browning gun in 1935. The Lewis remained in widespread use as a free mounted observer's weapon, and, mounted on the ingenious Foster rail mounting, it was also used as a pilot's gun. Large calibre shell firing weapons were tested for aircraft use, but very few were used operationally.

When hostilities ceased the vast stocks of weapons were put in store, and the Air Staff appointed a committee to investigate the most effective calibre of future aircraft guns. It had been found that aircraft hit by numerous rifle calibre bullets had survived without major damage, and many armament experts were advocating the use of 0.5 in cal. guns. Protracted trials were carried out to assess the merits of such a weapon, one problem being that no really suitable British heavy calibre gun was available at that time. In 1929 the findings of the committee were issued. It had been decided that although the range and hitting power of the heavier gun was superior to the rifle calibre weapon, the faster rate of fire and lighter weight of the small calibre gun was more suited to the needs of the service.

In 1933 it was decided to find a replacement for the 0.303 Vickers gun. It had proved to be an excellent weapon in many ways, but it was prone to stoppages and had reached the end of its development. After comparative trials it was replaced with the Colt Browning gun which was to provide the RAF with fast firing modern weapons in the coming conflict. There were many influential officers in the service who, remembering the success of the two seat Bristol fighter, championed the case for a heavy fighter. It was suggested that an aircraft armed with a heavy calibre gun could fly alongside but out of range of defending gunners, and decimate the massed formations of bombers expected in any future war. Consequently the Air Staff issued a series of specifications for such aircraft. Luckily the only gun available at the time was the rather ancient 37 mm COW gun, and the rate of climb of the aircraft submitted for trials was such that the bombers would have been long gone before the 'bomber destroyers' could attempt to shoot them down. Events were to prove that the whole concept of heavy fighters was badly flawed, the only multi seat turret armed fighter to attain production status was the ill-fated Boulton Paul Defiant which, like the

German BF110 was to prove no match for agile single seat interceptors.

Although the Air Staff had opted for the rifle calibre Browning gun as the main air weapon, it was realised that other air forces were introducing 20 mm shell firing guns to counter the armoured protection being built in to the latest types of aircraft.

In 1936 Major Thompson of the RAF Gun Section, and the Air Staff visited the Hispano Suiza factory in Paris, where they were given a demonstration of the newly designed Type 404 20 mm cannon. They were duly impressed, and after tests at the Aircraft and Armament Experimental Establishment at Martlesham Heath the gun was adopted for use in fighter aircraft, and possible use in gun turrets of bombers. Production facilities were established in the UK just before France was overrun, and the foresight of the Air Staff was fully justified when the gun proved to be one of the most successful air weapons of the war. As in the 1914–18 war, 40 and 57 mm guns were used for ground attack operations, but the advent of rocket weapons gave the service tremendous striking power, and these guns were phased out. After hostilities ceased many very advanced German developments in air weaponry were revealed. One of these, the Mauser 213c was adopted by virtually every major air force. There were others which seemed to be very potent designs. Among these were the recoilless guns of the Rheinmetall and HG companies. A development contract was issued to the Fort Halstead establishment to produce a weapon using the principles of these guns as a basis of the design. Work was at an advanced stage on this project when it was decided to concentrate all facilities on the development of guided missiles. The British version of the Mauser 213c was the ADEN gun, which has remained in service to the present day, when in its latest form it is specified for the latest RAF fighter aircraft. The gun section of the Royal Aircraft and Armament Establishment at Farnborough have carried out many aircraft armament projects over the post-war years. Among these have been the development and testing of the ADEN, the US Vulcan and Chain gun, the FN Mag 60, the Mauser 27 and many other weapons. The closing of this establishment will bring to an end an era of great achievement at Farnborough, where vital armament and aviation development has been successfully carried out for over 80 years. Most of the guns described on the following pages have been tested and improved by the talented staff at this historic site, which has been at the forefront of British aircraft development since its inception as a balloon factory.

The Lewis Gun

The Lewis gun was one of the most successful automatic weapons ever produced. Its advantages included light weight and a compact ammunition feed which worked in any attitude. These features made it an ideal infantry weapon, but it was most famous as a free-mounted aircraft gun. It was adopted by the Western Allies as almost the standard observer's gun, and was widely used as a forward-firing pilot's gun. The Lewis remained in service with the RAF from 1915 to at least 1940, being gradually replaced by the Vickers VGO.

Colonel Isaac Lewis was a leading ordnance expert in the United States Army. He graduated from West Point in 1884, and in 1911 became Commander of the Artillery School at Fort Munroe, where he gained international recognition as an expert in mechanical and electrical engineering. With his retirement from the army approaching, he accepted a consultancy with the Automatic Arms Company of Ohio. For some years he had been working on a design for a gas-operated light

Colonel Isaac N. Lewis, US Navy.

THE GUNS

A sectioned Mk I Lewis gun showing the fluted aluminium cooling vanes enclosed in the large diameter steel jacket. It was found that neither were necessary for air firing, as air flow over the barrel was sufficient for normal use. The wooden butt was replaced with a spade grip for air use.

machine-gun, and AAC suggested that he should develop the weapon further. AAC had acquired the rights to a gun patented by Dr Samuel McClean, which had some features which would improve Lewis's design. In return for the rights to manufacture the new gun, Lewis received major stockholdings in AAC, and assumed control of production and promotion.

Operating Sequence
After two years, Lewis had produced an air-cooled gun which embodied several unique features. When the trigger was pulled, a bolt unit consisting of a striker pin fixed to the rear of a piston was driven forward under the influence of a flat, coiled 'fuzee' spring under the gun. When the bolt reached its forward limit the striker fired the round. Some of the high-pressure gas behind the bullet was diverted through a small hole at the end of the barrel into a cylinder under the barrel, driving the piston back. As it travelled to the rear, the bolt ejected the spent cartridge, reloaded another from the circular magazine, and tensioned the return spring. When the piston reached its rear limit, the return spring rotated a pinion which engaged a rack under the bolt unit, and drove it forward until the pin fired the new round.

One problem with automatic guns is the intense heat generated during long bursts of firing. Lewis conceived a unique cooling system for his gun. The barrel was enclosed in a large cylindrical sleeve, at the rear of which was a fluted cast-aluminium radiator block. When the gun was fired, the high-velocity gas from the muzzle drew air over the radiator and into the space between the outer sleeve and barrel. Although this cooled the gun to a certain extent, bursts of fire of more than 20 rounds still tended to overheat the barrel. The magazine held 47 rounds, providing only 6 seconds' firing time. This was enough for skirmishes and, since it weighed little, the magazine could be quickly changed.

During his Service career Lewis had had several inventions rejected by the Service, most of which were then developed privately and sold back to the Army. He knew that this inbuilt prejudice against Service inventors would probably work against him, so he decided to adopt a rather devious approach. An acquaintance of his, Captain C. De Forrest Chandler, was Commanding Officer of the Army Field at College Park, Maryland, where the Signal Corps operated a few Wright biplanes. Lewis explained his new weapon to him, and suggested that Chandler should become the first man to carry out an aerial machine-gun test. Chandler was very enthusiastic and arranged to conduct the trial himself, the pilot being Lt. T. De Witt Milling. Lewis had informed various interested officers of the project, and on 2 June 1912 the lumbering biplane came puttering over the field. Chandler fired a burst into a sheet pegged out on the grass in front of a hangar.

The Wright then flew over some ponds, where Chandler fired the rest of the magazine into the water. This historic flight was reported enthusiastically in the press, but the General Staff were furious that Lewis had not officially notified them of the event.

However, due to the wide publicity the test had attracted, Lewis was invited to submit his gun for official evaluation. It would appear that the subsequent rejection of the gun was influenced by the fact that the Ordnance Board had already made a commitment to adopt the French Benet-Mercier gun, a version of the Hotchkiss. This weapon proved inferior to the Lewis – it used a strip magazine which projected sideways and was not popular with American troops. Lewis had by this time retired from the Army, and regretfully decided to leave America for Europe, where he knew his gun would be assessed on its merits. His offer of a demonstration was eagerly taken up by a group of Belgian businessmen, who had already heard of his gun. The gun performed well, and a mutual agreement was signed in which a new company, called Armes Automatic Lewis, was to produce the gun for European use. However, the only manufacturer able to offer production facilities of the required standard was the Birmingham Small Arms Company in England, so an agreement was drawn up and the Birmingham factory prepared for production.

With the help of BSA's publicity department, Lewis arranged for the weapon to be fired from the air at Bisley on 27 November 1913 before the press and Service officers. A Grahame White box kite was fitted with a wicker seat under the pilot's position for the gunner, a Belgian, Lt. Stellingwerf, the pilot being Marcus Manton of Hendon. At the appropriate time the biplane approached the range at 400 ft and Stellingwerf opened fire at a target 25 ft square. The results of this and other firings later in the day seemed to bear out Lewis's claims – in one burst, 28 of the shots hit the target.

BSA duly received orders for trial batches of guns from the War Office and from some overseas governments. These were promptly dispatched, and ordnance experts at Woolwich made a thorough evaluation. Weaknesses which had been played down at Bisley were soon found by the Woolwich team. The chief faults were overheating (prolonged firing soon 'blued' the barrel), and the drum-type magazine could be fouled by the ingress of dirt and grass. However, the gun was generally given a favourable report, and it was recommended for Service use.

The most significant trials took place when the RFC and RNAS carried out experimental air firing, for it soon became obvious that the Lewis would be an ideal air weapon. However, the War Office showed little enthusiasm, and BSA began to deliver the first guns off the line to fulfil orders from Russia and Belgium. With the prospect of war in Europe, the company decided to expand their production facilities, and ordered many machine tools from America. After urgent requests from the flying services, the War Office and Admiralty ordered ten guns in July 1914, and within two weeks a further order for 45 was placed. When war was declared, BSA received contracts for 200 Lewis guns, to be delivered at the rate of 25 per week. The initial price was £175 less magazines, tools and spares. When the true value of the guns was realized by front-line officers, the War Office placed further substantial orders. It soon became clear that in spite of increased production facilities, BSA could not keep pace with demand, so a joint Canadian–British order was placed with the Savage Arms Co. of Utica, New York, for 12,000 guns.

When the new BSA factory came on stream, production increased dramatically, until in 1915 over 300 guns were being delivered per week. The

Lt. Stellingwerf in his wicker seat preparing for the firing demonstration at Bisley.

Right *A Mk I Lewis mounted on an FE2B. The large cylindrical cooling jacket was removed in the later Air Service models.*

Middle right *Lewis guns fore and aft: a DH4 at Bacton RNAS station.* (Bruce/Leslie)

Bottom right *A BE.2c gunship of No. 8 Squadron. Mk I Lewises are ready for the fray.* (Bruce/Leslie)

factory in Liège was soon overrun by the German advance, proving the wisdom of Lewis's move to England; however, the original Belgian consortium was still paid substantial royalties on every gun produced.

The Lewis guns issued to the British Army remained virtually unaltered until the end of the war, but once air firing tests got under way the gun was progressively modified. The first alteration concerned the rifle-type stock: a small spade grip was much better suited for air use, where the gun had to be free to be pointed in any direction. The cylindrical barrel casing made it difficult to aim the gun to either beam: the slipstream would try to whip the gun to the rear if pointed forwards, the open end acting like a scoop. It was found that the barrel suffered much less from overheating in the air, so the bulky cylinder was removed, but the radiator was retained to support the gas cylinder and piston. It was also found that empty cases could be troublesome in the slipstream, and squadron workshops devised various ducts and bags to collect them. After receiving official complaints, BSA produced canvas collector bags holding 94 cases clipped to a small duct fitted over the ejector slot. These proved too small, so a bag was produced which held 330 cases. Known as the Mk II deflector bag, it was held rigid by an internal wire cage which could be emptied by unclipping a flap at the base. The 47-round magazine was inadequate for air firing and changing magazines with cold, heavily gloved hands was not easy, so a 97-round drum was developed for air use, with a canvas handle to help the gunner change drums. The larger magazine, known as the No. 5, was introduced in mid-1916, and soon became standard on guns used by the Air Services.

The Mk II

Following reports of damage to the gas cylinder under the barrel, a light metal tube 2.5 in in diameter was fitted to protect it. This and other modifications for air use were produced by BSA and adopted in November 1915. The modified Lewis

Twin Lewis mounting on a Scarff ring, giving increased fire power. Gunners commented 'they were rather a handful'.

Mk II became available in mid-1916. As described in the second section of this book, the original tangent sights were modified until compensating Norman Vane sights became standard on observers' guns.

The action of the Lewis was such that it could not be adapted for use with interrupter or synchronizing gear. Consequently, when used as a fixed pilot's gun it was mounted to fire outside the propeller arc, usually on the top wing of the aircraft. When used in this manner a special rail mounting was fitted, giving the pilot a means to change empty magazines. Known as the Foster mounting, it was invented by Sgt. R. G. Foster of No. 11 Sqn RFC. When the magazine was empty, the pilot could unclip the gun and pull it back down to the rear end of the rail, where the magazine could be changed easily.

By 1916 the Lewis had become the standard free-mounted weapon of the RFC and RNAS and, as supplies increased, it also came into widespread use

Left *The 97-round ammunition drum was loaded by inserting the rounds between the pegs into a spiral groove.*

Below *The Mk II Lewis, showing the modified barrel casing and cartridge bag. (MOD Pattern Room, Nottingham)*

by French airmen. It was also prized by German fliers, who retrieved the guns off crashed Allied planes for their own use.

In common with the Vickers, stoppages did occur: faulty or badly loaded cartridges would jam the mechanism, oil thickened in the cold upper air, and if the gas cylinder was not cleaned every 600 rounds, the piston would seize. Gunners were aware of the overheating which could occur if long bursts were fired, but when faced with a German fighter approaching head-on with guns blazing, such considerations were forgotten.

The Mk III
The RNAS used a slightly different version to the RFC. Many squadrons cut away the radiator casings, although this was strictly against orders. RNAS squadrons tended to strip the Mk II down to its absolute basic parts. Some attempt was made to protect the gas cylinder by light metal side guards, but the radiator was dispensed with, and the gun was still found to operate satisfactorily.

Right *The over-wing Foster mounting on an SE.5A, which enabled the gun to be fired forward. The drum was changed by lowering it down the rail. The pilot could also fire upwards at whatever angle was required.*

Below *Sectional drawing of the Mk III.*

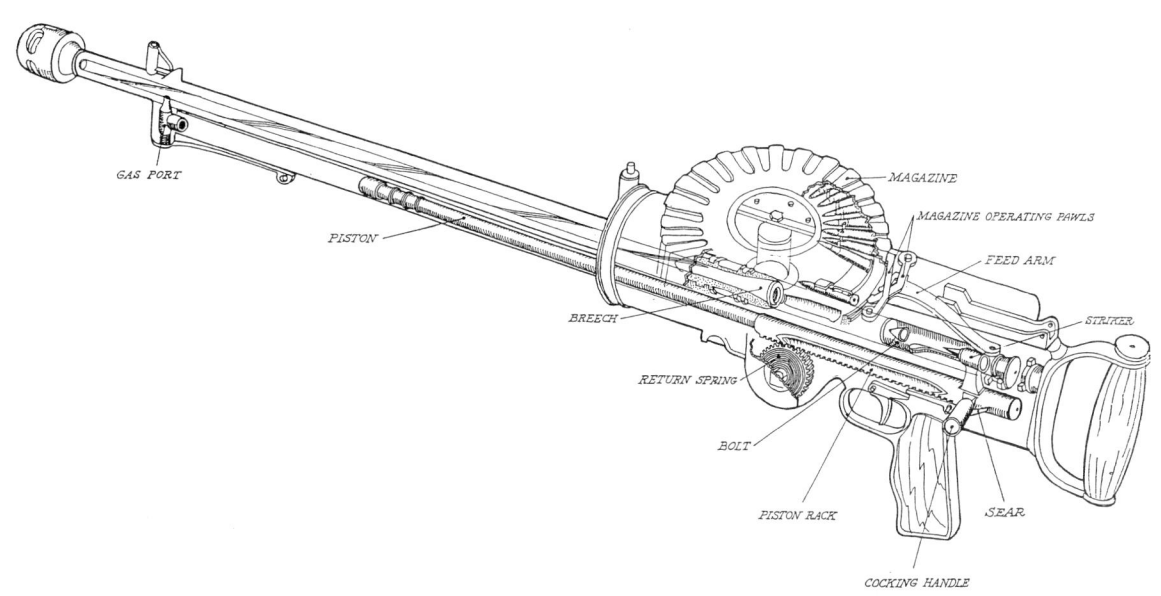

In 1917 it was realized that the naval version was lighter, and presented less air resistance than the Mk II. The absence of the radiator seemed to make little difference to barrel heating, and an official decision was made to include these features on a new version, the Mk III, which was the final version produced. BSA turned over to the Mk III, the Savage Arms Co. followed suit, and this model was used by many of the world's air forces for the next two decades. There were modifications to increase the rate of fire, muzzle chokes speeded up the movement, and recoil springs were strengthened. These alterations did speed up the action, but stoppages and wear tended to increase. As new aircraft types came into service on the Western Front, operating heights increased, bringing new problems caused by low temperatures. These made the Lewis guns sluggish and they sometimes failed to fire, whereas Vickers guns mounted on the engine cowlings were unaffected. Armament experts at Orfordness devised various modifications to solve the problem. A 36 watt heater was issued which clamped onto the body of the gun, and a conversion set was introduced which increased the gas pressure. These measures improved performance in low temperatures and increased the cyclic rate of the gun.

As mentioned in Part 2, gunners using Scarff ring-mounted Lewis guns had many allowances to make before hitting an enemy fighter. Once a pilot was behind his quarry he had a relatively stable target, but gunners had a more complex problem. However, some Lewis gunners had an uncanny gift for accurate shooting. Perhaps the most successful RFC squadron in this respect was No. 20 Sqn, operating Bristol fighters. Sgt. E. A. Deighton was credited with ten victories, and Sgt. A. Newland was awarded the DFC after shooting down six enemy aircraft. The first few weeks were usually the critical period: once a gunner established a liaison with his pilot and mastered drum changing and accurate sighting, he stood more chance of survival.

Although officially replaced by the Vickers K from 1936, the Lewis saw service during the Second World War, mainly as an anti-aircraft weapon. It was also used by gunners of Vickers Wellesleys in East Africa 1940. Faced with agile Fiat CR.42 fighters of

The Mk III Lewis was the final sub-type, remaining in service until 1941. (MOD Pattern Room, Nottingham)

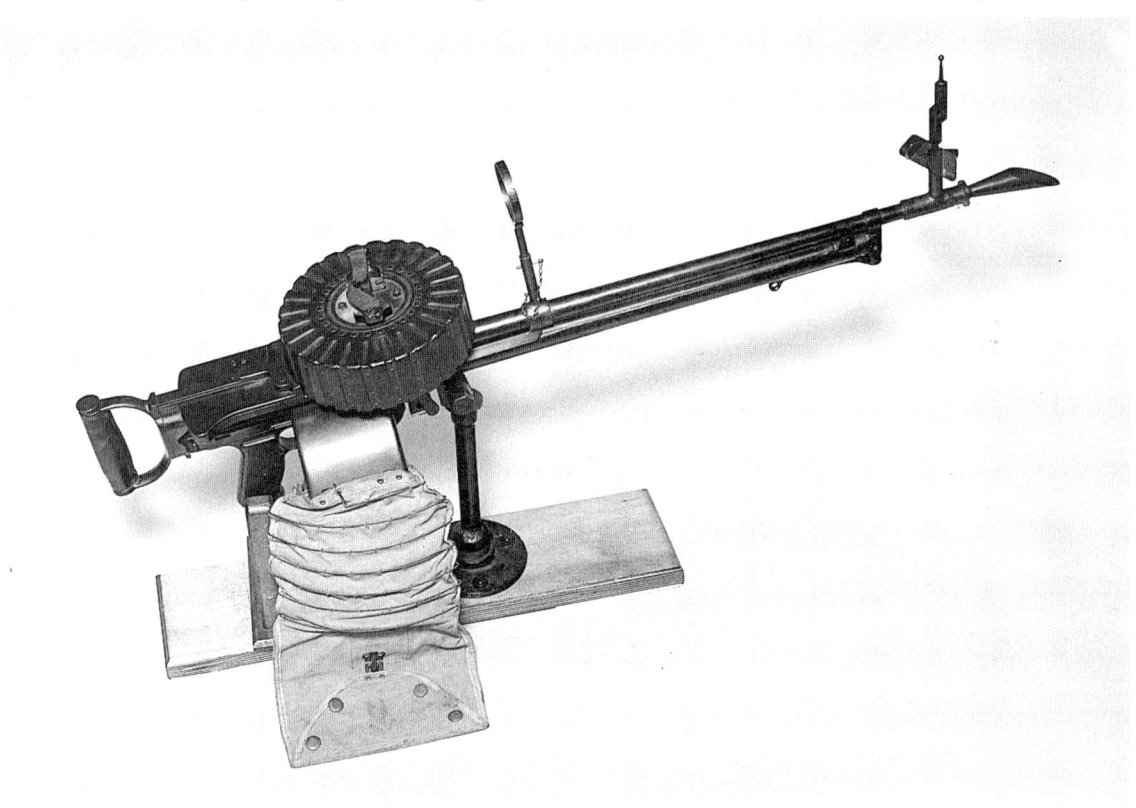

Prototype of the composite twin Lewis made by BSA in the late 1920s. The action was redesigned, but it was not accepted for Service use. (MOD Pattern Room, Nottingham)

the Regia Aeronautica, aircraft of No. 14 Sqn fitted three Lewis guns to fire from makeshift mountings covering the lower rear zone. The guns were loaded with 80 per cent tracer ammunition, and fired by cables from the observer's cockpit. During the first operation against Massawa, Fiats approaching from the lower blind zone suddenly found themselves flying through a snowstorm of tracer. One observer said they made off like scalded cats! This was probably the last time the Lewis was used in the air in any numbers by the RAF.

Details of the Lewis gun

Bore	0.303 in (7.7 mm)
Action	Gas-operated
Cyclic rate	Mk II 600 rpm, Mk III 750–1,000 rpm
Weight of gun	Mk I 26 lb (11.8 kg), Mk II 18.5 lb (8.39 kg), Mk III 17 lb (7.71 kg)
Weight of 97-round drum	8 lb 4 oz (3.74 kg)
Muzzle velocity	2,240 ft/sec (744 m/s)
Cooling	Airflow over barrel
Sighting	Ring and bead, Norman vane, Hutton
Rifling	4 grooves, right-hand twist

The Vickers 1½-pounder

This naval deck gun was first fired from the air in a Sopwith seaplane, No. 127, in May 1914, when 'excellent practice was done firing at targets both in the air and on the sea'. It was then fitted to a Short S81 with a specially strengthened nacelle. One report at the time stated that when the gun was fired the aircraft stopped in the air and fell 500 ft. It was a credit to the makers of these aircraft that they survived the fierce recoil of the gun.

A strong recoil spring was enclosed under the barrel, the action ejected the empty case to the rear (in the Short installation the case was retained in a leather bag fitted over the back plate of the gun) and the breech-block was held to the rear ready for the next round to be hand-loaded.

The gun, mounting and ammunition were far too heavy to be of any practical operational use, and after the first experimental firings the gun was not used by the Air Services.

Details off the Vickers 1½-pounder

Bore	1.443 in (37 mm)
Action	Recoil semi-automatic, hand-loaded
Cyclic rate	Single shot manual loading
Weight of gun	265 lb (120 kg)
Weight of round	1.5 lb (.68 kg)
Muzzle velocity	1,300 ft/sec (395 m/sec)

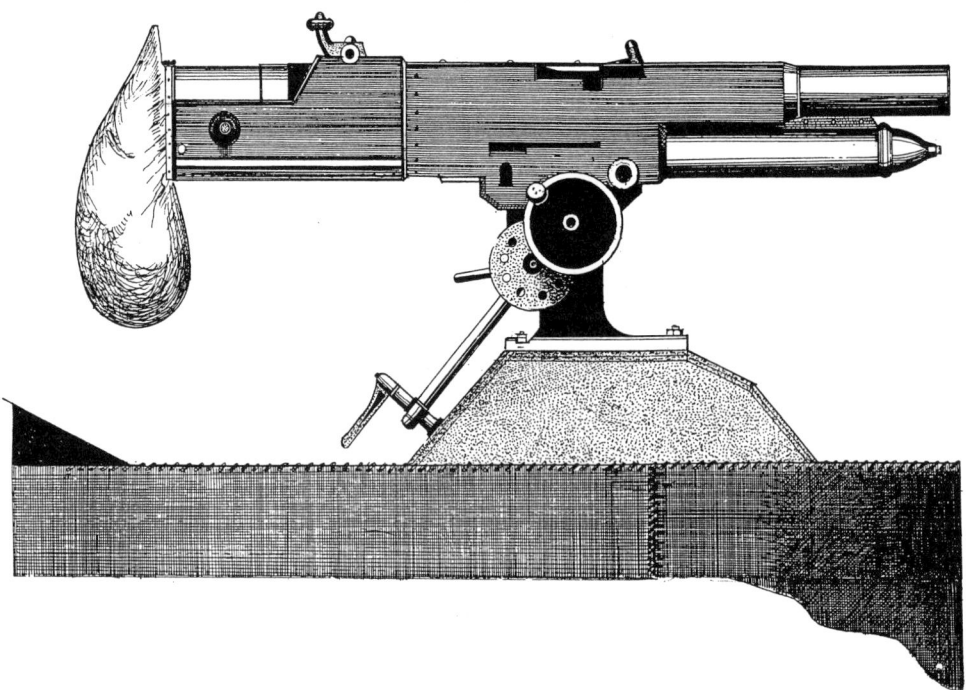

The Vickers 1½-pounder mounted on the strengthened nacelle of a Short 'gun carrier'. The fierce recoil proved too severe for Service use.

The Vickers Mk I*

In 1885 Albert Vickers and Son Ltd, steel producers with premises at Crayford in Kent, was approached by an American, Hiram Maxim, who suggested they manufacture an automatic gun of his design which was, he said, far superior to any similar weapon. Maxim had already demonstrated the gun to Government officials at Enfield, where he had fired 1,000 rounds in 1½ minutes, and he was able to convince Vickers that a Government order would soon be forthcoming. After various modifications had been carried out to improve the gun's action, and the weight reduced by one third, a new company, Vickers Sons and Maxim Ltd, was formed to produce the gun in July 1888. Chambered for rimmed 0.303 in (7.7 mm) calibre, the standard rifle round, the Vickers Maxim was adopted by the British army in 1912. It was destined to serve British forces for the next 55 years.

The Maxim was water-cooled, and used the force of the recoil to power the moving parts. The rounds were fitted to a canvas belt which was introduced into the receiver and the safety catch raised. When the thumb piece was pressed the sear was released and the firing pin fired the first round. The recoil forced back the breech-block and barrel for 0.75 in (20 mm), the bolt then being separated from the barrel by a swinging toggle. The bolt now continued to the rear, rotating a crank to tension a driving spring and cock the firing pin until the sear engaged. At the first moment of recoil after unlocking, the bolt face simultaneously extracted the empty case and withdrew a fresh round from the belt. Continued rearward movement engaged cams in the receiver, forcing the sliding bolt face down, bringing the new round in line with the chamber and placing the empty case in line with the ejection tube. When the rearward movement was completed, the driving spring started the move forward, the cartridge to be fired being chambered, and the T slot on the bolt face slipped over the next round. When the bolt reached the forward limit the toggle joint locked it to the barrel, the sear then being released to fire the next round. The Vickers gun differed from others only in that the toggle moved upwards instead of downwards, and the trigger bar was fitted at the top of the casing so that the trigger was in a more natural position. (This is the reason why the trigger motors of synchronizing gears were fitted to the top of the casing.) The barrel was cooled by a water/steam system, the water being contained in a large cylindrical jacket, through

Above *The Vickers Maxim 0.45 in. cal. 1891 pattern (left), compared with the Vickers Mk I from which the Mk I* was developed.* (Vickers)

Below *The robust Vickers action can be seen in this Mk III.* (MOD Pattern Room, Nottingham)

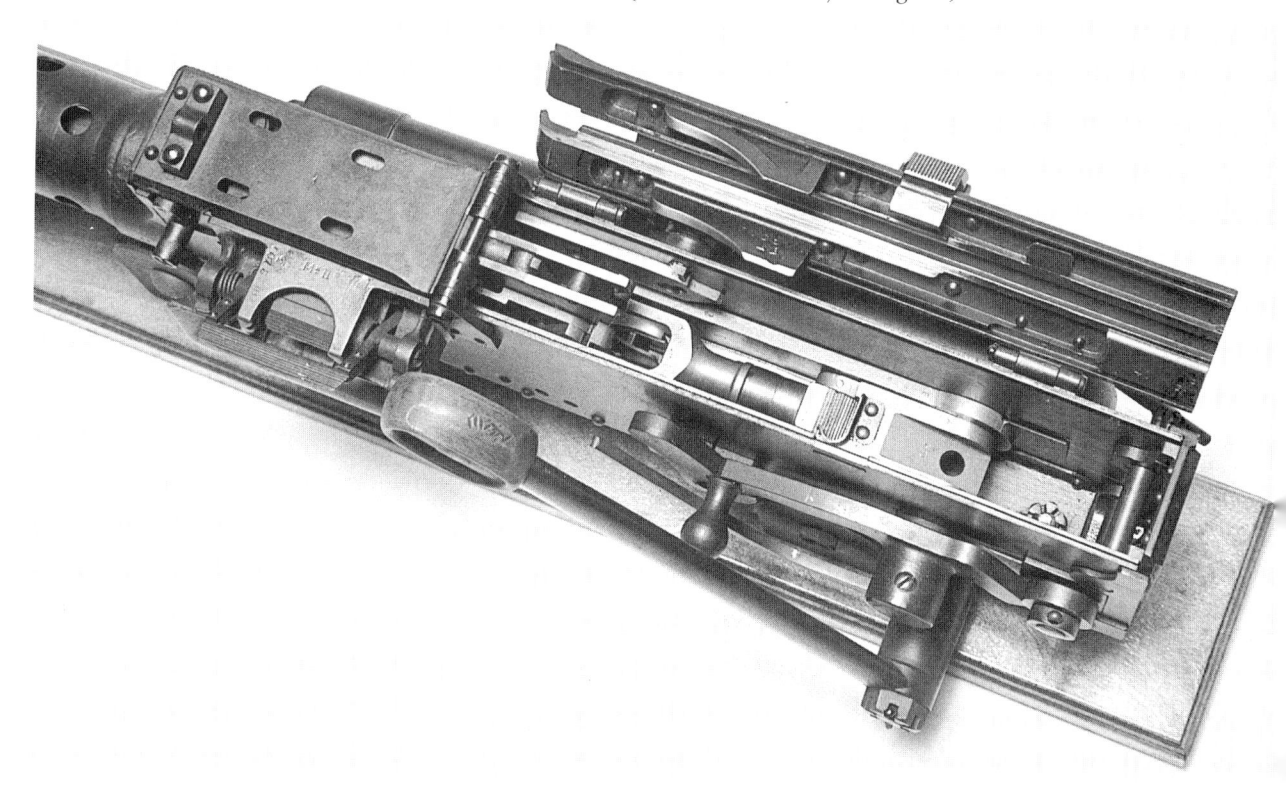

which the barrel was taken via watertight glands.

The War Office envisaged the gun being used by specialist units supporting the infantry, but several experiments were carried out to investigate its use as an aerial gun. Vickers were well aware of the possible extra sales this would mean, and in November 1912 exhibited a pusher biplane of their own design at the Olympia Aero Show in London. This, the Vickers EFB1 (Experimental Fighting Biplane), was armed with a Vickers Maxim on the front rim of the cockpit. To demonstrate the awesome power of the new weapon, the so-called 'Destroyer' took to the air after the show, but the pilot lost control and was killed. However, the Government team at Farnborough designed the FE2 (Farnborough Experimental), which carried out successful air firing trials with a Vickers Maxim on Salisbury Plain (see Part 2). The problem was that the aircraft were not really suited to the weapon, the ammunition belts flapped about in the slipstream, frequently jamming the action, and the weight of the gun (28.5 lb/12.9 kg) made it very difficult to manhandle. In 1913 Vickers developed the EFB1 into the FB.5 'Gun Bus', a two-seat pusher with a Vickers pivoted in the front cockpit. This was the first of a long line of production aircraft designed and built by the company, later to become Vickers Armstrongs Ltd. When war was declared in 1914 only a handful of pusher aircraft were fitted with Vickers guns. Aeroplanes were used only for reconnaissance, and clashes with German airmen were rare.

For aerial use the Vickers was stripped of its water-cooling pipes, openings were made in the front of the jacket (which had to be retained to support the barrel) and louvres were cut in the rear of the jacket to provide airflow over the barrel.

The appearance over the Western Front of the Fokker E1 monoplane gave the German Air Force a distinct advantage over Allied aircraft. It was designed specifically to engage observation machines over the German lines. Armed with a German version of the Maxim, the MG08/15, it was fitted with a cam-operated system which enabled the pilot to fire a fixed gun through the propeller arc. This enabled the pilot to approach his quarry from any direction, aiming the whole aircraft at it and thus setting a standard for all such interceptors. As losses from these attacks mounted, there were urgent requests from the front-line Allied squadrons for a similar system.

Ironically, two London brothers, William and George Edwards, had submitted a design for such a system in 1914, but had been politely told that 'no use could be found for such a device at present'.

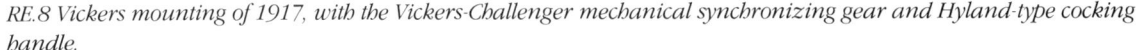

RE.8 Vickers mounting of 1917, with the Vickers-Challenger mechanical synchronizing gear and Hyland-type cocking handle.

THE GUNS

The first Allied synchronizing gear to be put to use was produced by the French Morane-Saulnier aircraft company. This was followed by the British ARSIAD system developed by Major A. V. Bellington of the Aeroplane Repair Section in France. Other early systems were the Scarff-Dibovsky, operated from cogs and teeth on the propeller shaft, and a similar design by Prof. George Challenger of Vickers, who had designed the EFB.

So urgent was the need to fit forward-firing guns that any available gear was fitted. The French and the RFC adopted the Vickers Challenger system, and naval machines used the Scarff-Dibovsky design. The first aircraft to be fitted with an official conversion by the makers was a Bristol Scout, on 25 March 1916. This was followed by six Sopwith 1½ Strutters, which were delivered to 70 Sqn on 24 May. Both these types were fitted with Sopwith-Kauper gear controlling a single Vickers gun. These systems relied on pushrods operated by cams on the propeller shaft. The rod alignment had to be very carefully set up and needed constant adjustment. Nevertheless, the balance between the powers was restored until the arrival of such new scouts as the Halberstat and Albatros.

Until this time, the Lewis gun had been regarded as the ideal aircraft gun, but when forward-firing fixed guns proved so effective, the Vickers came

A Sopwith Camel fitted with Sopwith-Kauper gear. This was soon replaced by the Constantinesco-Colley hydraulic synchronizing system.

into its own. The guns were usually mounted within reach of the pilot, who often had to clear jams and misfires. The first action if the gun stopped firing was to recock it. Various types of cocking handle were devised, the most effective being the Hyland type, which was adopted as standard.

Schematic drawing of the CC gear adopted by the British Air Services in 1917.

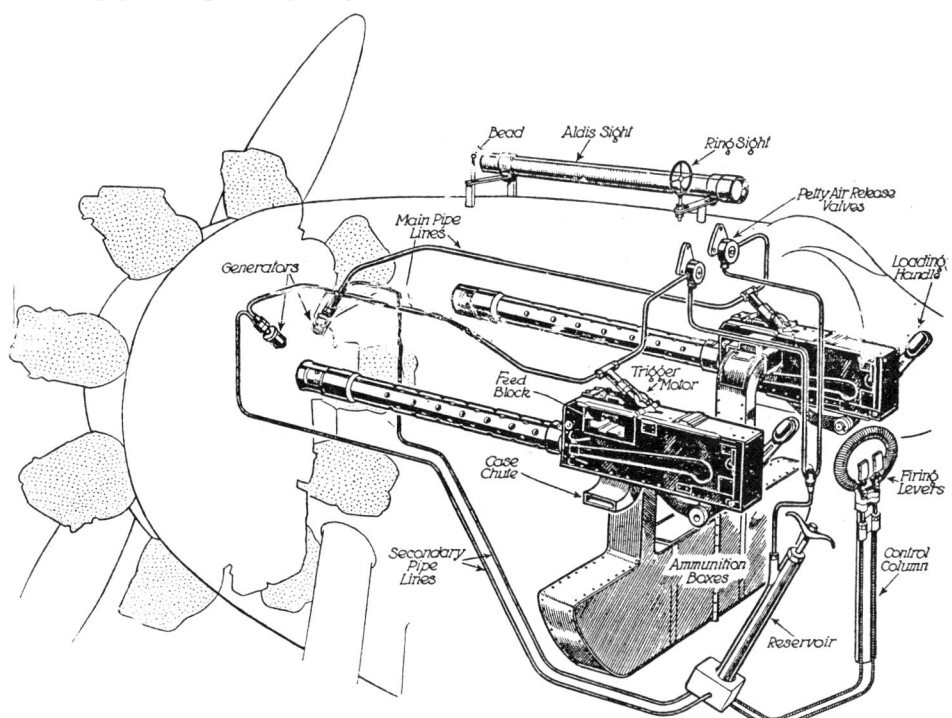

The Constantinesco-Colley (CC) synchronizing system was eventually adopted as standard by the RFC. This consisted of a hydraulic plunger actuated by a cam on the propeller shaft. A pipe was taken from this to release triggers on the pilot's control column, and from there to a second plunger, or trigger motor, on the Vickers gun. The plunger tripped the sear on the gun and fired a round if the action was ready to fire. With a two-bladed propeller, six blades would usually pass between one bullet and the next. Pressure was maintained by a hand pump, and a relief valve at the highest point in the installation released any air bubbles. The CC system could control more than one gun, and different aircraft could be fitted by cutting or lengthening the pipelines. The first firing trials took place in August 1916, when a BE.2c fired 150 rounds through the propeller arc. The Vickers company was given the contract to manufacture complete system kits, and CC gear was fitted to the new fighters as fast as they were built. Over 600 sets were installed on British aircraft between March and December 1917, and 2,000 in the first half of 1918. DH.4s and Bristol fighters were the first aircraft to use the gear on operations in early spring 1917. The 0.303 in Vickers then reigned supreme as the standard pilot's gun of the British flying services, and it was widely used by other Allied forces.

The gun was progressively modified: the back sight bridge was removed to accommodate new versions of the CC gear, the safety catch was removed, the cover over the fuzee spring was deleted, and various other alterations were made, many at the suggestion of squadron armourers. The gun was then officially designated the Vickers Mk I* (Aerial).

The next official modification was a device for increasing the rate of fire. Invented by Lt. Cdr. George Hazelton RN, it utilised the blast following the clearance of the bullet from the muzzle to add to the recoil forces, applying the surplus energy to accelerate the recoiling mechanism. A cyclic rate of 1,000 rpm was achieved, but this put undue strain on the other moving parts and the speed was adjusted to 850 rpm.

One problem with the Vickers was the fabric ammunition belt. During air firing the end of the belt would often blow back in the face of the pilot, or become entangled in some part of the aircraft. This was overcome by the adoption of the disintegrating link belt.

First used by the German Air Force, this system consisted of metal links held together by the cartridge cases. The RFC copied the idea until an improved design invented by a British dentist, William Prideaux and Maj. L. G. Hawker RFC, was adopted in 1918. As each round was extracted from the belt and fed into the gun, the link was separated from the rest of the belt, and was either collected in a container or ejected clear of the aircraft. The Vickers gun was ideally suited to such a feed system, as only two minor parts of the receiver needed to be modified. The disintegrating link system was soon

The Vickers Mk I with a CC trigger motor in place. The barrel casing was louvred at the sides and holes cut into the front to assist cooling.* (MOD Pattern Room, Nottingham)

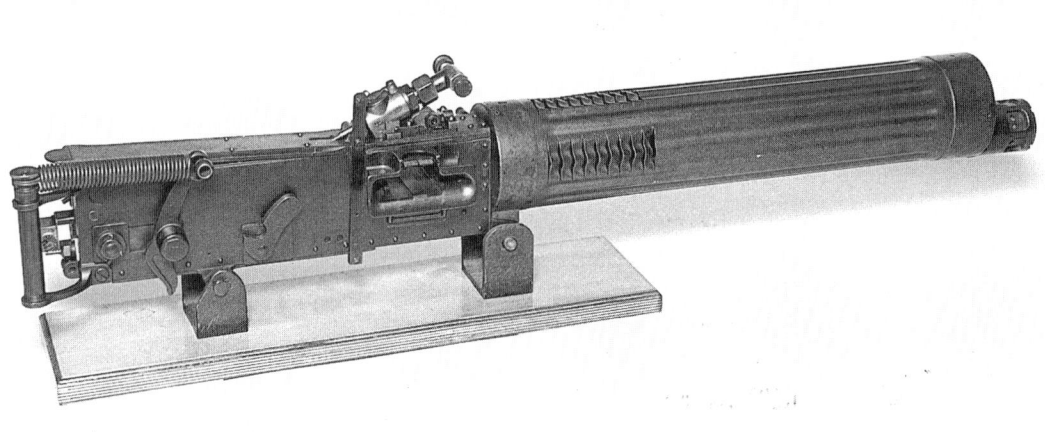

Prideaux steel cartridge links fell away as the rounds entered the breech. This system was a big improvement over the old canvas belts. (MOD Pattern Room, Nottingham)

adopted by nearly all belt-fed automatic guns, and is still in use today.

As fighting in the air took place over the front at even greater heights, the actions of the guns were affected by the sub-zero temperatures. Modifications to overcome this problem included a conversion set supplied to squadron armourers. This consisted of stronger return and buffer springs, a metal liner to the muzzle attachment, and electrical heaters. These modifications substantially increased the reliability of the gun at all heights. However, stoppages still occurred, the main causes being faulty ammunition and badly made-up belts. The jams were cleared by either a hooked clearance tool or, more usually, a well-directed blow from a leather-faced hammer carried in the cockpit. Thus aircraft designers had to install the guns within easy reach of the pilot.

As explained in the weapon aiming section of this book, gunsighting was progressively improved, the standard arrangement being an Aldis sight to the right, and a ring-and-bead on the left.

Heavy-calibre balloon guns

Both sides made use of captive balloons for artillery spotting and observation of the enemy's positions. These balloons were heavily defended by fighters and pre-ranged anti-aircraft guns. It was therefore with some trepidation that pilots briefed to attack these targets went about their task, and many were lost on these missions. Even when incendiary bullets hit the balloons, they often had no effect, because unless mixed with air, the hydrogen in the envelope did not ignite. It was thought that, if heavy-calibre bullets were used, they would rip bigger holes in the fabric.

The RFC decided to adapt Vickers guns to fire 11 mm French Desvignes incendiary ammunition. Some worn-out Vickers guns were rebored to take the heavy rounds, the feed slots and extractor slides being altered and the muzzle attachments enlarged. To evaluate the 'big Vickers', a series of trials took place at Orfordness on 6 March 1918. A number of balloons were tethered at various ranges, and a 0.303 in Vickers gun was loaded with Buckingham incendiary bullets for comparative trials. The results (Report C?195) were as follows:

	0.303 in Buckingham		
Range	Misses	Hits	Ignitions
200	0	8	3
400	6	21	1
700	13	29	1

	11 mm Desvignes		
Range	Misses	Hits	Ignitions
200	0	26	0
low	0	24	4
350	2	24	12

The modified guns were used against balloons in France, but the heavy recoil caused problems with the aircraft structure, and wear on the working parts of the gun.

In 1918 some British fighters were fitted with rounds counters made by the Veeder company. These gave the pilot an instant readout of his ammunition state, which served to discourage ammunition wastage on long-range targets.

After the Armistice of 11 November 1918, thousands of Vickers guns were scrapped, but most were put into store, many of these being used 20 years later in another 'war to end wars'. As new aircraft came into service in the early twenties, higher operating speeds demanded better streamlining. Aircraft designers found that the large-diameter barrel casings caused too much drag

when mounted outside the airframe, and they were impossible to accommodate inside. In 1917 a new version of the Vickers, the Mk II, had been introduced, but rarely used. This featured a small-diameter barrel sleeve perforated for cooling. This

Left *Veeder rounds counters gave pilots an indication of how many rounds were available.* (MOD Pattern Room, Nottingham)

Below *The Vickers Mk II with small-diameter cooling jacket, introduced for mounting inside the fuselage of post-war fighters.* (MOD Pattern Room, Nottingham)

Bottom *A Mk II Vickers with the breech cover opened.* (MOD Pattern Room, Nottingham)

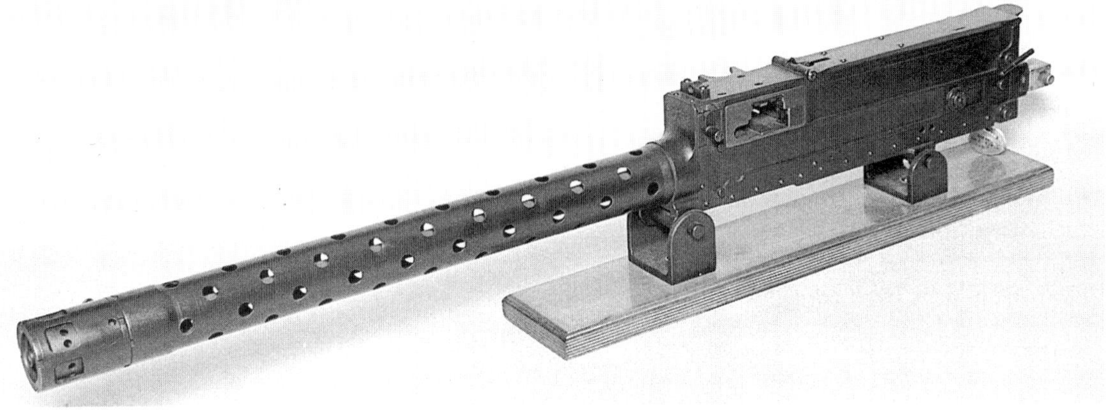

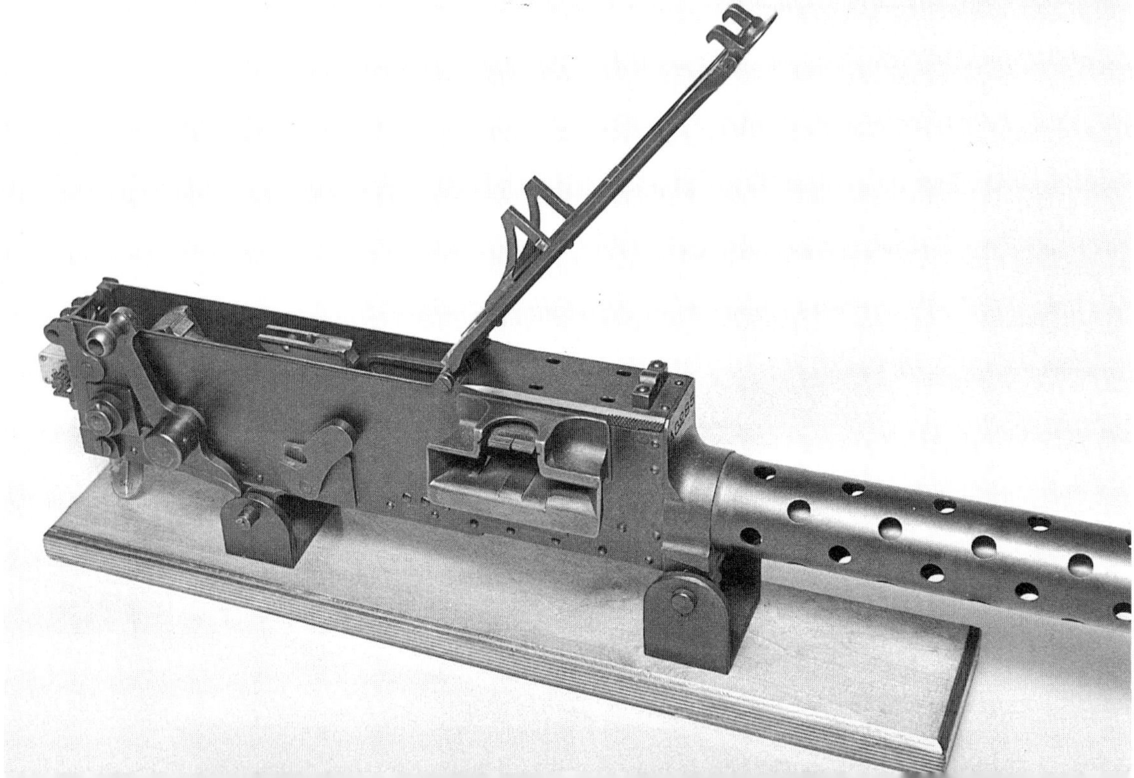

THE GUNS

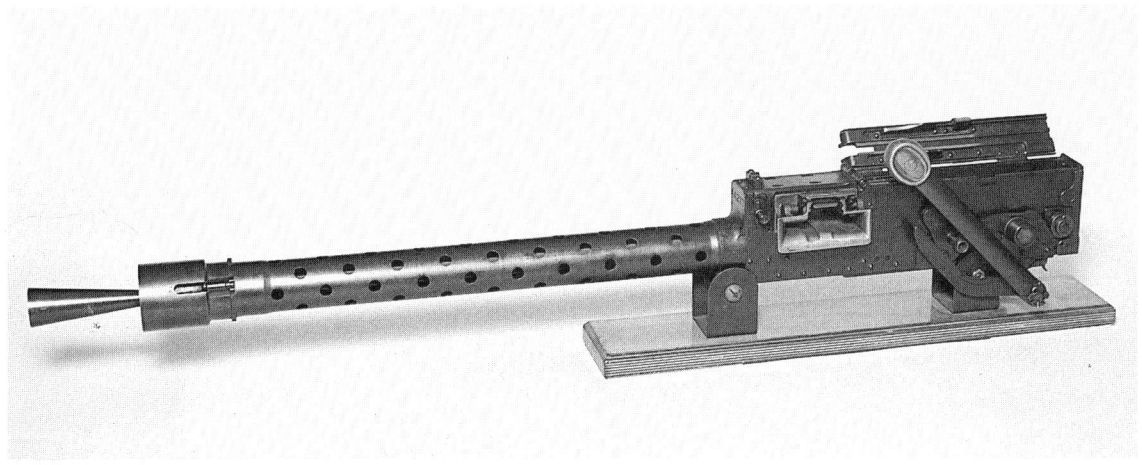

A Vickers Mk III with flash eliminator. (MOD Pattern Room, Nottingham)

gun, also known as the Class E, was adopted as standard for all RAF fighter aircraft until the introduction of the Colt Browning in the late thirties.

The first aircraft to be fitted with the Mk II were Armstrong-Whitworth Siskins for the Romanian Air Force. Any faults in the basic action had by this time been cured, and any stoppages that did occur were usually caused by rogue ammunition or case/link jams.

The biplanes of the inter-war years retained the CC synchronizing system, and the Vickers invariably remained accessible, enabling the pilot to clear any jams and recock. Ammunition was stored in boxes behind the engine firewall, the belts being lifted over rollers into the receiver by the gun feed units. After 1925, guns were usually mounted inside the fuselage, firing through troughs along the fuselage sides. Each gun was typically supplied with 500 rounds, enough for 35 seconds' firing.

Sighting in the twenties was similar to wartime,

Cut-away illustration of the Vickers class E.

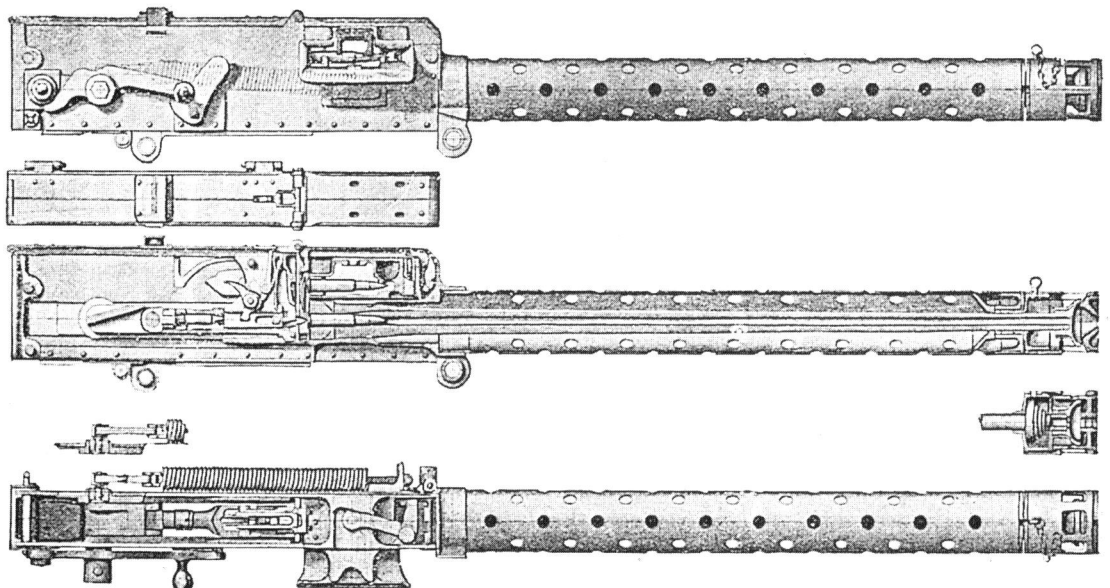

using an Aldis optical sight and a 4½ in ring and bead. In 1929 the first Barr & Stroud reflector sight was submitted for trials in an Armstrong Whitworth Siskin. This company produced several models for evaluation in the thirties, but the reflector sight did not come into general use until the Gloster Gladiator was introduced in 1937.

The next and final version of the Vickers gun was the Mk III. This differed only in having a side opening inspection cover and a flash eliminator.

The 1934 decision to adopt the Colt Browning brought the reign of the air service Vickers to an end. The ground service version remained in widespread use during the Second World War, its main advantage over the Bren gun being its ability to fire long-sustained bursts. The gun was in continual use by many nations for over 60 years, and Vickers were justifiably proud of the machine gun they had developed into one of the world's classic gun weapons.

Details of the Vickers Mk I*

Bore	0.303 in (7.7 mm)
Action	Recoil/fuzee spring
Cyclic rate	850–900 rpm
Weight	Mk I* 24.5 lb (11.1 kg)
Muzzle velocity	2,400 ft/sec (723 m/sec)
Ammunition feed	Mk 1*: canvas belt; Mk II and III; metal links
Cooling	Airflow over barrel
Rifling	Five grooves

The Davis gun

The fitting in 1942 of the Molins 6-pounder anti-tank gun to the Mosquito XVIII was a remarkable achievement, giving Coastal Command a valuable new strike weapon. It was even more remarkable that a 6-pounder was fired from some of the frail aircraft of the 1914–18 war.

Just before hostilities commenced in 1914, the major European powers were engaged in frantic efforts to provide their armed forces with modern weapons and equipment. Although a few recognized that the aeroplane might be useful for observation duties, and possibly machine-gun attacks on enemy troops, they knew that the frail machines would not be able to support any substantial offensive stores. Accordingly it is not surprising that an offer by an American naval officer to demonstrate a lightweight 6-pounder caused considerable interest at the British Admiralty.

This weapon was invented by Commander Cleland Davis USN, and was patented on 22 October 1911. Davis was a gunnery officer who had taken an interest in the fledgeling US Navy aeroplane section. He had designed the gun as an aerial weapon for use against air or surface targets. A small order was awarded in 1914, and arrangements made for manufacture at the General Ordnance Co. of Groton, Connecticut. Davis duly received an invitation from the British Admiralty to give a demonstration of his gun. Such was the rivalry which existed between the Service departments at this time that those chosen to attend the demonstration were told that 'The subject and nature of the demonstration should not be disclosed to any other branch of the Services, as this may adversely affect naval procurement of the weapon'.

It is easy to understand the interest taken by the Admiralty, for Davis's claims for the gun were considerable. According to him, it could be easily mounted on any of the aeroplanes being used by the air forces in England; its calibre was at least equal to that found suitable for field pieces and secondary batteries of warships; it could fire a projectile with sufficient velocity to secure long range and reasonable accuracy; and the weapon had no shock or recoil which could damage such frail structures as aeroplanes. If Davis's claims were correct, the more imaginative could envisage fleets of aircraft carrying such guns dominating any battlefield or naval engagement.

Davis duly set up a demonstration. He explained, 'The gun works on the principle that, if a mass is discharged to the rear of a gun at the same instant that a shell of equal mass is fired forward from the barrel, the recoil of one will balance out the other.' He then took the covers off a 6-pounder he had set up for the demonstration. The weapon appeared to be more of a long tube than a gun. It consisted of two barrels: one, which was rifled and chambered to receive a composite shell, faced forward, while the second, with a smooth bore, faced backwards and was joined to the first by an interrupted-thread mechanism operated by a central locking handle. The gun was loaded by releasing a catch on the handle, which was then pressed downwards, unlocking the barrel joint. This released the rear barrel, which rotated and swung downwards, leaving the breech open to receive the shell. The rear barrel was then snapped back into what was claimed to be a gas-tight joint.

The Davis round was longer than a normal shell,

because it comprised a projectile at the forward end, a cordite propellent charge in the centre, and a balance filling at the rear. The balance filling was slightly heavier than the projectile, being a mixture of fine lead shot lubricated with tallow or motor grease, sealed by a wooden plug at the rear. The round was fired by means of an electric primer mounted in the centre top of the case, activated by two contacts in the gun housing. The shell had to be carefully aligned into a guide to ensure the connections were made, safety switches being wired into the 12 volt firing circuit. In an early version the gunner held a spring-loaded double disc switch between his teeth which, when bitten, made the firing circuit!

Davis proceeded to fire the gun. The crack of the explosion was followed by a flash of flame from the breech joint, and a cloud of smoke from the front and rear of the tube. However, there was no discernible recoil movement, and the solid shot was seen to impact in the firing butts. After firing a few more rounds, Davis answered questions from the assembled experts.

It was generally agreed that the gun held much promise, and the Admiralty decided to order one 2-pounder and ten 6-pounders from the Groton factory. The weapons were delivered early in 1915 and were dispatched to Woolwich Arsenal for proof testing. Three 6-pounders were then sent to the RNAS Experimental Armament Depot at the Isle of Grain for Service trials, while the other guns were dispatched to the Marine Aircraft Establishment at Felixstowe, where it was hoped to develop suitable mountings for them in RNAS aircraft. It was at Felixstowe that, on 13 April 1915, Cdr Seddon took off in a Short S.81 (No. *126*) fitted with a 6-pounder Davis gun for the first air test of the weapon in the UK.

A twin-engined Cauldron G6 was also tested at the base with a Davis gun mounted to fire forward in the narrow gap between the propeller tips. The Felixstowe team's main project, however, was to provide a practical mounting for anti-submarine work. The gun was fitted to the front cockpit of a

The prototype Davis with the rear barrel down ready for loading.

Curtiss H1 (No. *951*) in February 1916, and various flying-boats, including the huge Porte Baby three-engined boat, in which firing tests were carried out later in the year.

The trial teams at the Isle of Grain were more concerned with the technical merit of the gun for service than its suitability to be mounted and fired from aircraft, as at Felixstowe. Several problems were encountered at the Isle of Grain, but none seemed serious. By October 1915 the Admiralty had ordered 293 Davis guns, with a large amount of HE and incendiary ammunition, and on 3 February 1916 Cdr Clark-Hall presided over a committee

The Davis composite shell.

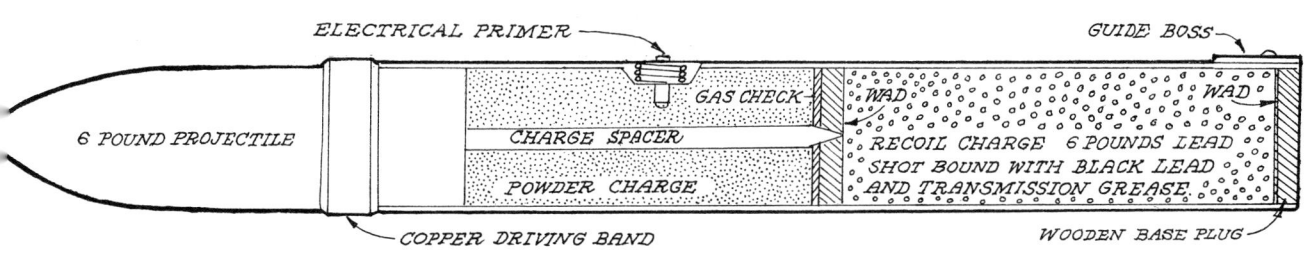

Davis 6-pounder on test rig with twin frame sights. (Bruce/Leslie)

consisting of four RNAS officers and Harrold Bolas of the Air Department. They were to examine the various trials reports to decide whether the Davis was a practical proposition for use on RNAS aircraft. In spite of the reservations from the Isle of Grain, it was decided to continue work. Robey & Company of Lincoln, one of the contractors building Short 184 seaplanes, were asked to produce an aircraft to carry two Davis guns, to counter the Zeppelin airships which were ranging over England. These frightening monsters could easily outclimb any aircraft then in service. The Davis gun could, it was hoped, engage them from below at far greater range than Lewis guns could achieve. The committee also recommended that a 6-pounder should be installed in a Breguet pusher aircraft at the RNAS station at Dunkerque for air firing trials.

Meanwhile, the Ordnance experts at the Isle of Grain were having further problems. Leakage of flame and gas from the barrel joint was not easily overcome, and forces on the interrupted-thread seal were giving cause for concern. When serious air firing began, 'interesting' events occurred. When the gun was fired in the stop butts, a safety area of 50 yd (45.7 m) had to be observed behind the gun, as the shot blast was similar to an enormous shotgun firing backwards, complete with a grease-laden cloud of black smoke. When fired from an aircraft, the whirling slipstream often deflected the shot and smoke back at the airframe, causing considerable anxiety to the crew. Test aircraft sometimes landed with damaged fabric flapping in the breeze, the crew shakily disembarking with blackened faces and tempers to match. The director of the establishment sent an urgent report to his superiors listing the faults, and requesting additional funds for modifications. On the other hand, the Davis was found to

Isle of Grain demonstration to RFC gunnery experts. (Bruce/Leslie)

Davis gun with twin Aldis sights fitted to a Handley Page O/400 at Redcar Naval Air Station. (Bruce/Leslie)

be very accurate up to a range of 2,000 yd (1,829 m), but some shells were seen to tumble uselessly end over end, causing blinds. The Admiralty would brook no talk of abandoning the project; indeed, great things were expected of the weapon. Besides widespread aircraft use it was planned to arm motor boats and armoured cars with the gun. Some motor boats were fitted out at Eastchurch, and took part in sea firing trials against moored targets.

It was not until February 1916 that the RFC was able to evaluate the gun, a 6-pounder being provided by the Admiralty to the machine-gun school at Hythe. It was mounted on a BE.2c in such a way that it could fire forward with either 45° depression or elevation, the fuselage being strengthened near the rear end of the gun. The RFC intended at first to use the gun in the air-to-air role, old Sopwith and BE.8 airframes being used as targets to assess the effect (one would have thought this would have been obvious). As a result of the trials at Hythe, the War Office sent details of both sizes of the gun to their main aircraft constructors, asking them to submit designs for special Davis gun positions on suitable aircraft. Armstrong Whitworth (FK.12), Vickers (FB.11), Aircraft Manufacturing Co. (DH.3) and Martinsyde (G.100) were some of the companies which carried out extensive design work. Perhaps the most bizarre mounting was that of the Dyott Battleplane. It was planned to fix a 12-pounder Davis to fire through a side porthole of the large biplane. When a ground target was sighted it was intended to circle round, pumping shells into it at a downward angle of 30°, exactly like the AC-47 and AC-119 in Vietnam many years later.

In early 1917 the Admiralty decided that the gun

Modified breech with strengthened barrel joint. (Bruce/Leslie)

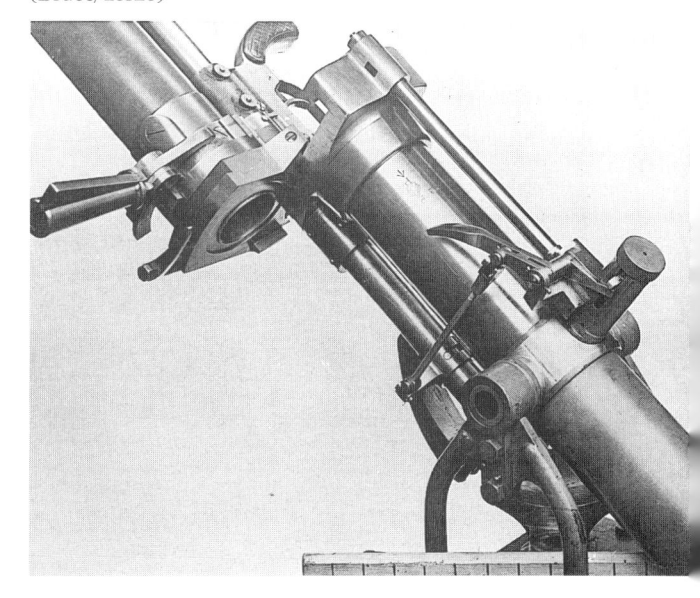

should be used operationally – probably to justify the large amount of time and money expended. Rather than use any of the special gun carriers, it was agreed that a Handley Page 0/100 should be fitted with a 6-pounder. A stout tubular mounting manufactured at the Battersea workshops in July was taken to Manston, where it was fixed into the front of the forward gunner's cockpit. After successful air firing tests three more 0/100s were fitted (Nos. *1459*, *1461* and *1462*). An 0/400 (*3127*) was armed and sent with 50 rounds of ammunition to the RN air station at Redcar on 7 September 1917 for use on anti-submarine patrols. Gunners were issued with caps fitted with thick canvas curtains to protect them from the flame and the hot gases which escaped from the breech joint. During the working-up trials on the Handley Pages, the blast stripped the fabric from the upper wing, and the pilot and other members of the crew were subjected to a barrage of lead shot and grease-laden smoke. However, the wings were protected by aluminium plates, and the Handley Pages were sent to France, where they were used operationally against ground targets by Nos. 7 and 7a Sqns in October and November 1917.

The measure of the gun's success can perhaps be gauged by a plaintive request to the Air Department from RNAS Dunkerque dated 7 December 1917: 'Regarding the Davis recoilless gun, no more ammunition will be required at present'. It appears that, after some hair-raising experiences, crews made it clear that they were unhappy with it. On 4 February 1918, following further disparaging reports, the guns were withdrawn from service in France. The project was finally terminated at the Isle of Grain. It was with mixed feelings of regret and relief that Major Thompson submitted the final trials summary. He stated that, although the principle was sound, the gun needed substantial redesign before it could be recommended for Service use. The decision caused dismay at various aircraft manufacturers, where work had reached an advanced stage on the specially designed gun carriers. Some considered replacing the gun with heavy-calibre conventional weapons, such as the Coventry Ordnance Works 1½-pounder or the Vickers 1-pounder, but none of the gun carriers saw service. What had been seen as a brilliant new air weapon had failed, like so many others before and after it, to stand up to the rigours of Service trials.

There is an intriguing postscript. A few 2-pounder Davis guns were dispatched to Mesopotamia where, in January 1918, 30 squadron armourers fitted one to an RE.8, a small aircraft. It was clamped to the starboard side of the fuselage

A Davis 12-pounder mounted on a US Navy Curtiss flying-boat. The Lewis was used as a sighting aid.

Details of the Davis gun

Bores	2-pdr: 40 mm (1.57 in); 6-pdr: 57 mm (2.24 in); 12-pdr: 75 mm (2.95 in)
Action	Opposing charge non-recoil
Weights	2-pdr: 45 lb (20.4 kg); 6-pdr: 165 lb (74.8 kg); 12-pdr: 220 lb (99.8 kg)
Muzzle velocity	1,200 ft/sec (366 m/s)
Projectile	HE, AP, incendiary
Firing system	12 volt electrical primer
Sighting	Frame, ring and bead, Aldis, Lewis sighting burst

The Robey Davis gun carrier with wing positions. (John Walls)

alongside the observer's cockpit to fire at 45° forward and downward. The installation gave excellent service on ground attack operations and, after a few modifications in the squadron workshop, no trouble was encountered with the gun's action or discomfort to the crew.

Aircraft fitted with Davis guns

Aircraft	Position
Norman Thompson NT.4	Above front cockpit (2-pdr)
Port-Victoria	Front cockpit (2-pdr)
Supermarine Nighthawk	Above upper wing (2-pdr)
Handley Page 0/100, 0/400	Front cockpit (2-pdr, 6-pdr)
Short 184	(2-pdr)
Curtiss H.12	Front cockpit (2-pdr)
Robey Peters	2 wing-mounted cockpits (2-pdr)
DH.4	Rear cockpit (2-pdr)
RE.8	Rear cockpit (2-pdr)
BE.2c	Rear cockpit (2-pdr)
Vickers FB.11	Cockpit above wing (2-pdr)
Martinside G.100	Above upper wing (2-pdr)
AW FK.12	2-pdr in each of two cockpits
BE.12	Rear cockpit (6-pdr)
Short 81	Front cockpit (6-pdr)
FE.2b	Front cockpit (6-pdr)
Voisin Canon	Front cockpit (12-pdr)
Breguet	Front cockpit (12-pdr)
Cauldron G6	Front cockpit (2-pdr)

There was no further interest shown in the opposing charge (recoilless) system until the armament of the Gloster Javelin was being considered in 1950. One of the options then was a new 114 mm recoilless gun, which would have been a novel departure from established aircraft armament. After several years of development the project was abandoned.

Above *A Vickers 1.59 showing breech opening, face shield and recoil spring.*

Below *A Vickers 1.59 in 'Rocket Gun' mounted on an FE.2D.*

The Vickers Crayford 1.59 in 'Rocket' gun

In 1915 Vickers submitted a design for a lightweight 1.59 in gun for infantry use on the Western Front. Relatively low-velocity projectiles using Balistite as a propellent were to be used. This enabled the designers to keep the weight and size of the breech down to a minimum. The gun was a single-shot, breech-loading weapon, the recoil being taken by a spring and buffer mechanism under the barrel.

The gun was not accepted for army use, but its light weight was seen to offer possibilities as an aerial anti-Zeppelin weapon. The prototype was delivered to Farnborough, where it was found to be remarkably accurate up to a range of 1,000 yd (914.4 m). Vickers supplied incendiary projectiles for use against airships. These shed a trail of sparks when fired, leading to the gun's unofficial nickname, the 'Rocket Gun'.

Various improvements were carried out, and in late 1916 it was test-fired on a Vickers FB.6. This aircraft, designed as a night fighter, crashed during the trials, so they were continued on another prospective night fighter, the NE1, designed at the Royal Aircraft Factory. Visiting pilots from the Western Front suggested that the gun could be used as an air-to-ground weapon, and Vickers developed a mounting for it for the front cockpit of the FE.2d. Six production guns were dispatched to No. 2 Air-

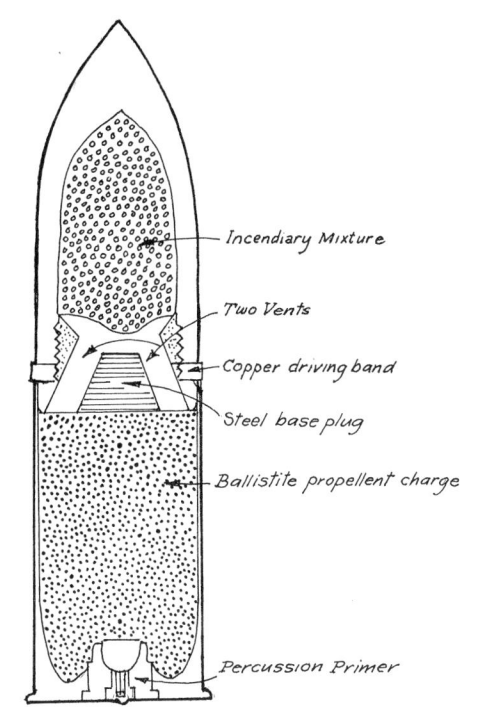

Above *The incendiary shell which gave the Rocket Gun its nickname. The incendiary mixture ignited on firing, leaving a trail of sparks behind it.*

Below *Interior of the FE.2D cockpit, showing the rocket gun mounting and the gunner's 'chair' and face shield.*

craft Depot in France, together with projectiles, HE, AP and incendiary.

In early 1917, 100 Sqn RFC were engaged in night intruder missions over enemy lines, and an FE.2d was fitted with a Rocket gun for operational testing. The unit reported: 'We have tested the Vickers 1.59 cal. gun. Of ten shells fired, three were misfires, one round did not fire, and the observer, thinking it was a dud, was about to open the breech when the shell suddenly launched itself amid a shower of sparks.' In contrast, Captain Martin of 102 Sqn sent a complimentary report to GHQ on the gun's performance. Reports show that both this gun and the Vickers 1-pounder were used on night operations during 1917, and stock lists from the Aircraft Depots show stocks of 1.59 cal. rounds for the 'Rocket Gun' at this time. It appears that the operational use of the gun was limited to a period in mid-1917 in France. As 0.303 in incendiary ammunition was found to be effective against Zeppelins, and bombs were used on night intruder operations, the various heavy-calibre projects for night use were dropped.

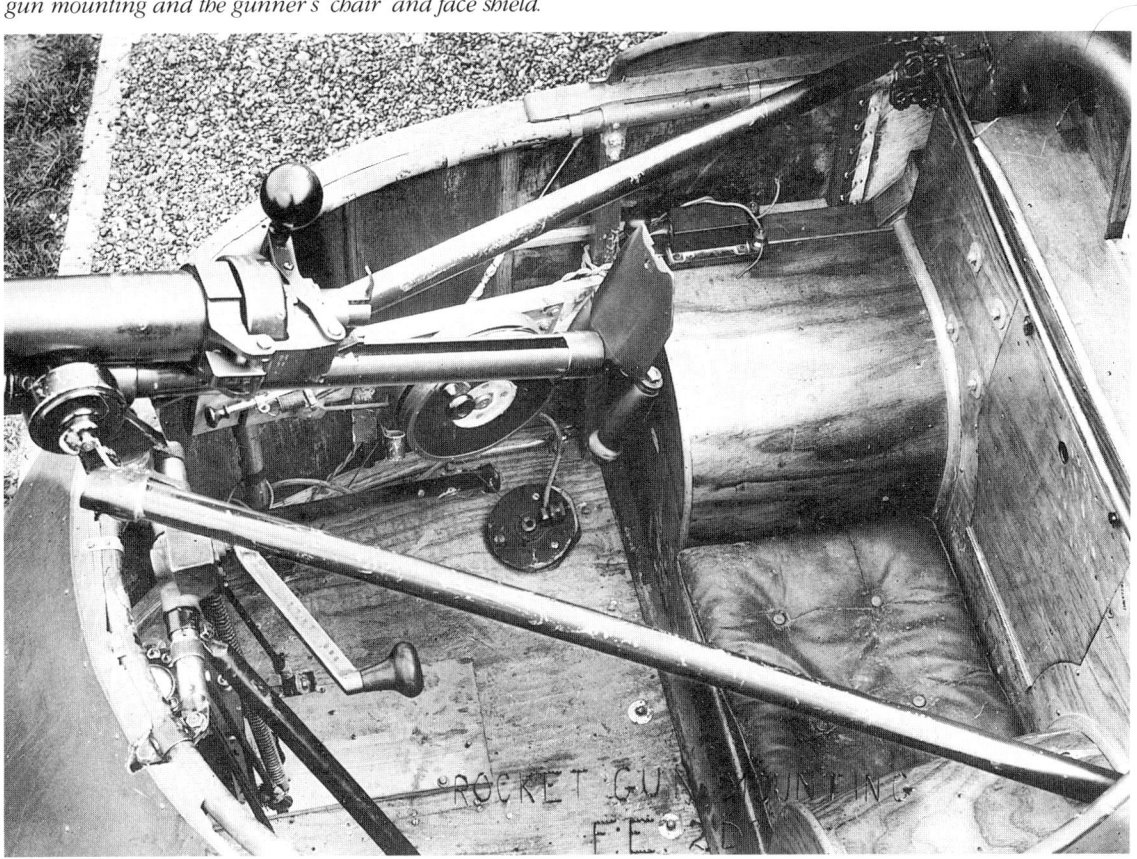

Operation

The gunner opened the breech by means of a large knob operating a locking lever. He then inserted a round and locked the breech. The gun was aimed by twin handles, over which was a face shield with a sighting aperture, cross wires in the shield being aligned with a ring foresight. On night operations Hutton illuminated sights were used, in which the foresight was illuminated by a small bulb shining through a 2 mm hole (see Part 2).

Details of the 'Rocket' Gun

Bore	1.59 in (40.4 mm)
Action	Single shot breech loading
Weight	47 lb (21.3 kg)
Muzzle velocity	1,050 ft/sec (320 m/sec)
Ammunition	HE, AP, incendiary

Vickers Mks III–V 1-pounder

In 1887 Hiram Maxim designed a compact 37 mm automatic gun. As explained later, this had been fixed as the minimum bore for any gun firing explosive ammunition. Most gun manufacturing concerns produced weapons of this size at the turn of the century. Maxim's gun was manufactured in small numbers at the Vickers works at Crayford, for promotional purposes. It was tested by the War Office but rejected because no use could be found for such a weapon. A small number were exported, including a few to France, which were then purchased by the Boer forces in South Africa. After several engagements when British forces suffered severe casualties from the quick-firing shells, the War Office and Admiralty arranged for further testing. A small quantity were ordered for use against torpedo boats, and the Army belatedly adopted the weapon as a highly mobile field gun on a wheeled carriage.

The Vickers 1-pounder mounted on an FE.2b as used by No. 51 (Home Defence) Squadron in March 1917. The two shells are (left) HE and (right) the AP round; the triple searchlight mounting was one of several tested. The gun was aimed by either an Aldis sight or the elongated ring and bead shown here.

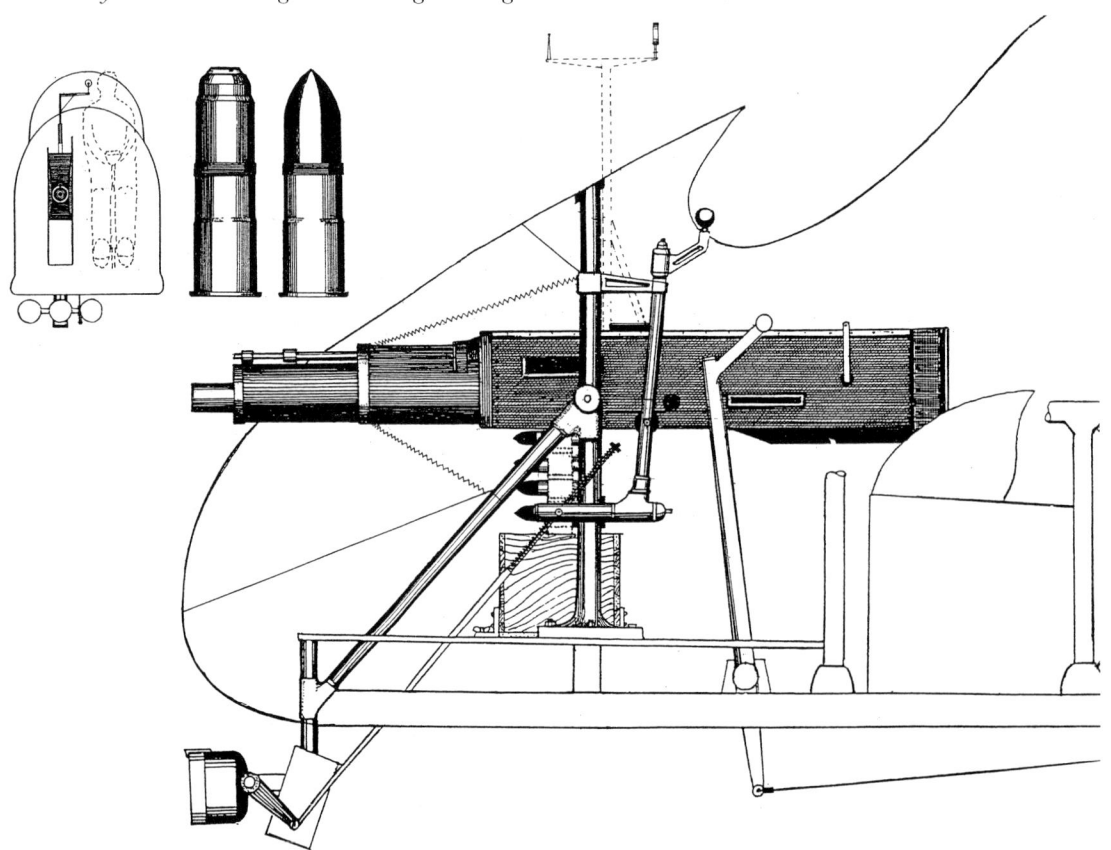

THE GUNS

When early experiments took place with heavy-calibre guns on the first aeroplanes, the Vickers 1-pounder was one of the few such guns available in Britain, and various trials took place with this and the heavier 1½-pounder Vickers gun. It was hoped to adapt the gun for anti-Zeppelin use, and a series of trials took place at Orfordness on an FE.2b, with a 1-pounder mounted to fire forward alongside the pilot. Several FEs were used operationally by Home Defence squadrons with Farnborough-produced mountings, but the gun was first used against German targets by FEs of 100 and 102 Sqns on night operations over the Western Front.

The gun used the Maxim recoil system, suitably enlarged and strengthened. Its worst features were the fierce recoil and flying empty cases. The ammunition was naval-type HE and armour-piercing rounds, fed from a 40-round canvas belt from a box on the right of the gun. The HE shell which was to have been used by Home Defence squadrons was fitted with a very sensitive fuse (no. 131) which was designed to detonate on contact with airship fabric. The gun was the only fully automatic shell-firing gun fired from British aircraft during the 1914–18 war.

Vickers 1-inch aircraft gun

This was an experimental aircraft gun designed by Vickers as yet another anti-Zeppelin weapon. It was a recoil-operated gun with a huge return spring mounted around the barrel. It was designed for the front cockpit of pusher aircraft, having movement in elevation only, this being controlled by hand wheel-turned gearing. A purpose-made illuminated prismatic sight was built into the mounting, giving the gunner an upward-firing field of fire. There are very few records concerning this weapon, but it was a practical concept given an aircraft strong and powerful enough to support the gun, mounting and ammunition, which was fed into the gun by a canvas belt from a 25-round box. The rate of fire was only 150 rpm but, as was found with the Hispano in the Second World War, the solid projectiles would have been very effective against the airframes of the period. No records exist of air firing, probably because no aircraft could have lifted the weight of a gunner plus the heavy mounting in the nose.

Details of the Vickers 1-pounder

Bore	37 mm (1.443 in)
Action	Recoil/fuzee spring
Cyclic rate	300 rpm
Weight	279 lb (126 kg)
Muzzle velocity	1,800 ft/sec (549 m/sec)
Ammunition feed	40-round belt
Cooling	Air
Length	74 in (1,880 mm)

Details of the Vickers 1-inch aircraft gun

Bore	1 in (25 mm)
Action	Recoil
Cyclic rate	150 rpm
Weight	110 lb (50 kg)
Weight with mounting	190 lb (86.2 kg)
Muzzle velocity	1,542 ft/sec (470 m/sec)
Ammunition feed	25-round box magazine with canvas belt
Length	54 in (1,372 mm)
Elevation	60°
Depression	30°

Ancient and modern: a Vickers 1-inch gun in front of an ADEN gun, from the MOD Pattern Room in Nottingham.

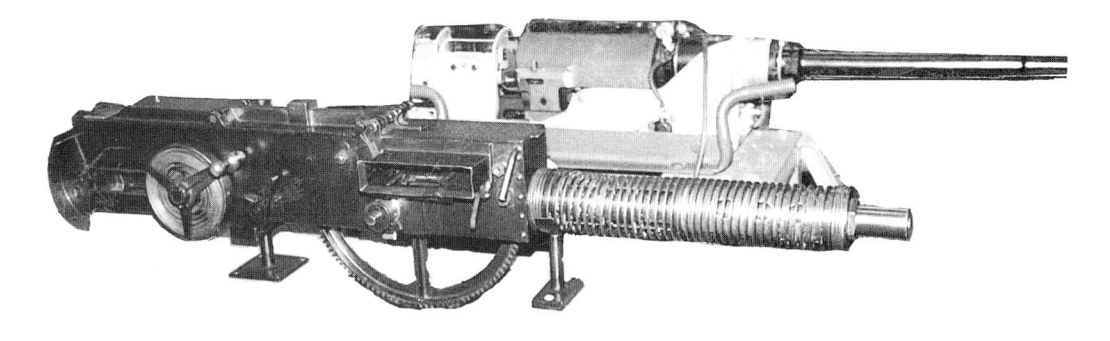

The COW gun

Following an international conference at St Petersburg in 1868, the major powers agreed that the use of explosive or deliberately deformed bullets against individual soldiers would be outlawed in any future conflict. A treaty was signed which limited explosive projectiles to a minimum weight of 450 g (15.9 oz). It was found that the smallest calibre of explosive shell would be 37 mm (1.46 in). Arms companies thereupon designed quick-firing guns of this calibre. In England, Vickers produced a naval deck gun of this size, but it was a design by the Coventry Ordnance Works that aroused most interest.

In response to a Royal Ordnance Board request in May 1912, the company submitted a semi automatic 1-pdr. the M1913. This was a magazine fed gun designed to fire the 37 mm round.

After hearing of experiments in France when a 37 mm Hotchkiss was fired from a Voisin, Geoffrey de Havilland at Farnborough decided to mount a 1-pdr. C.O.W. gun in the front cockpit of an FE.2. About this time news came through that the recoil force had caused the wings to fall off the Voisin in one of the French trials. It was known that the 1-pdr C.O.W. gun had a recoil force of 1,500 lb (680.4 kg), and de Havilland decided to suspend the FE.2 from a gantry and fire the gun into the stop butts. No damage was done to the airframe, and as a result of

Coventry Ordnance Works photograph of the 1½-pounder with twin Aldis sights, hopper magazine and mounting.

S81 Guncarrier built in 1914 with strengthened nacelle. The Vickers 1½-pounder gun was 'fired' in 1914, when excellent practice was had in firing at targets in the air and on the sea.

this experiment in 1914 the Royal Aircraft Factory produced the FE.6, to be armed with a C.O.W. gun (by this time the gun had inevitably been dubbed the COW gun, much to the chagrin of the makers). The FE.6 was not a success, and trouble with the engine led to the cancellation of the project.

In 1915 the Coventry company produced a $1\frac{1}{2}$-pounder version of the gun, and in the same year the FE4 appeared, a large twin-engined pusher with a $1\frac{1}{2}$-pounder in the front cockpit. This aircraft was designed by S. J. Waters and H. P. Folland as a ground attack aircraft, and plans were made for production at the Daimler factory, but the production order was cancelled. In 1917 the $1\frac{1}{2}$-pounder was evaluated at the Naval Experimental Air Station at the Isle of Grain, mounted on a Tellier flying-boat. Air firing against moored targets went well, and it was thought that the gun could be a useful anti-submarine weapon.

In March 1917 C.O.W. developed experimental models with 47 mm and 75 mm bores for the French Government, but no production order materialized. This was the time when Zeppelins were operating over England, and one of the Home Defence fighters, a DH.4 (*A2168*) was fitted with an upward-firing C.O.W. gun at Orfordness. The big airships could be detected more easily from below, silhouetted against the grey night sky. Lewis guns on Foster mountings were proving effective, and it was thought that a big gun might be another answer. The weapon was fixed to fire at an elevation of 45°, with the breech close to the cockpit floor and the barrel protruding through the wing centre section. The gun was sighted by the pilot, who told the gunner in the rear cockpit when to fire. This aircraft took part in several abortive anti-Zeppelin patrols, and on 12 August 1917 took off in search of Gotha bombers. Two similarly armed DH.4s were sent to France as bomber destroyers, but there are no reports of them being successful.

Vickers Vimy *F9146* at Brooklands was one of several other aircraft which took part in trials with the gun, but it was never used in any numbers during the First World War, the main reason being that its length and recoil made it too unwieldy for the aircraft available.

After the war, two large flying-boats, the Vickers Valentia and the Short Cromarty, were designed to carry COW guns. Large bow cockpits were provided with trunnions for the heavy cannon. The boats were to be used for anti-submarine and shipping patrols, but money ran out after the prototypes had been evaluated.

In the early twenties Vickers took over the rights

DH.4 A2168 *with upward-firing COW gun used on anti-Zeppelin patrols.*

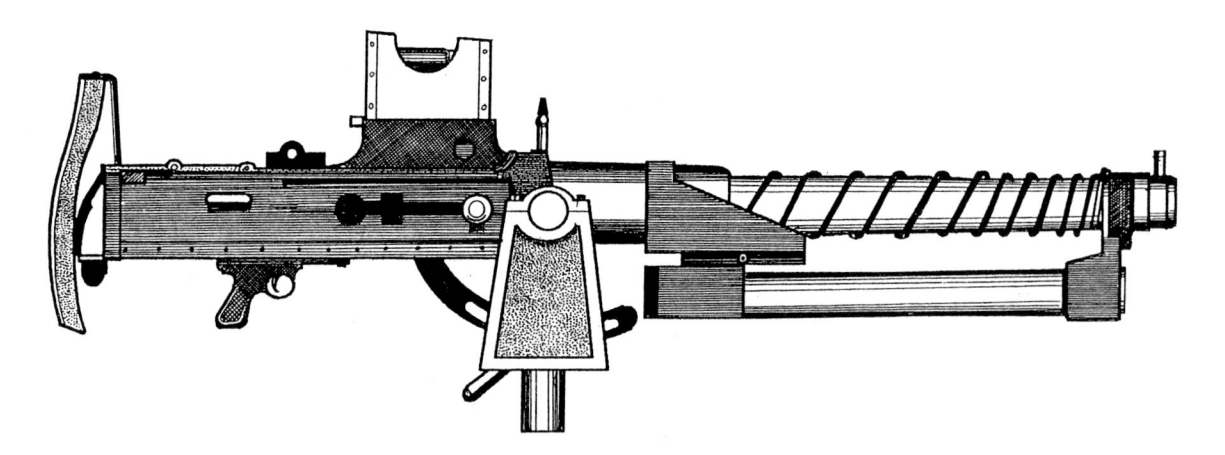

The C.O.W. 1-pdr. designed in 1913 to meet Ord. Board request. Experimental only.

from C.O.W., hoping the gun would be adopted by the RAF as aircraft became stronger. Experienced airmen expressed doubts about the continued use of rifle-calibre guns, having seen aircraft riddled with enemy fire still able to fly home. It was thought that heavy shell-firing guns would be much more certain to destroy an aircraft. The Air Staff reasoned that heavy fighters would be needed to break up enemy bomber formations, picking off the attackers with shell-fire.

In 1924 a specification was issued calling for a twin-engined machine capable of mounting two COW guns. Two companies submitted designs. The Bristol Bagshot was a high-wing monoplane with two spacious cockpits for gunners, one in the bows,

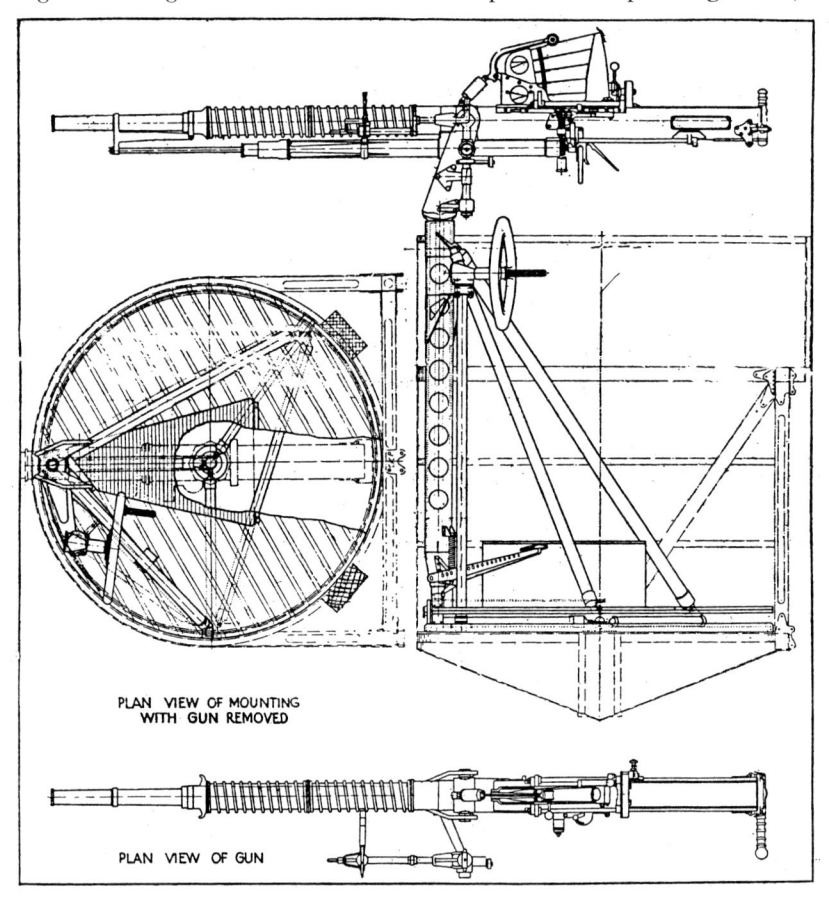

The Westland COW gun mounting.

the other behind the trailing edge of the wing, both gunners having an uninterrupted field of fire upwards and to the front. The Westland Westbury was a large biplane with similar provision for two cannon. Both had additional Lewis guns for defence. Companies were given every assistance by Vickers, who provided trials guns and assisted in the design of the mountings. The Westbury seemed to be the more efficient of the two (the Bagshot's wing was weak in torsion). Its bow gun could be traversed a full 360°, and could be locked in any position by means of a foot-operated brake. However, the poor speed, climb and ceiling of the heavy aircraft made it essential that they were already at operational altitude when the raiders arrived. As their only warning system consisted of lookouts and sound detectors, the Air Staff were forced to issue a new specification calling for a high-performance single-seater capable of climbing to intercept an enemy force approaching at 150 mph at 20,000 ft. This specification, F29/27, specified an upward-firing COW gun with an adequate supply of ammunition. Westland and Vickers were invited to tender.

The first design submitted was the Vickers Type 161 COW gun fighter, a pusher biplane in which great care had been taken to provide the pilot with a steady gun platform and an uninterrupted view to the front. He sat on the left of the gun, which he loaded with oversize 10-round clips. The front 4 ft (1.2 m) of the barrel protruded upwards from the nacelle at 45°. No fairing was added as it was thought that this would affect the cooling of the barrel. The Westland contender, powered, like its rival, by a Bristol Jupiter engine, was developed from the F.20/29 monoplane fighter. The gun was mounted in similar fashion to the Vickers machine, the pilot sitting on the left with an identical periscopic sight. The ammunition feed made the Westland fighter unique. Originally developed for the Westbury by V. S. Gaunt, a 39-round magazine provided the pilot with a continuous supply of shells, a 'rounds left' indicator telling him his ammunition status.

Both aircraft were evaluated at Martlesham Heath in late 1931, where tests were carried out against drogue targets. It seemed that either aircraft would provide an adequate defence, but the Air Staff kept changing their minds and the idea was abandoned. The COW gun fighters never hit the headlines, but an experimental mounting on a Perth flying-boat certainly did. In 1933 this big three-engined flying-boat took part in some highly publicized trials. International aviation journalists were invited by Vickers, who saw this as their last chance to promote the weapon, which by this time was certainly not the last word in weapon technology. The Perth was fitted with a mounting similar to that of the Westbury. An American magazine of 1934 spoke of 'squadrons of Perth flying-boats armed with heavy-calibre guns' but in fact only two boats were fitted with the gun.

Following a short-sighted Air Staff decision to abandon all work on heavy-calibre weapons, Vickers Armstrongs decided to stop further development of the gun. Shortly after this the Vickers K machine-gun was adopted by the RAF as an observer's gun, and this helped keep the Crayford works busy until the expansion schemes of the late 1930s.

In the critical early months of the Second World War, several makeshift proposals were put forward to arm aircraft with the small number of existing COW guns. One of these was to mount two to fire vertically through the hulls of Sunderland flying-boats on anti-submarine and shipping missions.

The COW gun spent a remarkable 30 years on the verge of adoption, but with hindsight none of the

Vickers Type 161 of 1931, built to specification E.29/27 for a bomber destroyer.

:he rotating bolt then locked behind the round, and :he weapon was loaded and ready to fire.

There are still some surviving examples of the gun, one of which is on show in the Royal Air Force Museum at Hendon. Another is held in the MOD Pattern Room at the Royal Ordnance Factory at Nottingham.

Details of the COW gun

Bore	37 mm (1.46 in)
Action	Long recoil, rotating bolt
Cyclic rate	60 rpm
Weight	140 lb (63.5 kg)
Weight of shell	1 lb (early); 1.5 lb (.68 kg)
Muzzle velocity	2,000 ft/sec (610 m/sec)
Ammunition feed	5- or 10-round clips
Recoil force	1,500 lb (681 kg)
Cooling	Air
Range	5,000 ft (1,524 m)

A Vickers-Armstrong 37 mm aircraft cannon mounted in a Blackburn Perth of 209 Squadron.

proposed installations would have been practical, and its ungainly length and primitive feed would have precluded any enclosed mounting.

Action

The gun operated by what is known as the long recoil system. The barrel and breech-block were held securely locked together until the distance of the recoil movement was greater than the overall length of the incoming round. The lock was then freed and remained held to the rear while the counter-recoiling barrel went into battery, pulling the chamber off the empty cartridge case. This type of action, which was first used on the Chauchat machine rifle, was commonly used on early large-bore automatic weapons, and was also adopted for the famous Vickers S gun. It was slow but reliable and solved such problems as the necessity of opening the breech under high pressure. The gun had an air-cooled barrel, and could be fired both semi and fully automatically. To fire the COW gun, the five-round clip was put in position with the bolt forward. With the aid of a crank, the gunner then pulled the operating mechanism to the rear until the holding device stopped the rammer and carrier frame. Upon release, the barrel released the carrier, this stripped the first round from the magazine and chambered it,

The Beardmore-Farquhar

The Beardmore-Farquhar was one of the lightest machine-guns ever produced, weighing only 16.25 lb with drum. It was the invention of Col. Moubrey G. Farquhar of Birmingham, and was manufactured by William Beardmore, also of Birmingham.

On 17 November 1919 the free-mounted version of the gun was submitted by Farquhar for an official test at Eastchurch, where it was fitted to the Scarff ring of a Bristol Fighter. The pilot was F/Lt. Rea, and the gunner F/Lt. Pynches. Twenty rounds were fired at 4,000 ft and 320 at 18,000 ft, where the gun was tested at all angles of elevation and depression. The gun proved easy to manipulate, and the magazine could be changed quickly and easily. The cyclic rate proved to be a little on the low side for an air weapon (430 rpm), but it had less kick than the Lewis, and cartridge cases were ejected 6 ft away into the slipstream. There was only one stoppage during the trial, caused by an oversized rim, and this was cleared by pulling the cocking handle. During the flight, loaded magazines had been placed on the floor of the cockpit, and vibration released the spring-loaded cut-off on one magazine, ejecting the contents around the feet of the gunner.

Though gas-operated, the piston did not directly return the action to the rear. A series of springs were tensioned by the piston, and these returned the bolt and chambered a fresh round, locking the

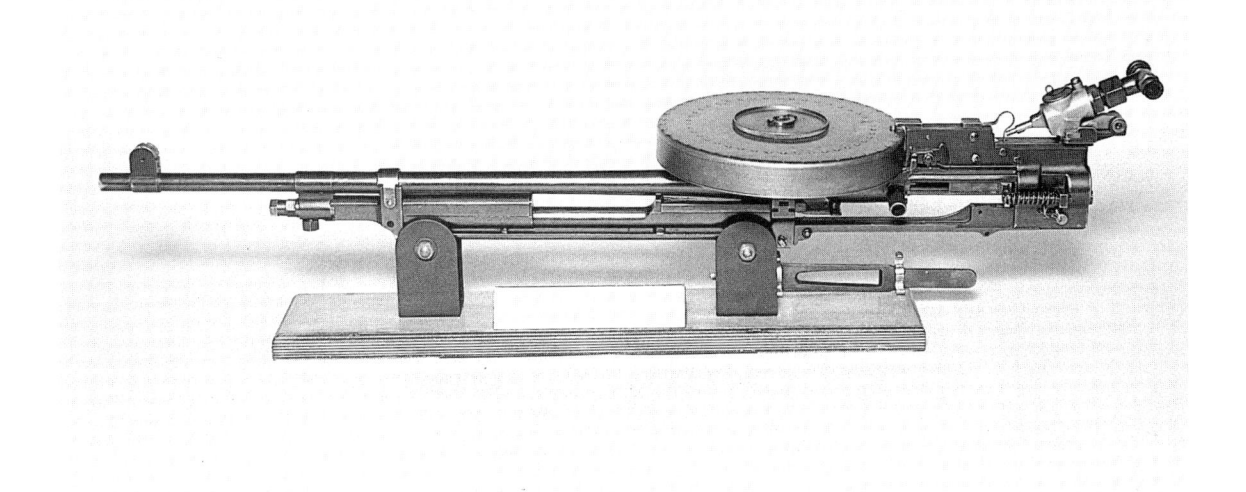

Beardmore Farquhar 0.303 in with Vickers trigger motor. (MOD Pattern Room, Nottingham)

bolt. The action depended on the precise balancing of the springs, but it was very smooth and free from vibration. Overall, the gun compared favourably with the Lewis, but not enough to justify its adoption by the RAF. The Finnish Air Force showed great interest in the gun in the early 1930s, but this did not result in a substantial order. The Beardmore company submitted the gun for further trials in 1940 as a fixed weapon, but again it was not accepted.

Details of the Beardmore-Farquhar

Bore	0.303 in (7.7 mm)
Action	Gas and springs
Cyclic rate	450–550 rpm
Weight	16.25 lb (7.37 kg)
Muzzle velocity	2,427 ft/sec (740 m/sec)
Ammunition feed	77-round drum
Cooling	Airflow over barrel
Rifling	Five grooves

The Adams-Willmott

Knowing that the Lewis gun had reached the limit of its development, and that the Air Staff were considering a replacement, the BSA Co. of Birmingham adopted a new design of observer's gun, the Adams-Willmott, which was the company's submission for the 1934 trials to find a new free-mounted gun for the RAF. Unfortunately, it was in the first batch to be rejected. It was evaluated by several foreign countries, and a small number were sold, but it was not adopted for widespread service use.

The Adams-Willmott was a design contemporary with the Vickers K, and was similar in many respects, being fed by a drum-type magazine with a quick-change mechanism. It was controlled by two handles, each with a trigger, to give more control when firing to the beam. A padded headrest was claimed to improve sighting and protect against injury. The trigger guards were large, for heavily gloved hands, and the cocking handle, low on the left of the gun body, was easily accessible. The action was gas-operated and bore many similarities to the Lewis, the cocking handle being easier to manipulate and the construction of the piston group more substantial.

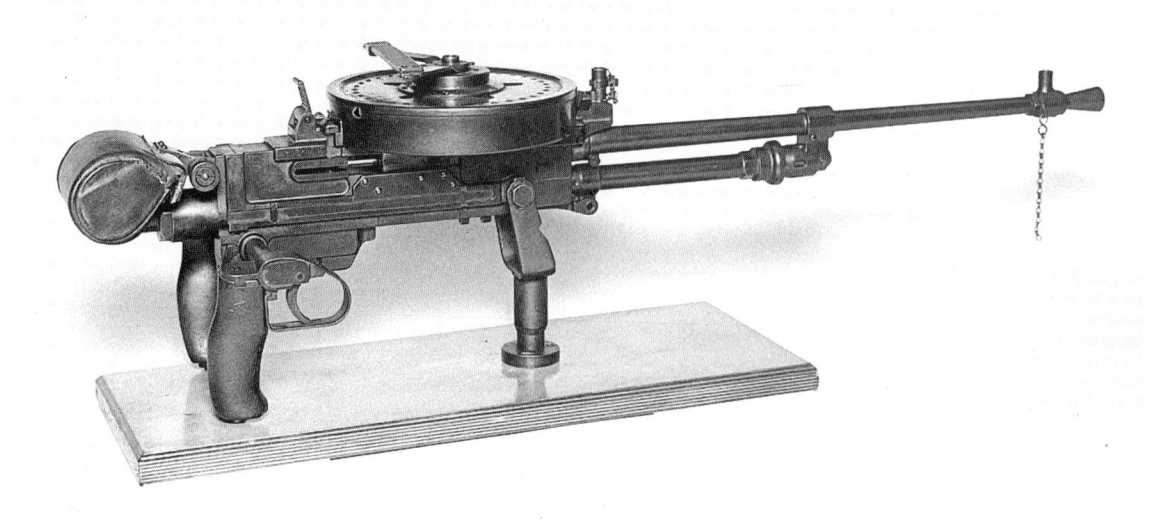

Above *The Adams-Willmott observer's gun, which was submitted for the RAF free-gun trials in 1934.* (MOD Pattern Room, Nottingham)

Below *Rear view of the Adams-Willmott showing the chin pad.* (MOD Pattern Room, Nottingham)

Details of the Adams-Willmott

Bore	0.303 in (7.7 mm)
Action	Gas-operated
Cyclic rate	720 rpm
Weight with magazine	23 lb (10.4 kg)
Muzzle velocity	2,400 ft/sec (732 m/sec)
Ammunition feed	100-round drum
Cooling	Airflow
Sighting	Norman Vane or ring-and-bead
Rifling	Five grooves
Length	46 in (1.4 m)

Vickers 0.5 in Class C

Listed by Vickers as the .5 H.V. aircraft gun, this weapon was originally designed for ground use, to fire special armour-piercing ammunition against armoured vehicles. It was evaluated by the RAF as a heavy-calibre fixed gun in 1928. The action was based on the Maxim system, with various parts strengthened to take increased recoil forces. The unusual side plates were originally intended to protect the barrel in the vehicle-mounted version; they could have been dispensed with to lighten the gun. It was found to be an efficient weapon, though it had a rather low cyclic rate. The Air Staff's decision to standardize on rifle-calibre weapons caused Vickers to discontinue development for air use.

THE GUNS

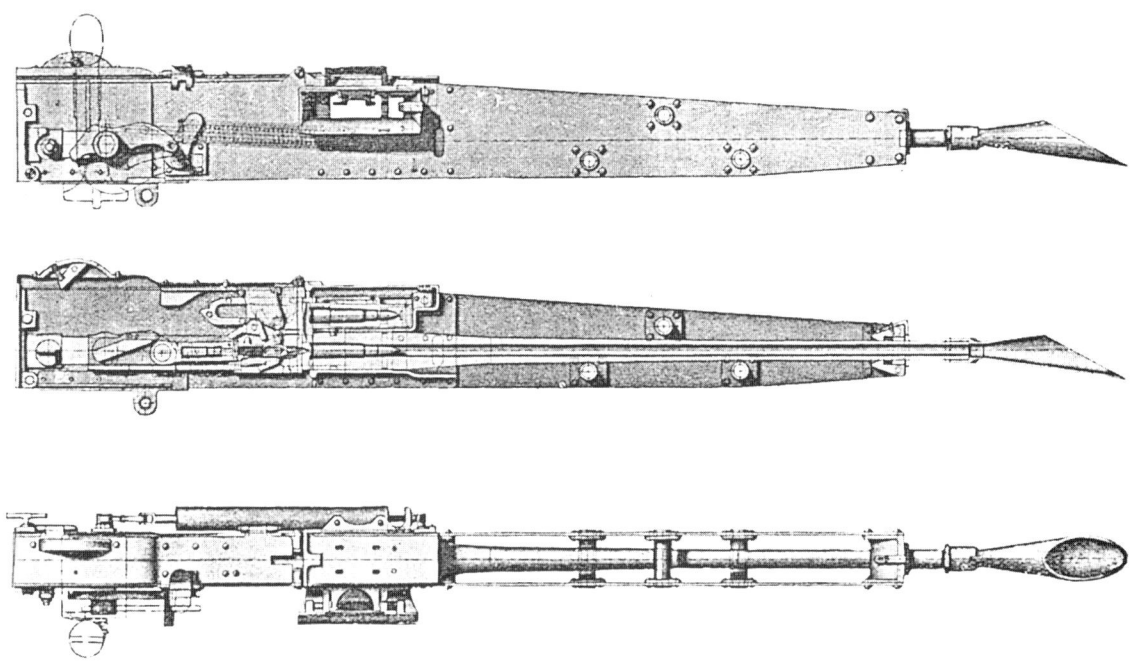

Vickers Class C 0.5 in automatic pilot's gun, evaluated as a fixed aircraft gun.

Details of the Vickers 0.5 in Class C

Bore	0.50 in (12.7 mm)
Action	Recoil/fuzee spring
Cyclic rate	340–600 rpm
Weight	52 lb (23.6 kg)
Muzzle velocity	2,550 ft/sec (777 m/sec)
Ammunition feed	Metal links
Cooling	Air
Length	45 in (1,143 mm)

Vickers Class F

Having excess capacity in their Crayford works in the 1920s, Vickers decided to produce an observer's gun as a private venture. With the Lewis beginning to show its age, the company adapted the robust action of the Mk III pilot's gun to take a Lewis drum magazine or a disintegrating link belt. Their main target was the export market, and after some problems with the ammunition feed system had been resolved, a sales drive resulted in some over

The Vickers Class F, produced at Crayford as a replacement for the Lewis.

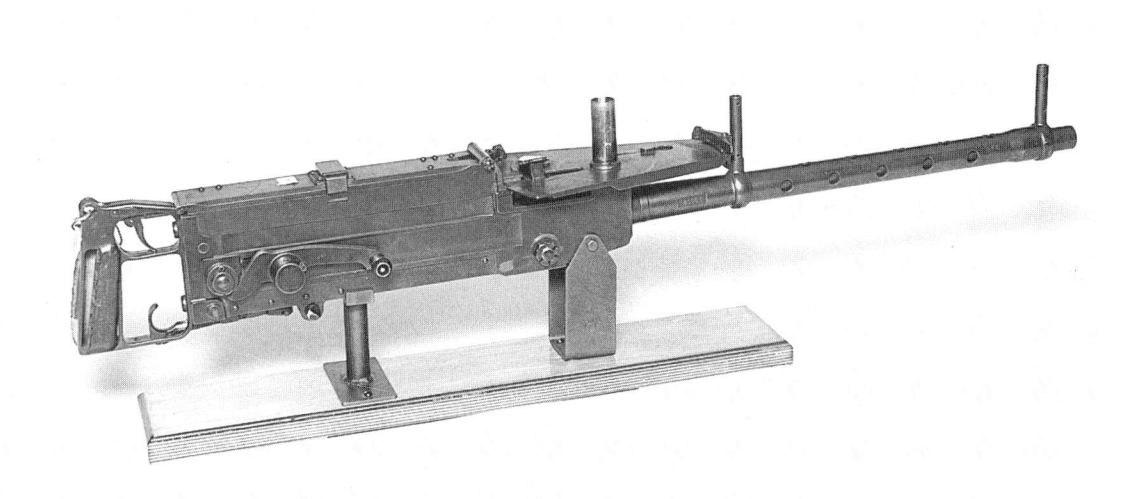

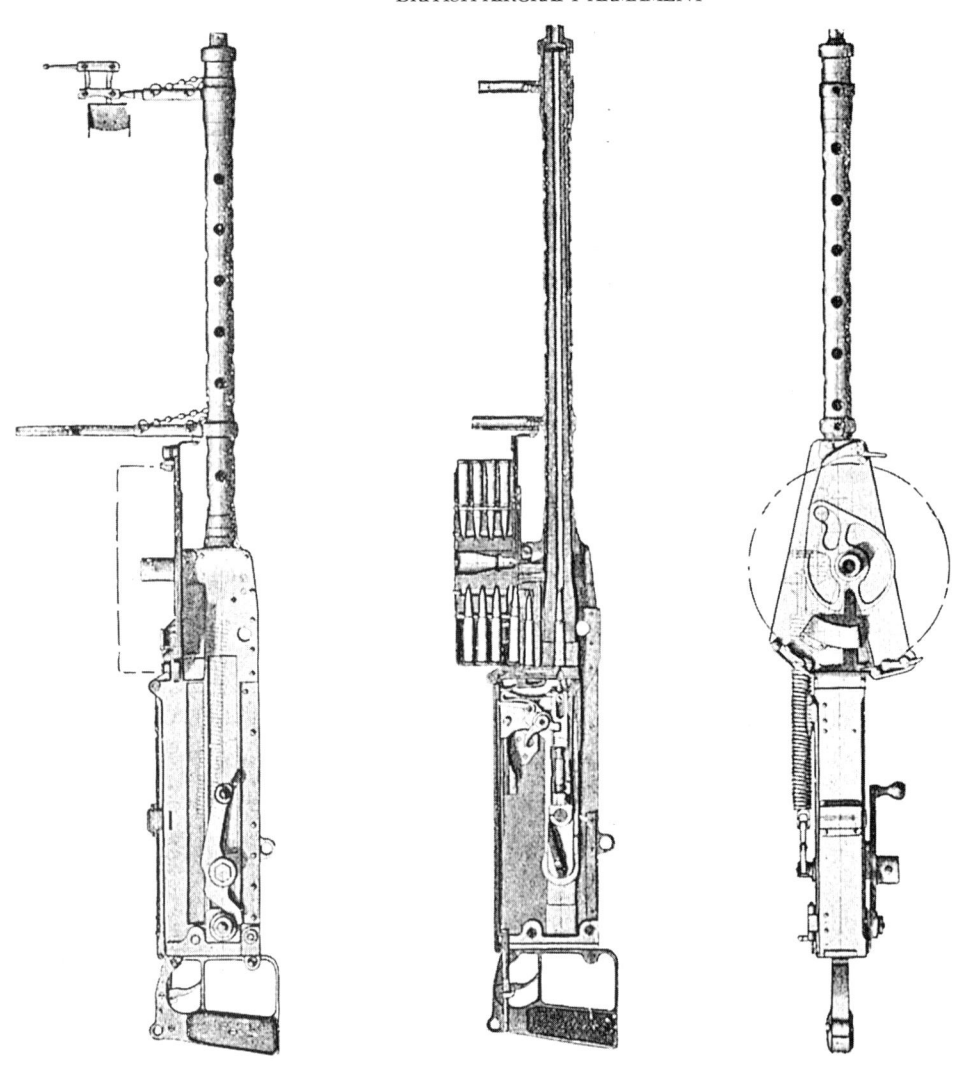

Above *The Class F with Vane sight and drum magazine.*

Left *A Class F Vickers gun mounted on the gun ring of an Avro Tutor gunnery training conversion.*

seas orders. A production line was set up and an order for 300 guns from the Polish Air Force completed. A few more orders materialized, and a small batch was supplied to the RAF, but when it was decided to embark on a new gas-operated design, resulting in the Class K, the F was discontinued.

Details of the Vickers Class F

Bore	0.303 in (7.7 mm)
Action	Recoil
Cyclic rate	600 rpm
Weight	24 lb (10.9 kg)
Muzzle velocity	2,200 ft/sec (671 m/sec)
Ammunition feed	97-round drum/ disintegrating belt
Cooling	Air
Length	46 in (1,168 mm)

The BSA 0.50 in

In 1918 the newly formed Air Staff believed it would probably be necessary to find a reliable heavy-calibre gun to counter the use of armour protection on aircraft. The trusty Vickers was rebored to 0.50 in, with questionable results, but the first British-designed 0.5 in (12.7 mm) gun was produced by BSA in 1924. It was a recoil-operated weapon which could fire single shots or automatically. It was simple, had robust components, and could be dismantled without tools. The drum-type magazine held only 37 rounds, but more would have made the gun very heavy. The gunner could replenish the magazine without taking it off the gun through slots in the top plate. Being designed primarily as an observer's gun, spent cases were deposited downwards onto the floor of the cockpit, not into the slipstream, which could have caused a possible hazard.

The gun was made in two versions: water-cooled, for naval use, and an air-cooled aerial version with radial fins on the barrel and large holes in the jacket. It also had large spade handles in place of a rifle stock. The ammunition was identical to that used on the Vickers 0.5 cal. gun, and was manufactured by Kynoch of Birmingham.

An official trial held at Hythe, proved very disappointing. After the low rate of fire, the most criticized feature was the 37-round magazine. It was suggested that this should be increased to 97 rounds, although this would have made the gun very difficult to move freely, and a full magazine would have been awkward to lift and clip into place. The BSA was used in comparative trials of rifle-

The BSA 0.5 in. The Air Service version had a spade grip and cooling holes in the barrel sleeve.

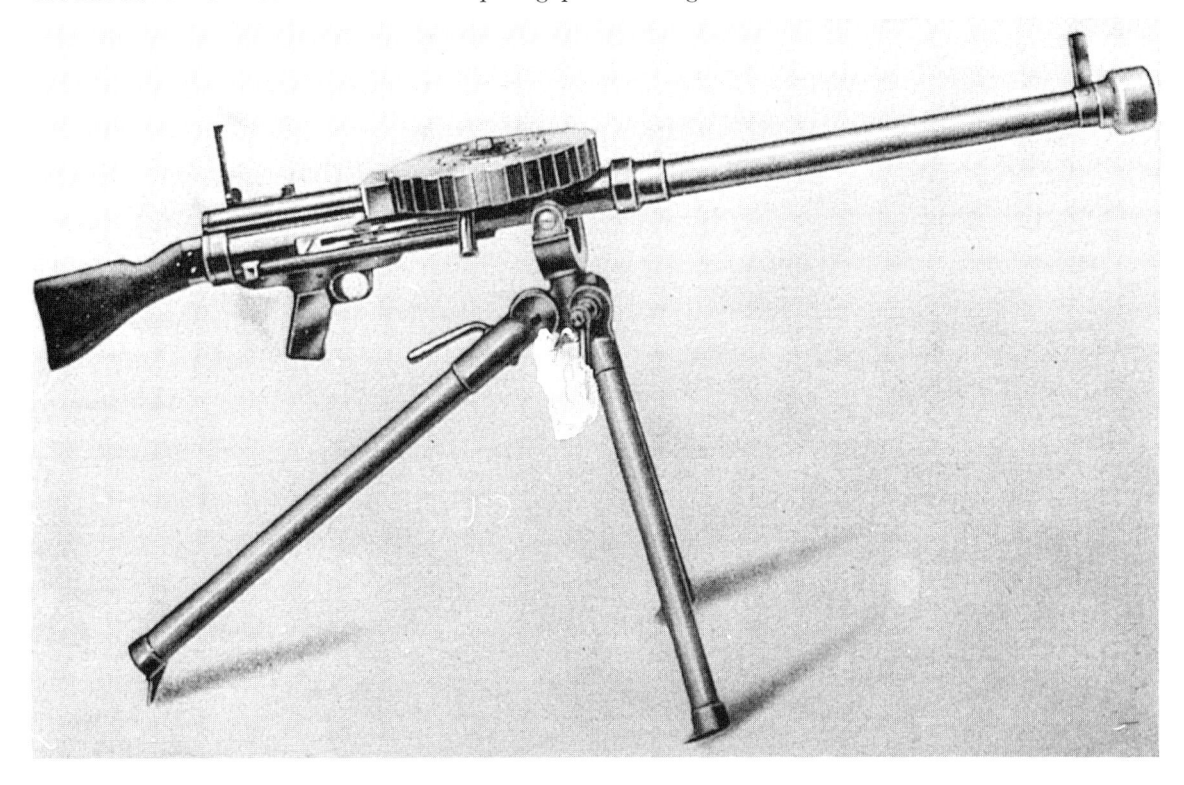

calibre and 0.5 in guns carried out by the RAF over six years. The result of the trials led to the rejection of the heavier-calibre guns. Given that they took place in the twenties, the result was probably inevitable, even with the excellent Browning 'fifty caliber' available, the US Air Corps also used mainly 0.30 calibre guns at this time.

Operation

To fire the gun, the magazine was placed on its post, with the first cartridge positioned at the cam mouth of the magazine centre. The cocking handle was then pulled to the rear until the bolt engaged the sear holding the recoiling mechanism in a spring-loaded position, and the selector was placed on automatic. When the trigger was pulled, the bolt went forward, a projection carried the round from the magazine into the breech, the firing pin was tripped by the rotation of the bolt sleeve, locking it to the barrel, and the pin was allowed to fly forward and fire the round. Recoil carried the barrel and barrel extension, bolt and extension rods backwards 73 mm (2.87 in). At this point the bolt was revolved by a cam, unlocking it from the extension and barrel, which then returned to its forward position. The bolt then ejected the spent case downwards, and the bolt returned ready for the next round. The gun compared poorly with the Browning, which could have been adopted had it been decided to proceed with heavy-calibre armament.

Details of the BSA 0.5 in

Bore	0.50 in (12.7 mm)
Action	Recoil
Cyclic rate	400 rpm
Weight	46 lb (20.9 kg)
Muzzle velocity	2,600 ft/sec (792 m/sec)
Ammunition feed	37-round drum
Cooling	Air
Length	54 in (1,372 mm)

The Vickers 0.303 in gas-operated Mk I No. 1

This gun had several official and unofficial titles. Its official service title was as shown above, Vickers called it the 'Class K', but in the RAF it was usually referred to as the VGO.

In the mid thirties, gunners in the open cockpits of RAF and FAA multi-seat aircraft were still armed with the Lewis gun, but it was becoming obvious that this would have to be replaced with a more up-to-date design. Trials had just been carried out for a fixed gun, which resulted in the selection of the Colt Browning, but it was decided that a second gun was needed for the free gun position of bomber aircraft, and also to provide a 'second string'. At Martlesham Heath in September 1935 six types of gun were submitted for what could have been highly lucrative production contracts. The final choice was narrowed down to between the French Darne and the Vickers Class K. The French Darne was a belt-fed,

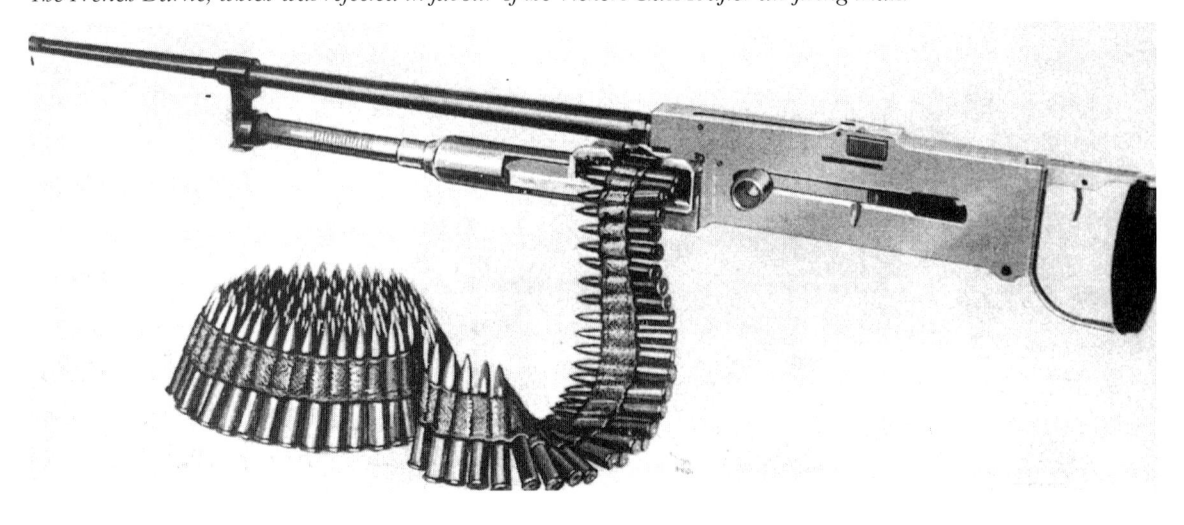

The French Darne, which was rejected in favour of the Vickers Class K after air firing trials.

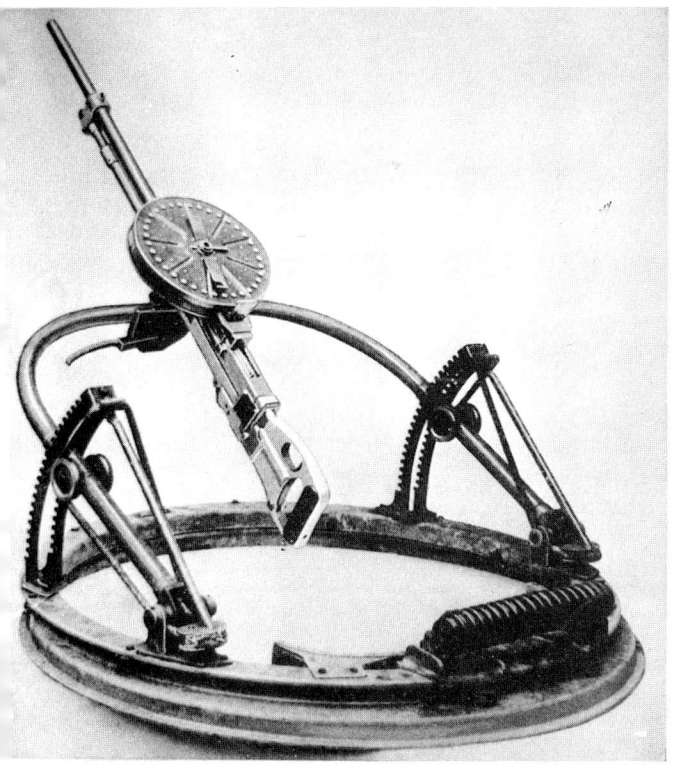

The Berthier gun from which the Class K was developed.

gas-operated gun which, because of its short bolt stroke, had a cyclic rate of 1,700 rpm. The Vickers was a sturdy drum-fed gun, ideal for pivoted mountings, and because it was a simpler design, it was easier to service. Owing to the disagreement among the evaluation experts, Air Vice Marshal Dowding decided to conduct air firing tests personally, and Vickers emerged the winner. The Crayford works were given an initial order for 3,000 guns, and the first 200 were delivered in 1937.

Design history and operation
Foreseeing the need for a light machine-gun for infantry, Vickers had acquired the rights to a design by the French lieutenant André Berthier in 1918. However, in 1934 the British Army selected the Bren gun instead, which was based on a Czech design. The Vickers weapon had been found to be light and easily manoeuvrable, but it was prone to component failure and had a rather slow rate of fire. Vickers decided to redesign it, and a team led by Percy Higson produced a better weapon with a cyclic rate of 1,050 rpm. As the Class K, it was then ready for the 1935 RAF trials, and was given a very favourable report by the two principal armament experts at Martlesham Heath, Major H. S. V. Thompson and Captain E. S. R. Adams.

The gun was cocked by pulling back a handle on the left side. The breech-block was retained at the rear of the receiver on a projection at the back of the piston rod. When the trigger was pressed, the rod was driven forward by the force of the main spring, carrying with it the breech-block, which pushed a cartridge from the magazine into the chamber. As the piston continued to move forward, the rear of the breech-block was engaged on a sloping projection on the rear of the piston, and was forced in front of a locking shoulder on the main body. The floating firing pin was then struck by a projection on the rear of the rod, firing the cartridge. The gas pressure impinged on the head of the piston housed beneath the barrel, driving it to the rear, compressing the main spring and unlocking and withdrawing the breech-block. The rearward movement of the block extracted the case and ejected it into a container at the side of the gun. A safety catch was incorporated into the hand grip which rendered the gun safe when not in use. After final assembly at Crayford, the guns were proof-fired with ammunition containing 25 per cent more charge than normal. They were then tested for automatic fire, first in a horizontal position, then at an angle of 90°.

The pistons in the early production guns tended to fail after firing 1,000 rounds or so, but Higson was able to correct this and other small faults. The modified gun proved to be highly efficient, the minimum life of any component being 10,000 rounds – a great improvement over the Lewis gun, for which replacement parts had to be carried in the aircraft. The 47-round magazine was dropped in favour of a 100-round drum, which gave a more realistic ammunition supply. Higson stressed the modest recoil, the absence of external moving parts, and the fact that the gun could be dismantled in a few seconds with no other tools than an empty cartridge case and a pen-knife.

Initially, armourers had trouble tensioning the spring of the 100-round magazine, but limiting the load to 97 rounds overcame the problem. Stoppages did occur, the most frequent cause being defective ammunition and badly filled magazines. Lack of maintenance could lead to short or broken firing pins, defective extractors and broken springs. Perhaps the most disturbing experience was not a gun which stopped firing, but a gun which would not stop! The following is an account by a trainee air gunner at Picton, Ontario, in 1941:

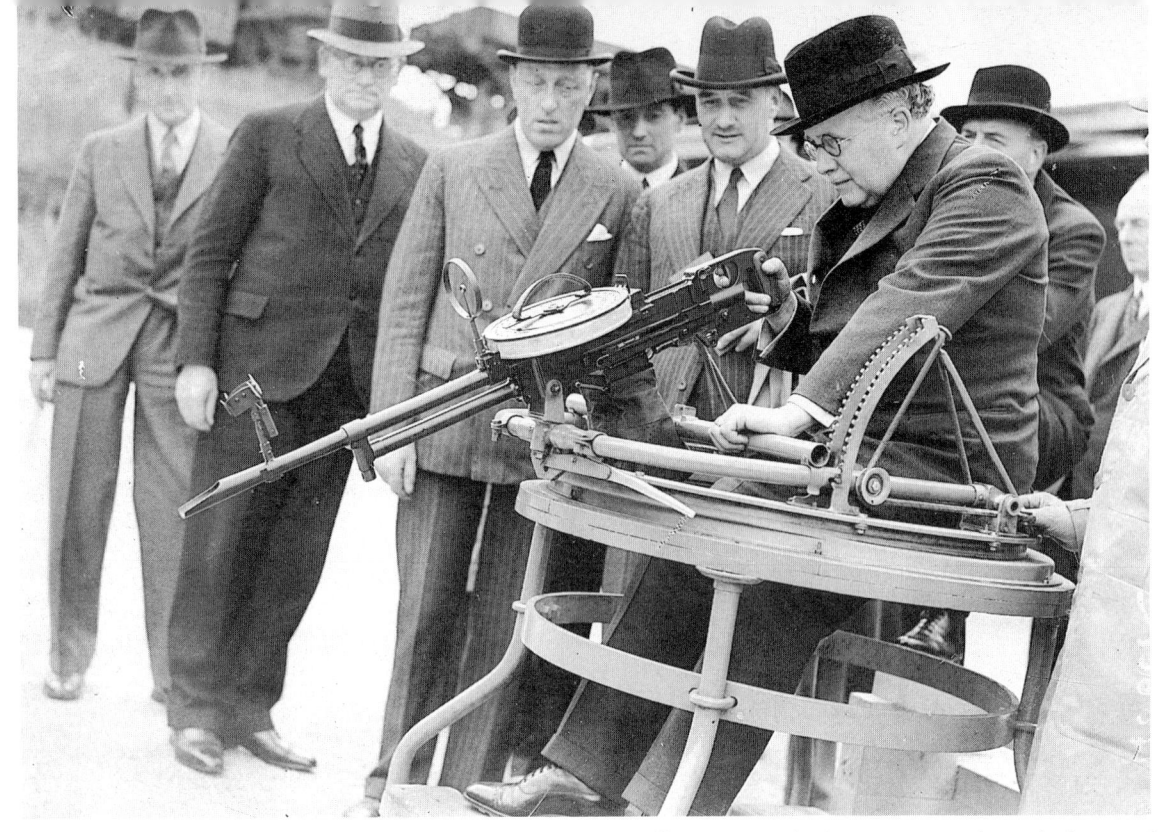

Sir Kingsley Wood, Air Minister, manipulating a Vickers on a Scarff ring at Crayford in 1938. Note the early 47-round ammunition drum. (Vickers)

The nearest thing I have ever experienced to pure flying was air firing from a Fairey Battle. The rear cockpit hinged up and gave some protection from the slipstream, but if you stuck your head out clear of the hood without goggles or helmet, the slipstream roared in your ears and blew the water from your eyes. Air firing took place over water, and as no inter-phones were fitted the pilot would waggle his wings when he turned towards the land; at this signal all firing had to stop. I had just put in a burst at the drogue when the wings waggled. I released the trigger, but the gun kept firing! I held on to the bucking VGO whilst it continued to blaze away the whole drum of ammunition. Meanwhile the pilot, not knowing what had happened, waggled his wings more

A cutaway Class K in the MOD Pattern Room at the Royal Ordnance Factory at Nottingham (formerly at Enfield).

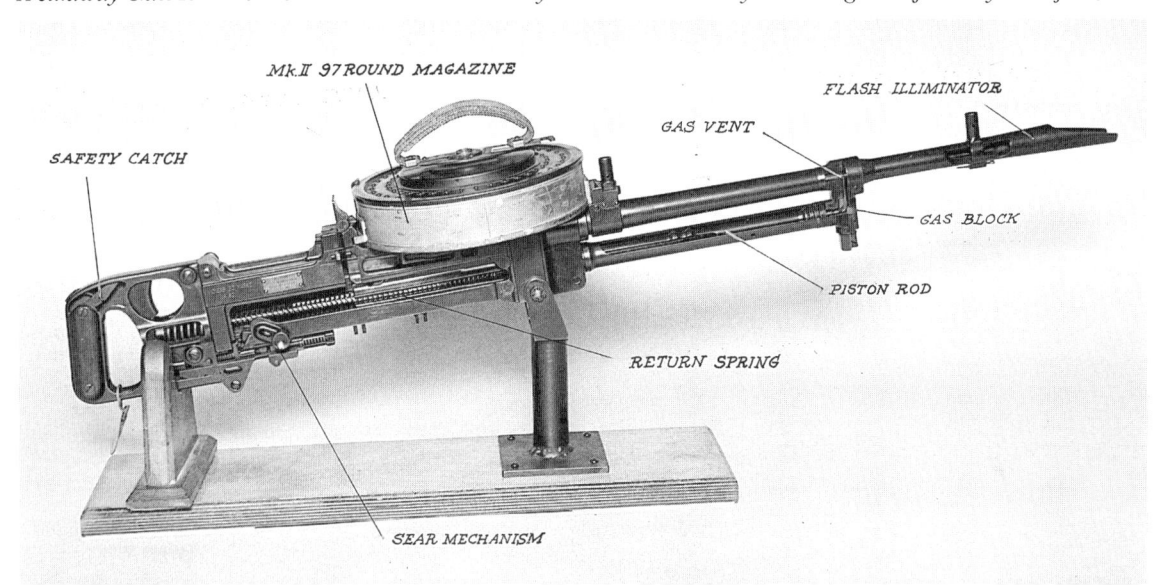

Typical mounting of the Vickers K, shown here in the cramped bomb aimer's position of a Handley Page Hampden.

violently, which sprayed lead dangerously near the towing aircraft. More experienced gunners would have removed the ammunition pan, but the experience left me shattered.

The gun proved so successful that a cable-operated version was specified for all turrets mounting a single gun and (in the case of Blenheims and Beauforts) twin guns. Vickers produced a belt fed fixed version, for which ammunition was supplied from 300- or 600-round tanks. Some early Blenheim 1 aircraft had the fixed version of this gun, but almost all were used on pivoted mountings and turrets. The Bristol Bombay nose and tail turrets each had a single gun, as did the front turrets of the Whitley, Sunderland and Lerwick flying-boats. When the gun was fired from these turrets the piston rods often broke. It was eventually found that the part of the gas cylinder outside the turret was chilled by the slipstream, whereas the section inside the turret soon reached a high temperature after a few bursts were fired, causing the piston to seize. The cylinder diameter was increased but, although this reduced the failure rate, gunners in No. 5 Group Wellingtons were advised to carry spares for the side hatch guns. Being self-contained and needing no belt boxes, the gun was used in the nose position of all later Halifax bombers, and Sunderland and Beauforts used it as additional armament to power turrets. Pivoted guns were usually sighted by 2 in ring-and-bead sights, but Mk III reflector sights were also used on both free-mounted and turret installations.

Many army units used the VGO. The Long Range Desert Group of the 8th Army found that it was far less liable to stoppage than the Bren gun, and it was also used by the Indian Army, who were reluctant to change over to the Bren. The VGO served with the RAF into the 1950s, being ideal in positions where ammunition belts and boxes could not be accommodated.

Details of the Vickers 0.303 in gas-operated

Bore	0.303 in (7.7 mm)
Action	Gas-operated
Cyclic rate	950–1,100 rpm
Weight	20.5 lb (9.3 kg)
Muzzle velocity	2,400 ft/sec (732 m/sec)
Ammunition feed	97-round drum
Cooling	Air
Rifling	Five grooves, left hand
Length	40 in (1,016 mm)

0.303 in Browning

When the Second World War began in 1939, the main gun fitted to RAF aircraft was the 0.303 in British version of the Colt Browning machine-gun. It was to prove reliable and efficient, justifying the Air Staff's decision to adopt it in 1934.

The history of this gun started in 1890, when two brothers, John and William Browning, of Ogden, Ohio, sent a letter to the Colt Fire Arms Manufacturing Co. of Hartford, Connecticut, asking if they would be interested in an automatic gun they had completed of their own design. It would, they said, 'shoot a government cartridge 0.45 about six times a second, and with mount weigh about 40 lb. It is entirely automatic, and can be made as cheaply as a common sporting rifle. If you are interested we would be pleased to show you how it works.'

Colt signed an agreement which was the start of a long association with Browning designs. In 1893 the gun was submitted for trials at the request of the US Navy Department. Three other guns were also being evaluated: the French Darne, the Maxim and the improved 16 mm Gatling. The Colt Browning was judged to be the superior weapon, and an initial order was issued for 50 guns, provided Colt 'could guarantee perfect operation with rimless cartridges and a maximum uninterrupted speed of 400 shots per minute'. The gun entered naval service as the Colt Browning Model 1895. It was soon dubbed the 'potato digger', owing to the action of the gas actuating arm which moved in an arc under the barrel. After modification in 1902, it was sold in great numbers in South America and Europe. In Britain, one was mounted in the Grahame-White Warplane exhibited at Olympia in 1913, and two or three were mounted on British military aeroplanes at the beginning of the First World War.

Constructed of less than 100 parts, the Colt Browning was gas-operated and the cartridges were fed from a canvas belt. There was so little vibration and recoil that it could be fired from the shoulder. One of the few faults to develop was that, after prolonged firing, the barrel and breech reached such a high temperature that the next round in the chamber might 'cook off'. When this happened the round detonated and the breech assembly was shat-

John Browning with his Model 1895 gun, the 'potato digger', which was used on some Allied aircraft in the early war years.

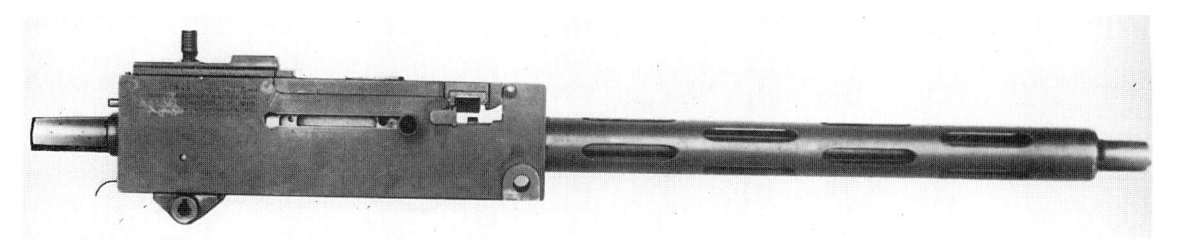

The Colt Model 1918, which was tested in a Bristol Fighter in 1918.

tered. A similar fault occurred when the British used cordite-filled cartridges in their quite different Browning guns 30 years later.

The US Navy were enthusiastic about their Model 1895 guns and increased their orders to several thousand, but the US Army rejected the weapon, preferring to keep their carriage Gatlings. It then became obvious that the unwieldy Gatlings would not be suitable for modern warfare. When the USA entered the war in 1917 the US Army had virtually no modern machine-guns: the inventory was 670 Benet-Mercies, 282 Model 1904 Maxims and 158 Brownings. This compared with the 12,500 Maxims with which the Imperial German Army had entered the war three years earlier.

In May 1917 a new design of water-cooled but recoil-operated Browning was tested at the Springfield Armoury. It fired 20,000 rounds of 0.30 in without a single stoppage, and was adopted by the US Army as its standard infantry weapon. Designated the Browning Model 1917, it was produced by Colt, Remington and Westinghouse. By the end of hostilities 43,000 had been made, but hardly any saw active service; in fact, 23,000 had a fault and were returned to the makers. An air-cooled version, the Browning Model 1918, was designed for use in aircraft.

British interest in the Browning began in July 1918, when a Model 1918 was fitted to a Bristol fighter and was found to be superior to the Vickers on several counts. The preliminary report (C186) by Major H. S. V. Thompson reads:

Subject: 0.300-cal. Browning gun Ref. APT/51/374, 16/7/1918.

This gun was brought here by Major Malony and Lt Wheeler of the US Army, and Major Hayley of the US Air Arm. It was fired on the ground on that date [16 July 1918] in the presence of General Salmond. About 400 rounds were fired, part of which were fired on automatic and part with CC gear fitted on a stand. The gun fired extremely well during the preliminary trials, having no stoppages or malfunctions.

The gun was then fitted to a Bristol fighter 4714 (Rolls-Royce Falcon) and an ammunition box and chute were later fitted to the gun. A blast tube of welded mild steel which projected through the cowling was also made up and the gun operated synchronously by trigger motor, mounted on the left side of the gun. This was connected up to the pipeline of the standard B-type CC gear of the aircraft.

The report was not acted upon. With the end of hostilities, money for new weapons was not available, and the huge stocks of Vickers guns still held precluded any thought of change. After studying various operational reports, the Air Staff decided to evaluate the merits of heavy-calibre guns, and because there was no great urgency, it was 1929 before the results were announced.

Meanwhile, in 1926 Armstrong Whitworth acquired the British rights to the Model 1918, and persuaded the Air Staff to evaluate six guns which had been produced in Coventry. The guns were not actually tested until 1929. The Air Staff decided that the advantage in range and hitting power of the 0.5 in gun was outweighed by the faster cyclic rate and lighter weight of the smaller weapon. Armstrong Whitworth therefore looked forward to a substantial order. In June 1931 two of the guns were mounted in an Armstrong Whitworth Siskin. The air firing trials confirmed the makers' claims, and thousands of rounds were fired without a single stoppage.

In 1933, at the request of the US Army Air Corps, two 0.300 in versions of the improved 0.5 in M2 gun were produced. The first was a pilot's gun, while the second, the M9/402, was designed as a pivoted observer's gun with a higher rate of fire and longer barrel length. Just after these guns had been produced, the RAF decided to hold competitive trials to

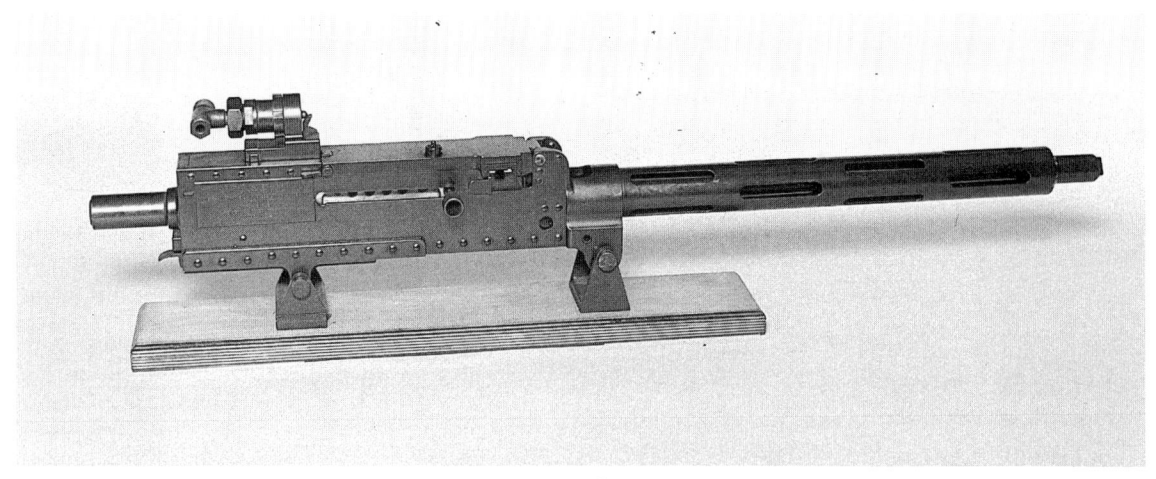

A Model 1918 made by Armstrong Whitworth. The gun was fitted to an Armstrong Whitworth Siskin for trials in 1931. Colt insisted that all licence-made guns should be stamped 'Invented in the U.S.A.'. (MOD Pattern Room, Nottingham)

select a modern automatic gun. The guns tested were the Vickers, Hotchkiss, Darne, Madsen, and the Colt MG40 and MG 40/2. The winner was the Colt 40/2, which proved to have the best all-round performance.

Once the gun had been selected, the Martlesham Heath gun section under Major Adams conducted Service trials. It was found that the cordite-filled 0.303 in cartridges used in Britain caused serious trouble (most countries used nitrocellulose propellent, which was less sensitive to heat than cordite). When a long burst was fired a round remained in the chamber, and the cordite then detonated as in the Model 1895. Major Adams redesigned the action to hold the breech-block to the rear with the chamber empty. The first trials of production guns from BSA showed a weakness in the feed. This meant a further extensive redesign, until the final

The two Colt Browning guns submitted for the RAF competitive trials of 1933. The successful weapon was the MG40/2, shown here in fixed and pivoted mounting patterns.

FIXED MACHINE GUN
(Including fixed back plate and operating slide group assembly)

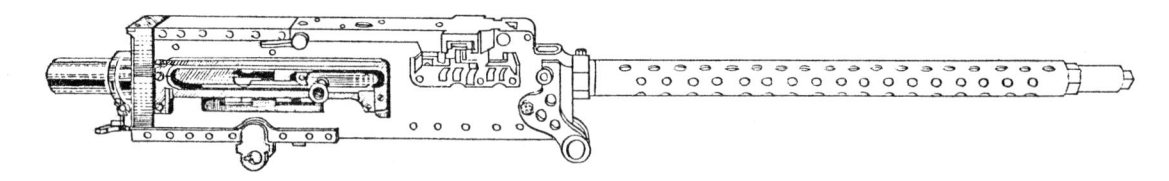

FLEXIBLE, WITH FIXED BACK PLATE, MACHINE GUN
(Including fixed back plate and retracting slide group assembly.) Used in turret installations.

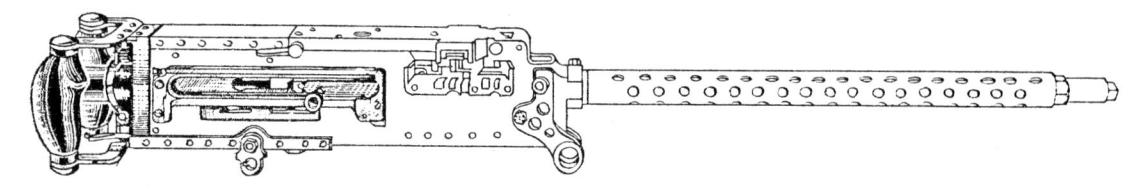

THE GUNS

gun was quite different from the MG40/2.

The Browning gun was the first in RAF use to have the facility of adjusting the barrel in relation to the breech-block. Some armourers adjusted the barrel too far forward, leaving too much of the case protruding from the barrel, so that the end of the round was blown off causing a 'separated case' stoppage. With experience this problem was overcome, and during the Battle of Britain, if a fighter returned from a sortie with a separated case stoppage the armourer responsible was put on a charge. Trouble was also caused by excessive fouling of the muzzle attachment, the guns seizing after about 200 rounds. A sharp pen-knife seemed the best way to clear the hard residue. In 1940 BSA redesigned the muzzle attachment by adding cooling fins and chromium-plating the bore of the unit. This modification caused a hold-up in production at a vital period, but the gun could then fire 300 and more rounds without fouling. After the troubles were rectified, production at BSA, Vickers Armstrongs and sub-contractors kept up with the demands of the Service (one Hurricane and Stirling needed 16 guns).

The Browning was a recoil-operated gun with safety features which ensured near trouble-free operation. The gun was fired when the rear sear was depressed. This was done by hand, or by pneumatic, hydraulic or electrical solenoid actuation, depending on the installation. With sufficient maintenance, malfunctions were minimal. The most usual stoppage was caused by rogue ammunition or badly made-up belts, though the links would also sometimes jam in the ejection outlet. Turret gun-

Above *Mk I Browning guns with the original muzzle unit (which proved unsatisfactory) in a Boulton Paul Type A turret.*

Below *The modified muzzle attachment developed by BSA. The bore was chromed and cooling vanes were added.*

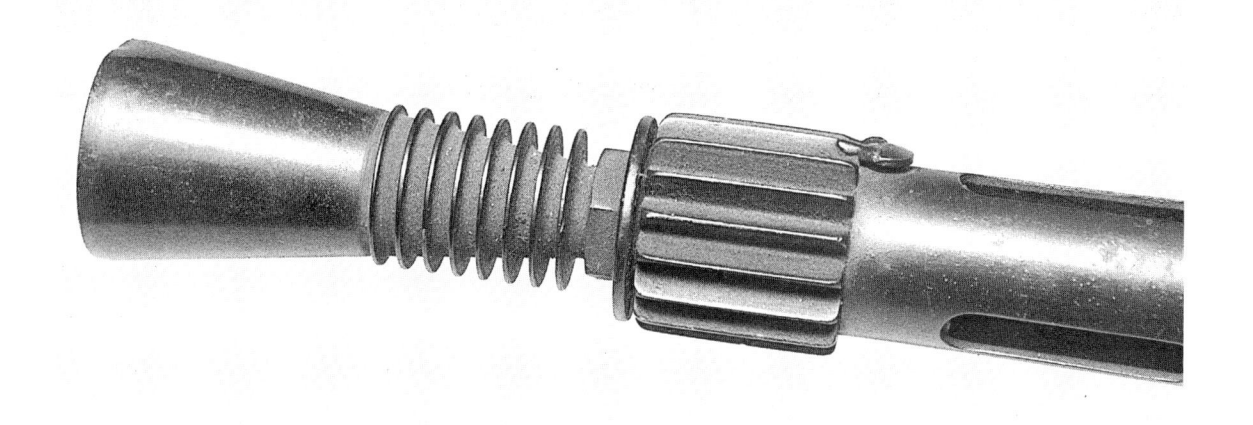

Brownings being fitted to the FN4 turret of a Whitley at Driffield in 1940.

ners could clear the stoppages with a hooked tool kept handy in the turret, and gunners also kept a looped wire or hooked metal cocking tool to clear the gun. Stoppages were reduced after special belt-making machines were introduced (see below).

Very cold conditions could also lead to problems. Heaters were provided in fighter gun bays, but oil spillage in some power turrets made heaters a safety hazard. Anti-freeze oil helped, and one Bomber Group fixed a paper seal over the cartridge ejection slot to stop the fierce draught which could enter the aperture. The 'fire and safe' unit mounted on the side of the gun body was operated by a pneumatic actuator on the same pressure line as the sear release unit on fighter aircraft. Turret guns were fired by hydraulic units or electrical solenoids controlled by triggers or push buttons on the turret control handle. The guns were made safe by press-

Cocking handle (top) and link clearance tool used by air gunners in turrets armed with Mk II Brownings.

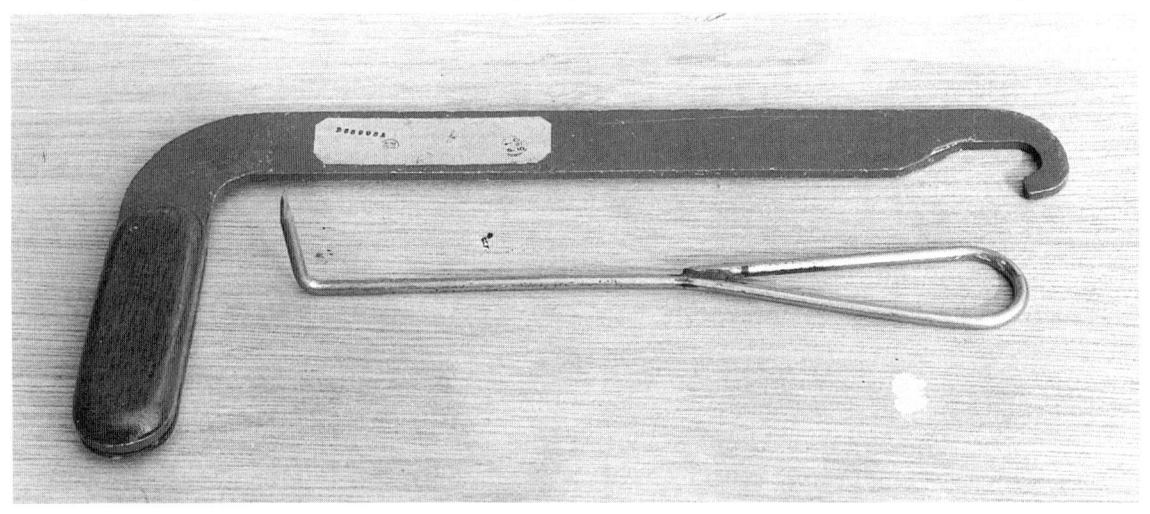

THE GUNS

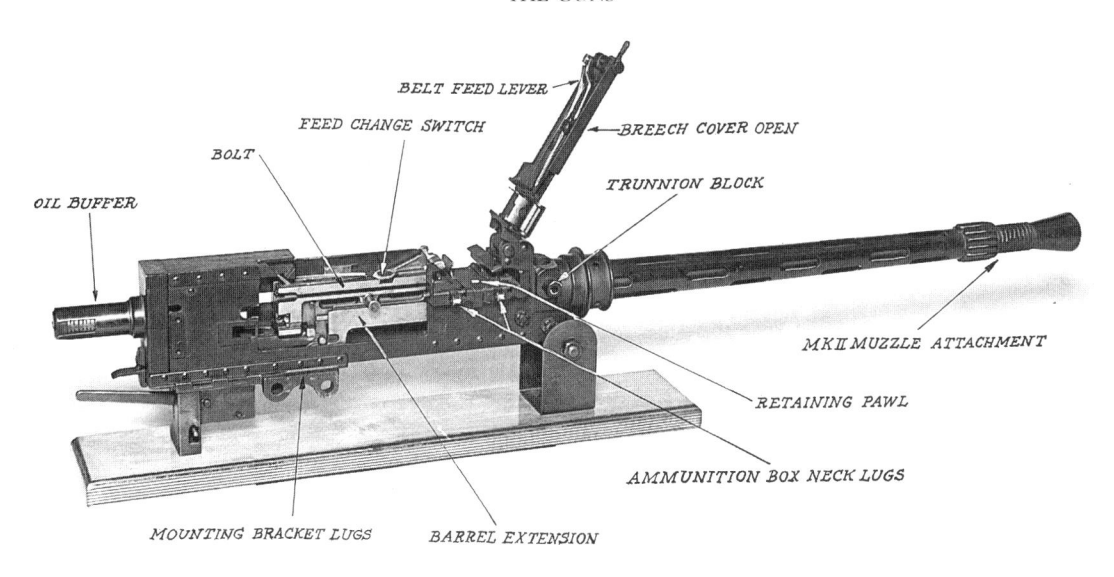

Cutaway of the Mk II Browning. (MOD Pattern Room, Nottingham)

ing a release pin at the back of the fire and safe unit.

Each gun was marked with its number and Mark designation. Marks I and II were almost identical, having the early muzzle attachment; Mk II* guns were fitted with the BSA modified unit.

The Browning was rarely used as a free-mounted gun, except in Beaufighter TFXs of Coastal Command, where the observer's cupola was so small that the ammunition drum of a Vickers could not be accommodated. Otherwise, when a free-mounted gun was needed, the Vickers K was used.

Ammunition belts

Browning ammunition belts were made up in the early war years by armourers using hand-wound belt-making machines. On some occasions the bullets were even inserted into the links by hand. The make-up of the belts varied with the unit: in most Bomber Command Groups one in five rounds was tracer, but AP and incendiary rounds were often included, especially in fighter units. When the Plessey electrically powered belt-making machine was introduced, the work of the armourer became a little less irksome. The bullets were fed into a hopper one end, and when the machine was switched on a continuous belt was produced from the other.

Browning gun production

A total of 460,000 complete Brownings were manufactured in the UK with spares for another 100,000. Most of these were manufactured by BSA but an extensive sub-contract scheme was set up in 1941, BSA supplying key personnel and supervising the work. Series production started in 1938, when 3,809 were produced, and production figures rose until in March 1942 16,300 were completed. Production contracts were completed by the end of 1944, and in 1945 the lines were closed down.

Make-up of the 0.303 in cal. disintegrating link ammunition belt. The belt is held together by the rounds.

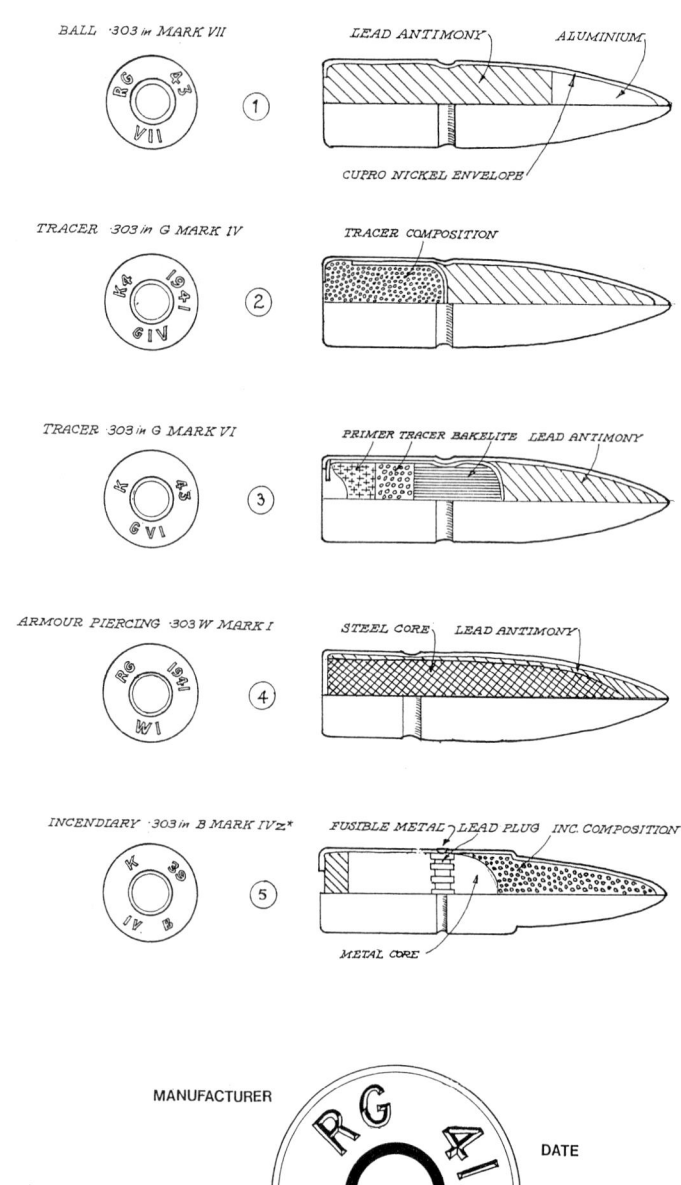

Construction of some British rifle-calibre bullets (The circles on the left show the base markings for each type of ammunition). 1. The Mk VII ball cartridge introduced in 1910 which remained in use until 0.303 in cal. weapons were phased out. As in all rounds, the nitrocellulose-filled rounds had the code letter Z added. 2. The Mark IV tracer was approved for service in 1940 as the G Mk IV. It was a short-range air-to-air tracer based on the earlier G Mk 2 round. The first model burned to 400 yd, later increased to 600 yd. 3. The Mk VI tracer was an air/day round which traced to 550 yd. 4. After many experiments with hardened cores, the basic W Mk 1 armour-piercing bullet was found to be more easily manufactured and gave good penetration. 5. The B Mk IVZ* incendiary was derived from the First World War Buckingham design. The incendiary mixture in the nose spurted from holes revealed when the lead plug melted. Other incendiary rounds were the Mks 6 and 7. Enlargement of typical ammunition base marking. Note: *German rifle-calibre aircraft ammunition belts contained no ball rounds.*

Details of the Browning 0.303 in

Bore	0.303 in (7.7 mm)
Action	Recoil
Cyclic rate	1,150 rpm
Weight	21 lb 14 oz (9.9 kg)
Muzzle velocity	2,660 ft/sec (811 m/sec)
Ammunition feed	Metal links
Cooling	Air
Rifling	Five grooves, left-hand twist, 1 turn/10 calibres
Length	3 ft 8 in (1,130 mm)

Identification of .303 ammunition used in the Browning and Vickers K

Service ammunition is identified in several ways, viz.: 1. Labels on the container; 2. A code stamped on the base of the cartridge case; 3. Coloured dye on the annulus of the round (centre of the base); and 4. Colouring of the bullet tip (1939–45).

The *base marking* gives the main details. These consist of: A. Code initials of manufacturer; B. Year of manufacture; C. Type of propellent (usually only Z for nitrocellulose); D. Mark of cartridge; and E. Type of bullet. The *annulus colour code* is found in the centre of the base, and signifies the following: Black: Ball until 1918; Purple: Ball after 1918; Blue: Incendiary; Orange: Explosive, including PSA, Red: Tracer.

During the 1939–45 war, station armourers needed a more instant way of identifying special ammunition. The method adopted was to colour the tips of the bullets, the code being as follows: Blue tip: some marks of incendiary; White tip: air-to-air short-range day tracer; Grey tip: air-to-air short-range night tracer; Black tip: observation bullet; No colour: ball.

Head Stamps

Each manufacturer was given a code to be used on the head stamp on the base of the round. The main makers were: BE – ROF Blackpole (1939–45), CP – Crompton Parkinson, E – Eley, FN – Fabrique Nationale, K – Kynock (ICI) – at various factories (K2, K3, K4, K5), RG – ROF Radway Green, RL and RG – Royal Laboratory Woolwich, RW – Rudge Whitworth, SR – ROF Spennymore. *Bullet types* were also shown on the headstamp, as: AA – PSA or Pomeroy, B – Buckingham explosive, F – semi-armour-piercing, G – SPG tracer, K – Brock incendiary, R – Explosive from 1918, T or G – Tracer, W – Armour-piercing, Z – Nitrocellulose propellant after 1917. Ball ammunition was stamped VII with no letter.

The Hispano Suiza

During the First World War it was recognized that, if large-calibre guns firing explosive ammunition could be mounted in aircraft, a single hit might destroy any target aircraft. Many experiments were carried out along these lines by the major powers. In Britain, the C.O.W. $1\frac{1}{2}$- and 1-pounder, the Vickers $1\frac{1}{2}$-pounder and the Davis gun were among those tested, but they were not adopted for general use. Such guns were too heavy for the frail airframes of the time – indeed when a Vickers $1\frac{1}{2}$-pounder was fired from an early Short floatplane, the aircraft literally stopped in the air and stalled. The French were probably the most successful in this field, a Swiss engineer, Marc Birkigt, founder of the Hispano Suiza Automobile Co. mounted a 37 mm (1.443 in) gun to fire through the hollow propeller shaft of one of the company's aero engines in 1917. Two French aces, René Fonck and Georges Guynemer, shot down several German aircraft using this weapon. The gun was loaded with canister shot containing 24 half-inch steel balls – one well-aimed shot was usually enough for a kill.

In 1930 the Swiss Oerlikon company designed a 20 mm (0.78 in) cannon to be mounted between the cylinder banks of geared vee-type aero-engines, the engine absorbing the recoil forces. The gun fired through the propeller shaft as in the HS (Hispano Suiza) wartime mounting. In 1932 Hispano Suiza purchased some Oerlikon guns for use in conjunction with their HS.12x aero-engine, but in spite of much development work the project was abandoned owing to problems with the gun. The company then decided to design a new 20 mm gun based on the action of the German Becker cannon of 1918, and after early design faults were sorted out the new gun, designated the Hispano-Suiza Type 404 'Moteur Canon', proved to be an outstanding success. A production contract was received from the French Government, and the first aircraft to be fitted was the new low-wing fighter, the D510, one Hispano being mounted in each wing outboard of the propeller disc. The performance of the gun was so good that the air arms of all the major powers evaluated examples for possible licensed production. In 1935 the British Air staff decided that sooner or later a heavy-calibre gun would have to be introduced to counter the introduction of armour in future aircraft. The Air Ministry Gun Section advised the adoption of the Hispano 'Moteur Canon' as no British design was available. The Air Staff visited the Hispano works in Paris, where a demonstration, accompanied by the head of the Air Ministry gun

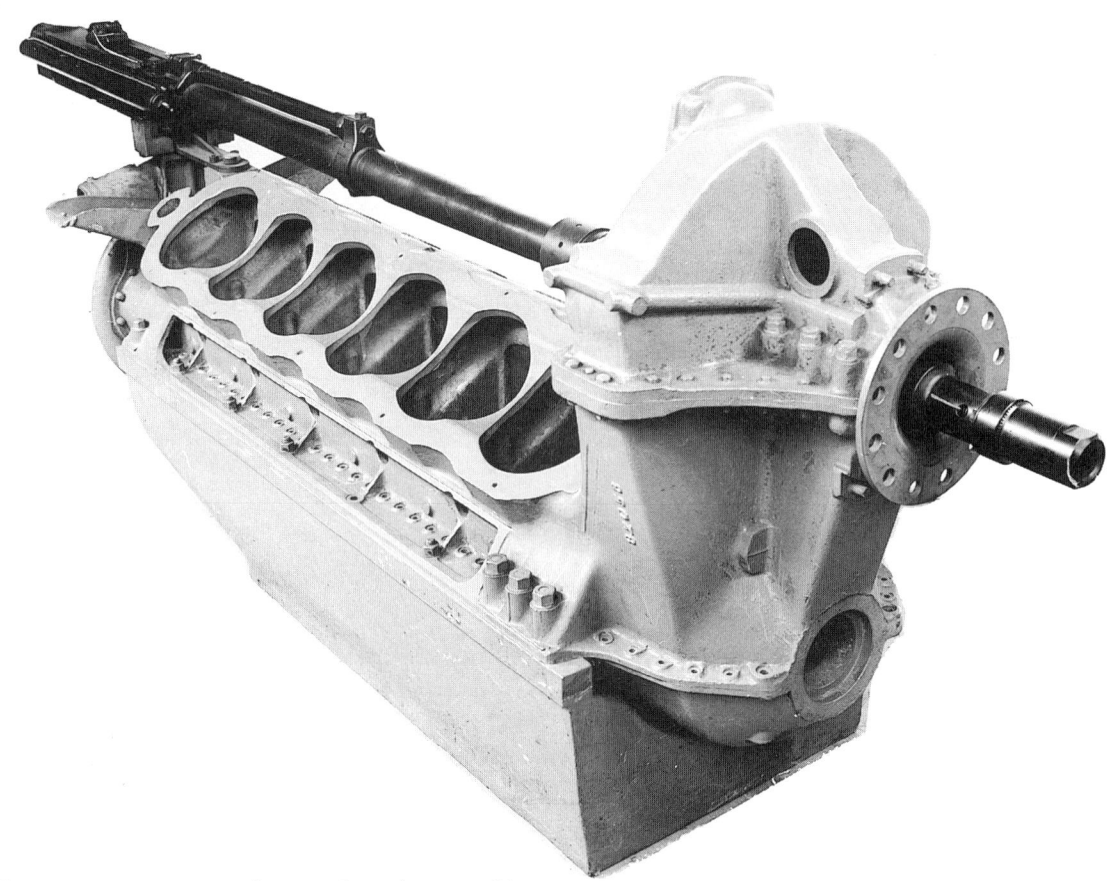

Above *A Hispano-Suiza 'Moteur Canon' mounted in the cylinder block of an HS12Y aero-engine.*

Below *The enormous increase in destructive power over the 0.303 round can be seen when the two projectiles are compared.*

section, Major H. S. V. Thompson, convinced all present that the Hispano should be adopted by the RAF, and an order was placed for six guns. The Section had conducted a series of tests on a specimen gun during the previous year, and several features, including the positive locking of the breech when the round was fired, had convinced them that the weapon was superior to other weapons with blowback actions.

Briefly, the action sequences was as follows: As the breech-block was released by the sear it travelled forward, feeding a round into the chamber. The breech was then locked and the round fired. Diverted gas pressure then acted on the piston, mounted over the barrel; this unlocked the breech, and gas thrust on the case forced the breech-block to the rear, the empty case being extracted from the chamber and ejected. A pneumatic recocking unit, operated from the aircraft's system, was built into the gun body. The gun was fitted with a recoil reducer and a heavy spring, damped front mounting unit. The firing sear was released by Bowden cable, pneumatic- or solenoid-operated unit under the rear of the gun body.

When the decision was taken to accept the gun as

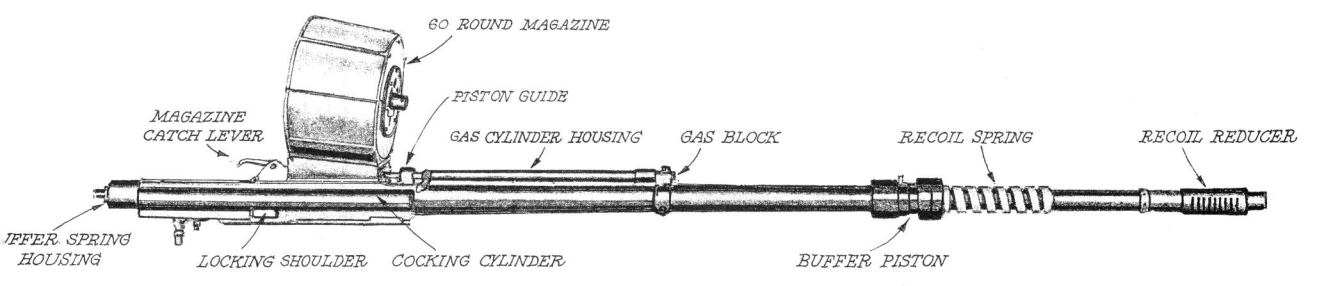

The Hispano Mk I fitted with a 60-round magazine.

a possible standard weapon for RAF fighters, the Air Staff realized that it would have to be manufactured in the UK. As the only gun-producing companies were fully committed to other weapons, it followed that new production facilities would have to be provided. After initial reservations, the Hispano Suiza company finally agreed to form a British subsidiary company to manufacture the gun in England. A factory was built at Grantham which was to be known as The British Manufacturing and Research Company (the title was intentionally vague for security reasons). All dimensions were metric, and engineers from France helped to train the operatives. The first guns were proof-fired in January 1938. Several faults were found, some due to the inexperienced workforce, but others due to problems with the design which had not shown up on the hand-built guns tested in Paris. The return spring was prone to breakage, the extractor spring had a very short life, and the breech-locking system needed redesigning. Whilst these problems were being investigated at the Chatellerault works of Hispano-Suiza, Captain E. S. R. Adams, Senior Technical Officer of the Directorate of Armament Development, and experts from BSA investigated all aspects of the project. It was agreed that it would be very difficult to manufacture the gun at a normal engineering works if the need arose, and the Air Staff set up a team to produce British drawings and simplify the manufacturing process. Hispano Suiza did not take kindly to this development, but after close co-operation between Captain Adams and Hispano engineers it was agreed to proceed.

Most of the defects had been sorted out by early 1940. Meanwhile, John North at Boulton Paul had made other improvements to the design, and prepared the gun for turret mounting. The Air Staff had originally planned to introduce the gun for the new fighters which were to succeed the Spitfire and Hurricane, but when the Munich crisis occurred the Air Staff suddenly directed that the new fighters coming off the production line should be fitted with the new gun. It was soon realized that the Grantham factory would not be able to produce the vastly increased number of guns required, and top priority was given for four new factories. A second, 'shadow' factory was built at Grantham, another, run by BSA, was constructed at Newcastle-under-Lyme, a new purpose-built factory was specially erected at Poole, and part of the Enfield Royal Small Arms factory was turned over to Hispano production. The first guns were delivered from these factories in early 1941, but the Grantham factory was able to provide several hundred guns for Spitfire Ibs in 1940. These aircraft took part in Service trials, and No. 19 Sqn took part in operations against German raiders with spectacular results, but the gun's début was marred by faults in the ammunition feed, and was withdrawn from squadron use.

Future development work was carried out at the BSA works, where the Air Ministry had set up a drawing office to produce the anglicized Hispano Mk II under the leadership of Captain Adams. The ammunition feed problem was mainly due to the use of a bulky 60-round drum, which could only be accommodated if the gun was mounted on its side. Just before the Germans overran the Hispano factory in Paris, Adams made a risky journey to the works and retrieved a newly designed belt-feed unit and a full set of working drawings. The production contract for the new feed units was given to the Molins Machine Company. This company progressively developed the feed unit until stoppages were brought down to an average of one every 1,500 rounds. The large 60-round magazines were still used on Beaufighter aircraft where the observer could change drums during operations. At last, RAF fighters were being armed with weapons second to none.

Most air forces opted for shell-firing guns in the 1930s because it was thought that explosive shells would destroy an aircraft with very few hits. This was not borne out by events; in fact, trials often showed that solid ball ammunition did as much damage as explosive rounds, which tended to

Final assembly of Mk I guns at the BSA works in Birmingham. (BSA)

explode directly on contact with armour or airframes without penetrating them. The heavy ball projectiles penetrated both, and was the main type of ammunition used in the early war years. However, it was then realized that an ideal round would first penetrate an airframe and then ignite the fuel or

oil inside it. Incendiary ammunition tended to break up, but a composite explosive/incendiary shell was found to be very effective. Known as HE/I, this new round was however less effective against armour. In July 1942 the semi-armour-piercing incendiary (SAP/I) was introduced, with a tungsten nose and a shell containing an incendiary composition. On impact, the tip penetrated the armour, and there was then a flash of flame which ignited anything

Reloading the early Mk I guns of a Westland Whirlwind.

THE GUNS

The Molins belt feed unit (right) increased the ammunition supply and reduced the height of the installation. The Bristol belt feed (left) was neater but was not adopted for Service use.

inflammable within a foot (300 mm) of it. From mid-1942 these two types of ammunition replaced other types, belts being made up equally of HE/I and SAP/I.

Tremendous skill and responsibility was required by squadron armourers. Considerable technical knowledge was essential, and very few stoppages were attributed to poor maintenance or servicing. Many modifications were introduced at squadron level. The recoil distance of the Hispano was critical

Armourers reloading the guns of a Hurricane IIC. The belt-feed units can be seen ready to be fitted into place.

for trouble-free firing. After one round was fired into the stop butts, the recoil distance had to be 20 mm (0.78 in) when cold. This proved to be very difficult to measure, until an armourer discovered that, if a piece of Plasticine was pressed on to the end of the piston so that it came into contact with the front face of the feed unit after firing, the recoil could be measured by using calipers on the resulting indentation.

Until the end of 1941, the British development of the Hispano was led by Captain Adams. It was then thought necessary to promote the younger generation of armament engineers, and the Hispano was passed over to Mr G. F. Wallace, who had co-ordinated the development of the belt-feed mechanism. The weapon was still not ideal for wing installations, the main problem being the length of the barrel. The gun had been designed to fit the Hispano engine, with the breech behind the engine and the barrel projecting through the propeller hub. The long barrel gave increased muzzle velocity, but when installed in fighter wings, some 2 ft (600 mm) projected in front of the leading edge. Also, in late 1941 the cannon turret projects had been belatedly revived, and the need for a shorter gun was urgent.

As a first step towards achieving this, Wallace shortened a MK II by 12 in (305 mm) and test-fired it at the Poole Ordnance Factory. Other than a slight reduction in muzzle velocity, performance was not affected. A small batch of Mk IIs with shortened barrels was supplied to Boulton Paul, Bristol and Parnall for use on experimental turrets. Wallace then made more fundamental modifications, increasing the cyclic rate and reducing the weight of the gun. Although the faster speed imposed more stress on the working parts, it was found that few guns fired as many as 1,000 rounds before the aircraft was lost in action, taken out of service or written off in an accident. Though the specification called for a service life of 10,000 rounds, it was agreed that this could safely be reduced to 2,000 in the interests of performance, the speed being increased from 650 to 750 rpm. Reduction in weight was achieved mainly by deleting the integral cocking cylinder. This gave the pilot a means of recharging the gun in the air, but was rarely used – with three other guns, the pilot was often unaware of the stoppage.

After these modifications the gun was faster, lighter and shorter and in May 1943 a batch of modified guns, known as the Mk V, were tested at Boscombe Down. It was found that the front gun mounting would not stand up to the increased recoil forces. The US Army Air Force had adopted the Hispano for the P38 and other aircraft, and had fitted a very efficient front mounting made by the Edgewater Co. An anglicized version was designed and produced which solved the recoil problem, and the Hispano Mk V was accepted for use on RAF aircraft for 30 years. During the last year of the war aircraft of the 2nd Tactical Air Force alone fired 13,500,000 rounds of 20 mm ammunition at a stoppage rate of one per 1,500 rounds fired, mostly due to badly-made-up belts.

After the war the Mk V continued as the standard gun of the RAF and Fleet Air Arm. The next major conflict was the Korean War, when Royal Navy carriers carried out operations against North Korean targets. When Hawker Sea Furies were returning to one of the carriers from one of these engagements, a serious accident happened. The breech-block in one of the guns had jammed under the base of a round. When the tail hook on the aircraft engaged, the jolt was enough to clear the jam, which freed the action and fired the round. The shell hit the fuel tank of one of the many aircraft on the flight deck, eight other aircraft were destroyed and there were many casualties. Similar stoppages had occurred during the Second World War, but only on airfields where the runway ahead was clear. As carrier flight decks were usually full, an urgent message was sent to the A&AEE Weapons Department at Farnborough to devise some way of preventing a recurrence of the

The Hispano Mk V barrel was 12 in (305 mm) shorter than the previous Marks, and the recocking unit was deleted.

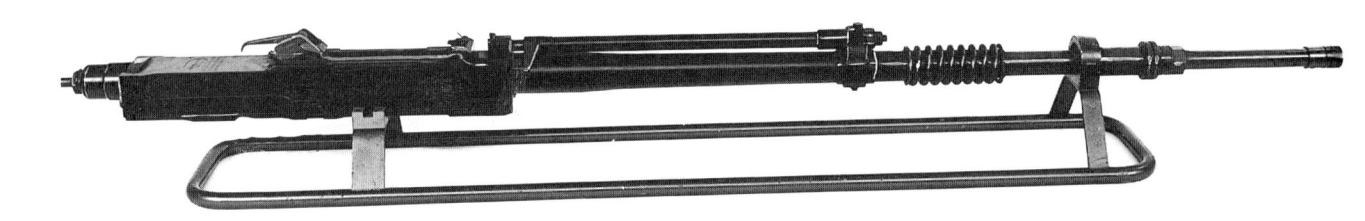

THE GUNS

The Bristol B17 mid-upper turret, which was used on the Lincoln and Shackleton MR1, was armed with twin Mk V Hispanos.

problem. Technicians Shaw and Ainley were told to sort it out – quickly. With the help of the Electrical Section a solenoid-operated pin was fitted to project into the breech-block when the gun button was not being pressed. The pin circuit had a delay of .1 seconds to allow the breech to close and fire the last round. When the action returned to the rear, the pin engaged. The device was completed in ten days, and modification kits were flown out to the carriers.

From 1942 to 1955 the Hispano was the standard fighter gun of the RAF and FAA. No aircraft armour was able to withstand hits from its high-velocity projectiles. Although it was fitted to the Bristol B17 turret and Shackleton nose mountings, it was never fired in anger from a turret.

Night firing tests on a Hawker Typhoon 1B.

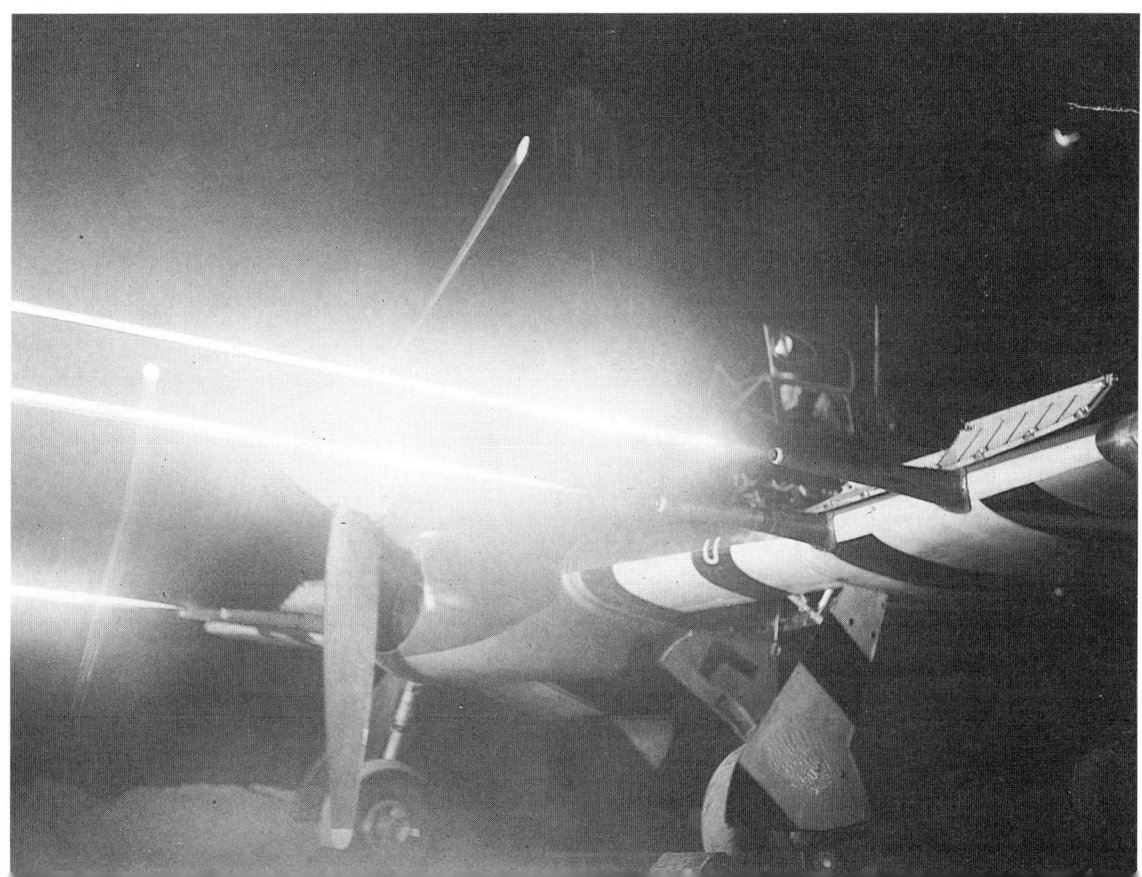

Summary of Hispano aircraft guns made in the UK

Mk I	First model, made to French drawings.
Mk II	Minor changes, made to British drawings.
Mk II*	Mk II with Mk III unlocking plates having no inertia blocks.
Mk III	Enfield designed, prototype only.
Mk IV	Mk II* with barrel shortened 12 in (305 mm).
Mk V	Wallace modification: lighter, with special locking plates, short barrel, cyclic rate 750 rpm.
Mk 6 (Arabic numerals adopted)	Modified to fit American cradle.
Mk 7	Electrically primed ammunition.
Mk 8*	Mk V modified to give 900 rpm.
Mk 9	Mk 8 with electrically primed ammunition.

Projectile identification

Projectile	Colour marking
Ball	Black
Gun functioning	White
Armour-piercing	Black with white tip or white nose rings
Tracer	Black with stencilled red T in inverted circle
Semi-AP/high-explosive/inc	Red with red tip
Semi-AP/high-explosive/inc	Red lower body, buff upper body, white tip
High-explosive	Buff
Incendiary	Red
High-explosive/incendiary	Green upper body, red lower body

Cartridge headstamp identification

BMARC	British Manufacturing and Research
GB	Greenwood and Batley
H	Halls Telephone Co. Ltd
K	ICI Birmingham (K3 and K2 indicate ICI Kynoch)
P&S	Plasters and Stampers.
RC	Raleigh Cycle Co.
RG	Royal Ordnance Factory, Radway Green
RH	Raleigh Cycle Co.
ST	Royal Ordnance Factory, Steeton and Thorpe Arch
JES	Post-war drill/inspection rounds

Details of the Hispano Suiza

	Mk II	Mk V
Bore	20 mm (0.78 in)	20 mm (0.78 in)
Action	Gas-operated	Gas-operated
Cyclic rate	650 rpm	750 rpm
Weight	109 lb (49.4 kg)	84 lb (38.1 kg)
Muzzle velocity	2,880 ft/sec (878 m/sec)	2,750 ft/sec (838 m/sec)
Ammunition feed	60-round drum/ dis. links	Belt-feed (dis. links.)
Cooling	Air	Air
Rifling	Right hand	Right hand
Cocking	Pneumatic charger	By ground crew only
Grooves	Nine at 7°	Nine at 7°

Vickers S gun

In 1936 the Air Staff decided to carry out a series of trials to find the minimum size of shell capable of destroying an aircraft with one hit. After various experiments it was found that any aircraft hit by a shell 40 mm (1.56 in) or over would probably not survive. Once again the 'big gun' lobby in the Air Staff suggested a fighter armed with such a weapon. The COW gun was now obsolete and had other drawbacks, so in 1938 a specification was issued for a 2-pounder gun suitable for aircraft. The obvious choice was Vickers Armstrongs, but at a meeting at the Air Ministry E. W. (later Lord) Hives of Rolls-Royce announced that his company could produce such a weapon in 18 months, so both Rolls-Royce and Vickers Armstrongs received development contracts.

The chief designer, at Crayford, Percy Higson, had foreseen the result of the projectile trials, and made sure by obtaining the conclusions long before they were officially announced (Vickers had informants both in the Air Staff and throughout the international armaments industry). Thus by the end of 1938, only months after receiving the order from the Air Ministry, Higson's gun was complete.

The gun used the long recoil system similar to the COW gun, but there all similarity ceased. The new gun was smaller, had a much faster rate of fire, and

THE GUNS

Vickers-Armstrongs Class S gun.

was fed by a magazine holding 15 of the big rounds. The gun fired the same 2-pounder shells as a much heavier naval gun also designed by Higson. In early 1939 Vickers submitted a scheme for mounting the gun in a large dorsal turret in a Wellington 'heavy fighter' with a predictor and rangefinder. Such an aircraft, it was claimed, could engage hostile formations at a range well beyond that of the fighters' defensive fire.

A prototype gun began testing in 1939, and suffered far fewer teething troubles than was usual with an entirely new gun. In early 1940 the gun was dispatched to Woolwich for Ordnance Board certification, where no faults occurred during extensive testing. A small production order soon followed for the gun, known at Vickers as the Class S. Vickers also went ahead at Brooklands with fitting the prototype Wellington II (*L4250*) with the big mushroom-shaped 40 mm emplacement (see Volume 1).

The company also submitted a fighter design to the F.22/39 specification, mounting an S gun with a limited angular movement in the nose. Another Wellington conversion (Type 439) was to have an S gun in the nose, the gunner having a sighting cupola similar to the turret. As described in Volume I, the S gun was also used in the Bristol B.16 nose turret installed in some Coastal Command Fortress II aircraft for anti-submarine operations.

After the fall of France in June 1940 it was obvious that some means would have to be devised to knock out German tanks. Ordnance experts suggested that, if a suitable armour-piercing projectile could be devised, the S gun might provide one answer. A warhead was produced which would

The Vickers S gun turret on the Wellington II prototype.

The Mustang used in S gun trials at Farnborough. (RAF Museum)

penetrate the German PzKw tank's frontal armour, and an S gun was tested in a Beaufighter. Vickers were given an immediate order for 100 more guns, and Hawker Aircraft were asked to make the necessary structural alterations to a Hurricane to take the

The main user of the S gun was the Hurricane Mk IID.

weight and recoil shock of an S gun under each wing. A mounting was devised by Higson, the big magazine proving difficult to accommodate. In the meantime a Mustang (*AM106*) was used to test the mounting and devise the best method of attack (the Mustang might have been a more suitable aircraft to use than the lower performance Hurricane).

The trials carried out from Boscombe Down proved very successful, and the first two production guns were fitted to a modified Hurricane. Known as the Mk IID, it was flown to Boscombe Down for assessment in September 1941. Attacks were conducted against a Valentine tank at the Lulworth range. The AP shells penetrated both the front and turret armour, and the go-ahead was given for a IID squadron to be sent to North Africa. As the guns were virtually hand-made, it was decided to air-test every gun fitted to the Hurricanes. In the first test the empty cases of both guns failed to eject and jammed. Why this should have happened after prolonged firing tests remained a mystery, until someone realized that hitherto, Vickers-made shells had been used. The Kynoch shells used in these tests had a slightly softer brass case, so that when fired they expanded fractionally more than the Vickers, and their rims were torn off by the extractor. As an interim measure, the rounds were slightly oiled (usually a punishable offence in the RAF, but accepted in this case as a stop-gap solution).

The officer in charge of the project was Wg Cdr. 'Dru' Drury, who was the driving force behind the IID programme. He nearly crashed during the early trials, when the two guns were first fired: the recoil

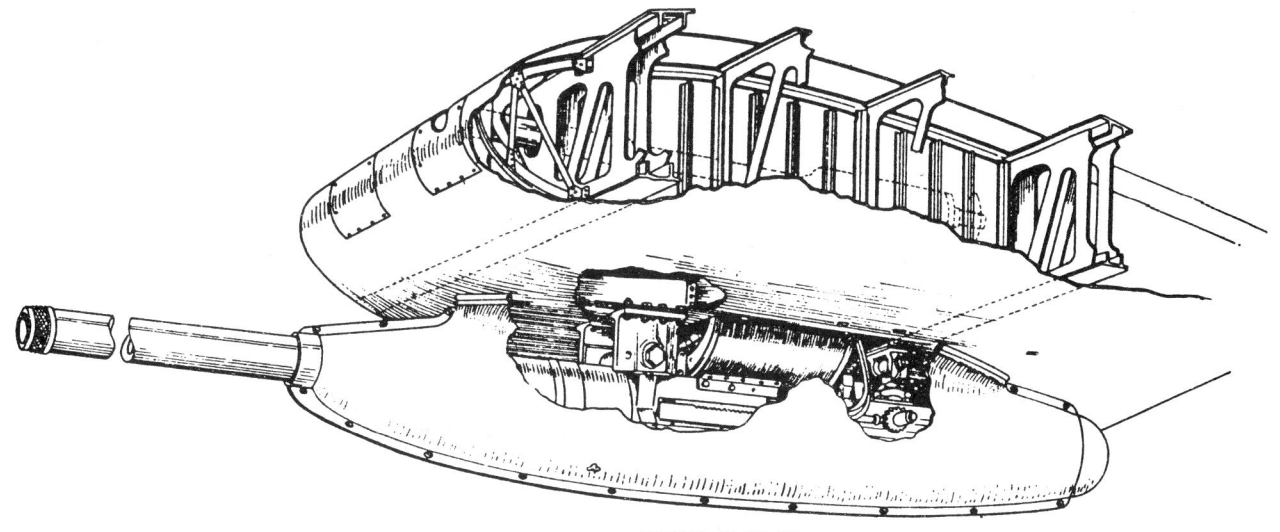

THE GUNS

VICKERS 40 MM GUN.

Detail of the under-wing mounting of the Hurricane.

caused the aircraft to dip nose down. He recovered just in time, but this remained a problem, and was countered by easing the nose up slightly at the moment of firing. The first squadron, No. 6, began training at Shandar in Egypt on 20 April 1942, Drury taking charge of the first period of training. The gun was aimed by the usual Mk II reflector sight, but two Brownings loaded with tracer ammunition were retained, and they gave a good indication of the impact point.

The first operation took place on 7 June, when two tanks and a number of trucks were destroyed. The squadron was in continuous action from this time. In early August two DFCs were awarded to 6 Sqn pilots, F/Lt. Hillier pressing home an attack so low that his tailplane struck the tank he had hit. A captured German tank commander described how his company of 12 PzKw IV tanks were attacked by No. 6 Sqn. Six tanks were knocked out, the other six managed to escape, though one of these had its turret pierced right through. On the other hand,

A Vickers S gun shell, weight 2.5 lb (1.13 kg).

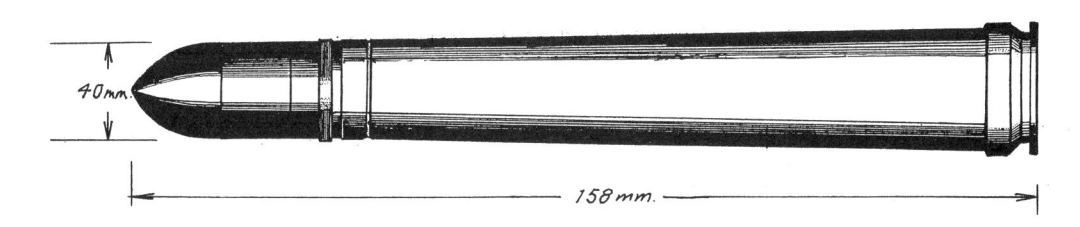

No. 6 Sqn suffered a high casualty rate: the guns slowed the aircraft by 40 mph (64 km/h), and even the fighter version was no match for the more agile and powerful Bf 109F. With the appearance of rocket projectiles, the Hurricane was withdrawn from service in North Africa, although a few were used on what were virtually suicide attacks on V1 launching sites in 1944. Most were despatched to the Far East, where they were very effectively used by No. 20 Sqn in Burma.

As the Vickers S gun was originally designed for air-to-air firing, the first shells used were HE. Although based on a naval projectile, the length of the round was increased to obtain the maximum explosive charge. In September 1941 Vickers designed the armour-piercing shell, known as the armour-piercing Mk I. Weighing $2\frac{1}{2}$ lb, it was a solid projectile with a tungsten nose which could penetrate 50 mm (1.95 in) armour, and was the ammunition used in North Africa. Vickers Armstrongs later produced a 3 lb shell for the gun which gave an increased penetration of 9 per cent. This was the final round, with the Service title AP Mark V. HE ammunition was used in Burma, where most targets were 'soft skinned'.

Details of the Vickers S gun

Bore	40 mm (1.575 in)
Action	Recoil
Cyclic rate	125 rpm
Weight	295 lb (134 kg)
Muzzle velocity	1,800 ft/sec (549 m/sec)
Ammunition feed	15-round spring-loaded drum
Cooling	Air
Effective range	2,500 yd (2,286 m)

The Browning 0.50 cal.

US general John J. Pershing realized whilst serving on the Western Front in 1917 that there was a need for a heavy machine-gun which would give range and hitting power against vehicles and aircraft. He made an urgent request to Washington for such a weapon, and John Browning was asked to develop a larger version of his machine-gun, retaining its simplicity. He produced a prototype for testing on 15 October 1918. The Winchester Repeating Arms Co. produced a rimmed cartridge for the gun, but this was rejected and a new rimless cartridge, based on a German anti-tank rifle round, was developed which, with minor modifications, was used on all future 0.50 in Brownings.

It was decided to produce the new gun, known as the Colt Browning Model 1921, in no less than seven versions: 1. Army and navy anti-aircraft (water-cooled); 2. Heavy barrel ground service (air-

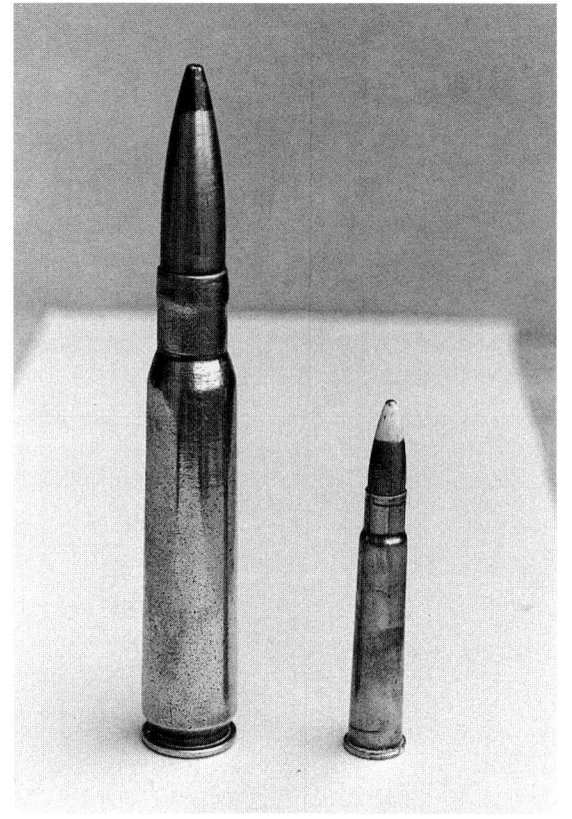

Above *The Browning 0.5 in rimless round (left) compared with the British rimmed 0.303 in.*

Below *A Browning 0.5 in gun with an E11 gun cradle and K6 swivel mounting fitted, as used for manual operation from enclosed side hatch positions.*

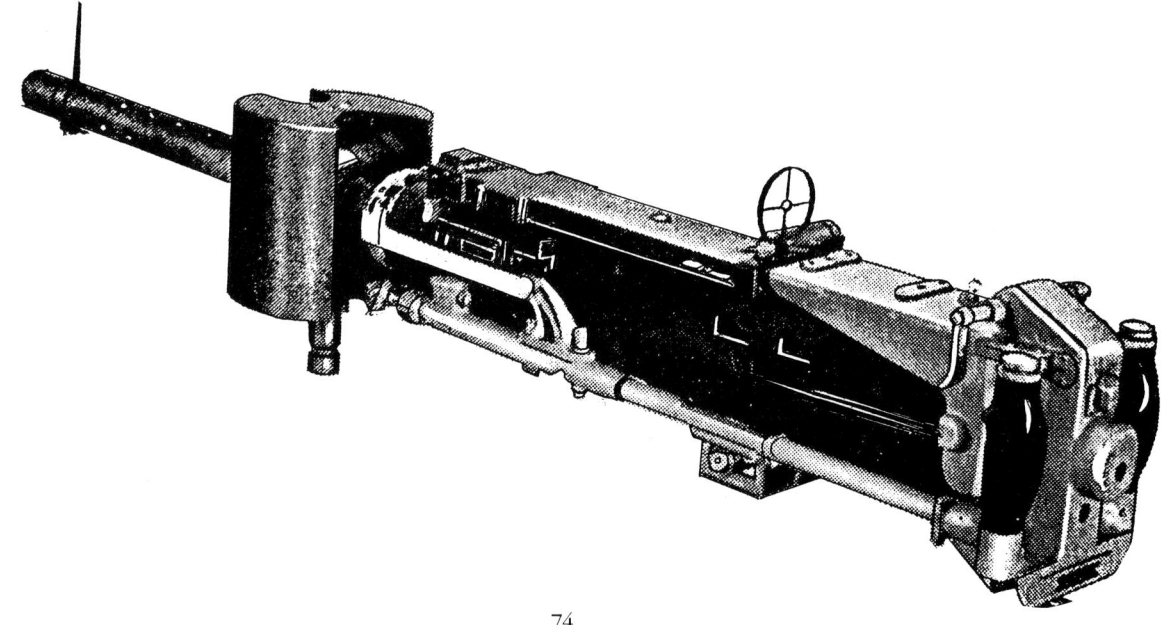

THE GUNS

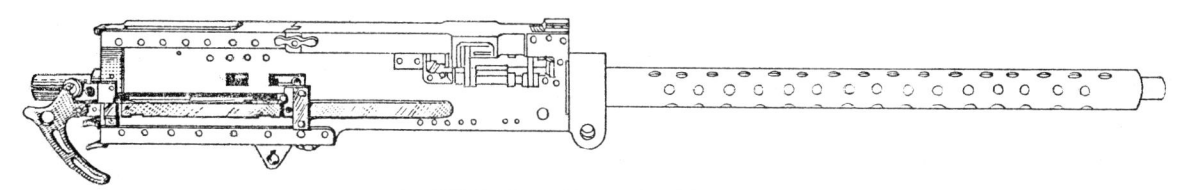

FIXED MACHINE GUN
(Including fixed back plate and operating slide group assembly)

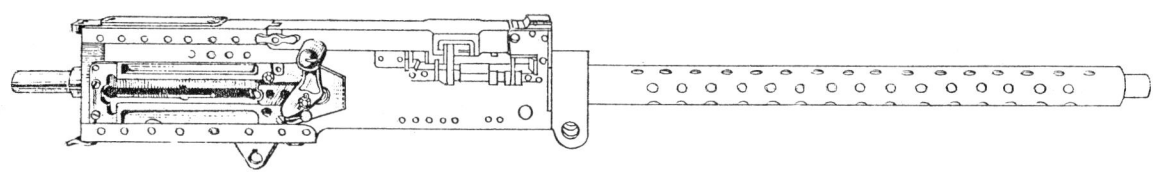

FLEXIBLE, WITH FIXED BACK PLATE, MACHINE GUN
(Including fixed back plate and retracting slide group
assembly.) Used in turret installations.

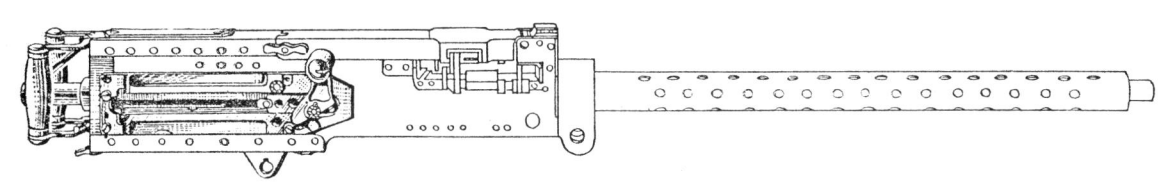

FLEXIBLE MACHINE GUN
(Including flexible back plate and retracting slide group assembly)

Above *Three models of the Browning 0.5 in gun.*

Below *The Browning E18/E2, with turret mounting equipment fitted* (P. Sanders)

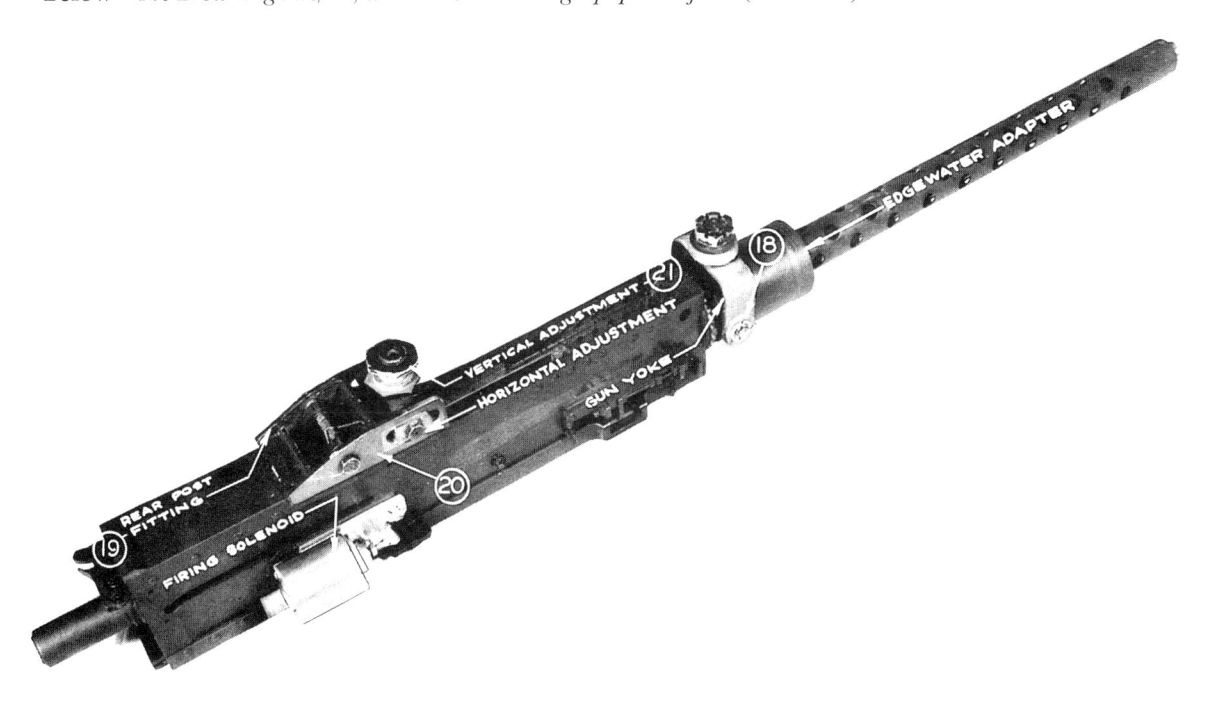

cooled); 3. Ground service (water-cooled); 4. Fixed air-cooled, aircraft; 5. Aircraft observer; 6. Air-cooled, turret, aircraft; 7. Air-cooled (navy).

The air-cooled versions suffered from various problems, incuding overheating and the fact that the receiver could not accept a left-hand feed. In 1927 Dr S. G. Green was asked to modify the weapon, and in late 1932 the new Model M2 was tested. A longer barrel improved the cooling of the air-cooled versions, and the receiver was modified to take right- or left-hand feed. With minor improvements, the M2 armed nearly all US warplanes in the Second World War. One of its drawbacks was short barrel life: the ammunition was not fitted with the usual driving band, and when armour-piercing rounds were fired, the hard metal soon wore down the rifling. However, this was not seen as a major problem, and the supply of barrels was always adequate.

As the war progressed, the Royal Air Force received many US warplanes, and the 'fifty caliber' Browning's firepower was appreciated by British aircrews. As described in Volume 1, all UK makers produced turrets armed with the gun in the last years of the war. The Spitfire Mk IX E was also equipped with these weapons, and other British fighters were fitted with them experimentally.

After 1945 the Browning 0.5 in gun remained the primary air weapon of the US services. In 1946 the final sub-type appeared, the M3, in which the cyclic rate was increased to 1,200 rpm. This was used in fighters until 1955, when fast-firing cannons were introduced. Post-war production in the USA was limited to Army versions and export contracts, and the Belgian licensee, Fabrique Nationale Herstal (FN), began mass production. This company devised a quick-change barrel, and remains a major supplier of both ground and naval models worldwide.

With the advent of armed helicopters the aircraft version is again in demand, and an American company, SACO, have developed a version 35 per cent lighter than the M3. This gun is offered as an alternative to the FN-built M3 pattern on the Lucas helicopter turret (see Volume 1).

Action

The gun can initially be charged manually or, in turret guns, by hydraulic or pneumatic units. Turret guns are fired by relay-controlled solenoids, free guns by twin triggers at the rear in an E ll cradle unit.

When the gun is fired, recoil carries the barrel, barrel extension and bolt backwards a short distance. This unlocks the bolt from the barrel extension and the bolt is thrown further to the rear against the main spring. The empty case is withdrawn by the bolt and the next round extracted from the belt. As the bolt travels forward, the case is ejected and the next round moves into the breech. The rearward motion of the barrel and extension is checked by the oil buffer and spring, which then drives them forward again. This locks the bolt to the barrel, and the final motion of the bolt causes the firing pin to fire the new round. This cycle continues as long as the sear is depressed and ammunition is available.

Sectional drawing showing the main working parts of the Browning 0.5 cal.

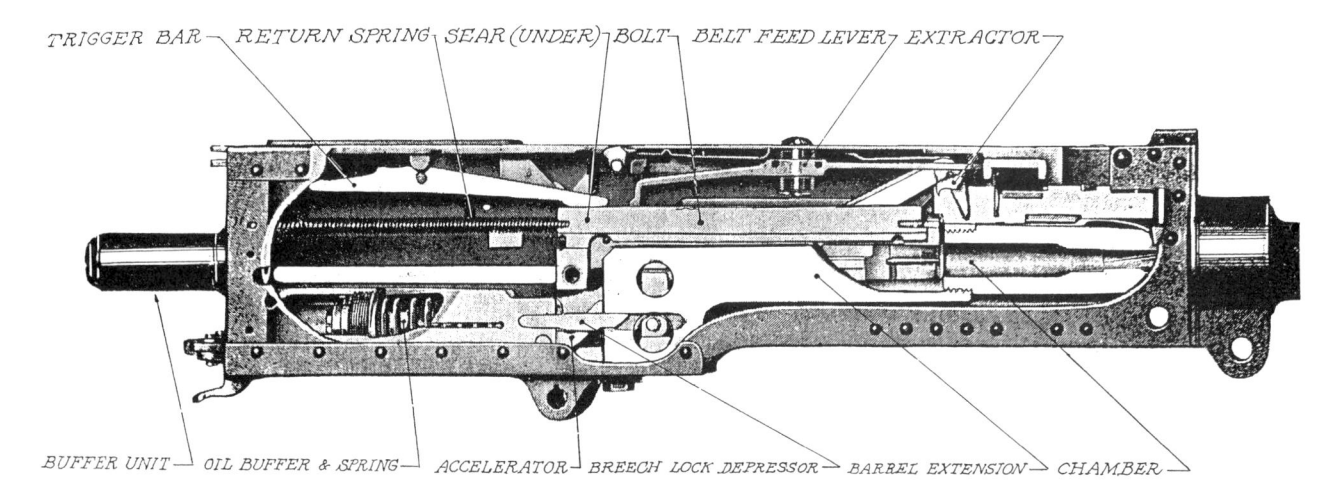

The Rolls-Royce 40 mm gun

In September 1938, at a meeting called by the Air Staff to discuss various aspects of aircraft armament manufacture and design, a member of the Air Staff asked how long it would take to produce a heavy-calibre aircraft gun. The general opinion of those present was that it would probably take three to four years, but E. W. Hives, General Manager of Rolls-Royce, said that he could form a design team and, with the engineering resources of his company, produce one in 18 months. The Air Staff decided to take him at his word and issued a development order for a 40 mm automatic gun. Accordingly, a gun design department was set up at the experimental works at Derby headed by Dr Mario Spirito Viale, a naturalized Italian who had previously been employed by Armstrong Siddeley Aircraft on engine design. The Air Ministry provided a 37 mm Coventry Ordnance Works gun of First World War vintage which it was thought could provide a design basis, but Viale had other ideas. He used the Degtyarev breech-locking system, modified to incorporate cams to force the locking struts apart. It was recoil-operated, the return of the barrel and breech-block being pneumatic. RR50 aluminium alloy was used for the gun cradle and parts of the recoil mechanism; this proved very successful and was used in two types of .5 in (12.5 mm) cal machine-guns designed later by the team. Rounds were clipped four at a time onto an aluminium charger plate which could be replenished whilst the gun was being fired, but this rather basic system gave trouble and was replaced by an eight-round hopper in later versions.

Design work started in late 1938, and in February 1939 Viale was granted patents for some of the new features of the design. The prototype was completed in December of that year, and was despatched to Woolwich for proof firing, mounted on a Victorian gun carriage which, strange as it may seem, provided an ideal mounting, being both steady and transportable. The initial test was completed without mishap, and work proceeded on the magazine-feed system, which was to prove troublesome during test firing at Melton Mowbray in April 1940. During the trials it was found that stoppages were in part due to the pre-First World War design of the 2-pounder naval ammunition used, which was unable to withstand the high pressures involved. The gun was returned to Derby where

Twin 0.5 in guns with British-pattern flash eliminators in the Rose turret of a Lancaster.

Cartridge identification

Ball	Gilded metal (copper-coloured)
Armour-piercing	Black tip
Tracer	Red tip
Incendiary	Light blue tip
Dummy	Hole in case

Details of the Browning 0.50 cal.

Bore	0.50 in (12.7 mm)
Action	Recoil
Cyclic rate	M2: 750/850 rpm; M3: 1,200 rpm
Weight	64 lb (29 kg)
Muzzle velocity	2,750 ft/sec (838 m/sec)
Cooling	(Aircraft) Air
Rifling	Eight grooves, right hand
Length	4 ft 8 in (1.42 m)

Above *The prototype Roll-Royce 40 mm gun fitted to a Victorian gun carriage for test firing.* (Rolls-Royce)

Below *Rolls-Royce engineers prepared a 40 mm gun for a filmed test firing at Sinfin. This gun is fitted with an experimental belt feed.* (Rolls-Royce)

further modification work was undertaken.

Whilst this was in progress the company was informed that operational experience had shown that damage caused by enemy 40 mm anti-aircraft guns had been far less than expected, and it had been decided to treat the project with less urgency. However, the Admiralty meanwhile had found that the defensive guns on their coastal motor gunboats were completely inadequate, and the Rolls-Royce gun seemed a possible answer to this situation. So great was the need that the Navy requested a gun for trials without the troublesome automatic feed, adapted for single-round firing. A gun was duly despatched to HMS *Excellent*, and in July 1940 the Rolls-Royce 40 mm was introduced into service as the 2-pounder Mk XIV. Production was entrusted to the British United Shoe Machinery Co. of Leicester.

When early production guns were mounted on patrol craft, several fatal accidents occurred during gun drill. It was found that the action of the gun was so fast that when the round was fed into the chamber, some over-sensitive primers were set off before the action was locked, firing the round in an open breech a few inches from the gun layer's head. Another cause for concern was that standards of maintenance on small craft could not approach those of RAF armouries, and the aluminium parts suffered from salt-water corrosion. The open breech problem was partially overcome by slowing down the action, and when a modified feed was introduced the Navy ordered a further batch of the weapons.

In the same month that the modified gun underwent sea firing trials, the Air Staff issued a requirement for an airfield defence system using a mobile heavy-calibre gun. Mr W. A. Rowbotham of Rolls-Royce's Experimental Department mounted one of the guns on a 15 cwt Chevrolet truck, with hydraulic rotation and elevation control. When test firing commenced, the vehicle started to vibrate and rock violently, and when further improvements were specified, such as armour for the crew, the project was sidelined. Meanwhile, a second and more urgent order had been issued for an airborne anti-armour weapon to be mounted in Hurricane aircraft, and competitive trials were to take place between the Rolls-Royce gun and the Vickers Type S 40 mm weapon. A modified version of the Rolls-Royce was prepared, and during the test firing Walter Hampton, who was in charge, upset the LMS Railway Company. The test gun was being fired in the usual manner – from behind a brick wall, using a lanyard tied to the firing mechanism, with the gun rigidly secured. A long burst was fired, and the shells all struck the same spot, the later shells hitting the mass of solid shot and ricocheting. A

The troublesome hopper feed prior to test firing. Note the boxes of practice ammunition stored close by. (Rolls-Royce)

goods train at Sinfin Moor suddenly screeched to a halt and its driver came running across the fields to the butts. When he got to within shouting distance, he brandished a missile and screamed, 'How about this, then?' It had come through the roof of his cab and narrowly missed him and his fireman! Walter told him that if nobody claimed it within seven days he could keep it! A strong protest was sent to the company and Hampton was told to be more diplomatic in future.

Mr G. N. Wallace of the RAF Gun Section supervised the test firing of the 40 mm guns for the A&AEE at the Pendine Sands test facility, where the Vickers weapon was found to be much superior. The Rolls-Royce gun was found to be adversely affected by altitude and temperature, and would not fire in an attitude of 80° depression as required in the specification. Wallace duly recommended the adoption of the Vickers gun, which was fitted to the Hurricane IID and proved to be an excellent weapon against all but the heaviest tanks in the Western desert. When Vickers had difficulty in supplying the required number of weapons, Rolls-Royce were given a production order for 1,000 of their guns, to be manufactured at BUSM, but during acceptance trials two guns had breech explosions, after which deliveries were stopped. It was found that the cause of the explosions was the failure of a case in the ejector slot, so the slot was modified and trials continued. Extensive firing tests proved that the gun was now much improved. Considering the relatively short development period. Viale's team had been remarkably successful in bringing a gun incorporating many new and untried features to a standard where it was passed as fit for RAF service by the Boscombe Down evaluation team. However, production was limited to parts for 200 guns, after which production ceased at Leicester. No further orders were received for either the Rolls-Royce or the Vickers Type S, as the emergence of rocket weapons promised to give the RAF a weapon which, when fired, left the host aircraft with no heavy firing mechanism.

During the period when the gun was being improved, Viale had been working on another large-calibre design, a 47 mm weapon using a more conventional rising breech action. This was 90 per cent complete when the project was terminated, and the team dispersed to other work. The frustration of the team can be well imagined, but they were by no means the only design group to see promising projects cancelled as Service needs changed. This gun consisted of a barrel, breech ring, and a body in which a rectangular breech-block could reciprocate. Attached to the barrel was a hydro-

Dr Viale with the 40 mm gun. (Rolls-Royce)

pneumatic recoil and recuperator mechanism which was used to absorb the energy of the recoil, return the barrel to the firing position, and close the breech. The entire assembly was free to slide in its cradle, to which the main ram of the recoil system was anchored. The cradle also carried an eight-round hopper-type magazine at the rear. A pneumatic recocking mechanism was incorporated in the body to recock the striker.

Details of the Rolls-Royce 40 mm gun

Bore	40 mm (1.56 in)
Action	Recoil-operated
Cyclic rate	120 rpm
Weight	328 lb (148 kg)
Muzzle velocity	2,440 ft/sec (744 m/sec)
Ammunition feed	Eight-round hopper
Number of grooves	12
Length of barrel	2,030 mm (80 in)
Length overall	2,870 mm (113.5 in)

Below *The Rolls-Royce gas-operated gun was the second, and less successful, 0.50 in design of Viale's team.*

Bottom *If the supply of American 'fifty caliber' guns had stopped, the Rolls-Royce recoil-operated gun would have been used. It had a faster cyclic rate and was lighter.*

Rolls-Royce 0.50 in (experimental)

As well as the 40 mm airborne weapon, the Rolls-Royce gun design department designed two heavy machine-guns which had several new features. The first design was gas-operated, with a cyclic rate of 650 rpm, but it was decided to concentrate on a recoil-operated weapon. Viale again decided to construct the body and breech cover of RR50 aluminium alloy. The barrel was shorter than the Browning and this, combined with the alloy construction, reduced the weight. The gun fired from a locked breech. As the breech-block recoiled, a pair of accelerators carried back a wedge-shaped balance-piece and retracted the striker pin.

The gun was tested at the Proof and Experimental Establishment at Pendine Sands in March 1941. It was found that the short barrel caused an abnormally large flash, and a long flash eliminator was added. The trials revealed minor snags, and Rolls-Royce decided to redesign the gun for 0.55 in Boys anti-tank ammunition. This development was showing great promise when the company decided to cancel all armament work.

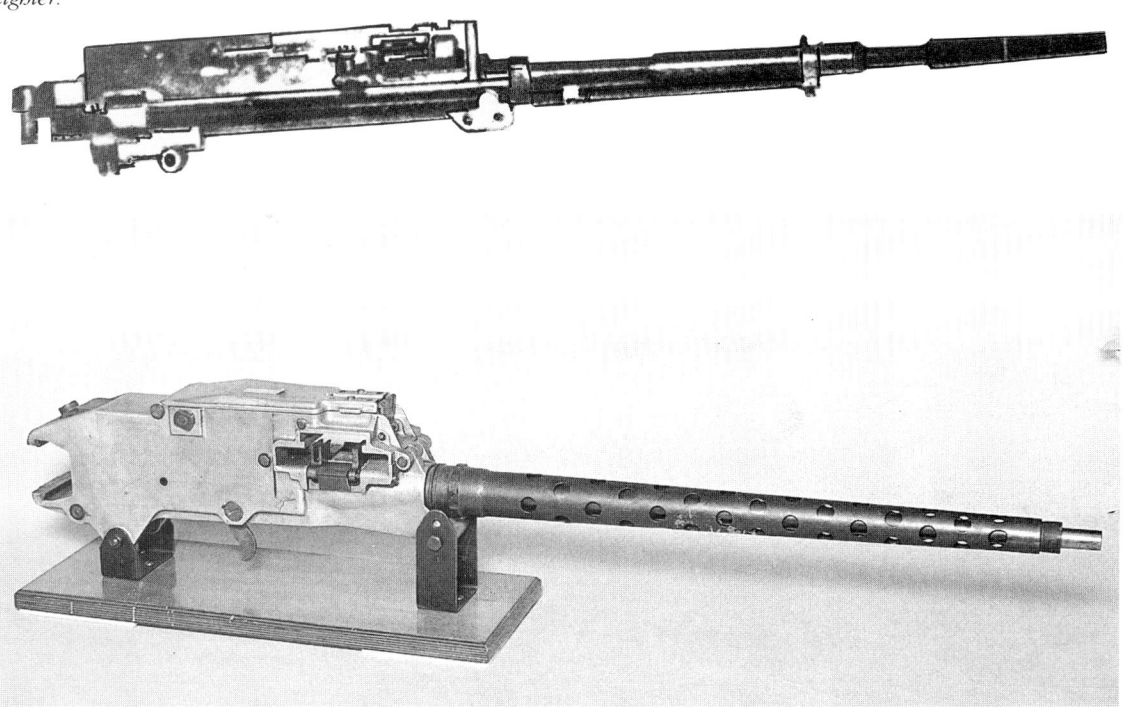

Details of the Rolls-Royce 0.50 in experimental gun

Bore	0.5 in (12.7 mm)
Action	Recoil
Cyclic rate	1,000 rpm
Weight	49 lb (22.2 kg)
Muzzle velocity	2,340 ft/sec (713 m/sec)
Ammunition feed	Disintegrating belt
Cooling	Air
Rifling	Four grooves, right-hand twist
Length	50 in (1,270 m)

The Molins 6-pounder No. 1

The de Havilland Mosquito Mk XVIII was armed with a 6-pounder anti-tank gun. This weapon was adapted for aerial use by the Molins Machine Co. of Deptford, London, the leading British manufacturer of tobacco packaging machinery. Apprentices at the present works at Peterborough have rebuilt one of these historic guns, which can be seen in the Mosquito Museum at Salisbury Hall. This weapon gave the German Navy problems out of all proportion to the very small number actually used on operations.

The swift advance of the Panzer armoured units in May 1940 revealed a lack of mobility as well as firepower in the British 2-pounder anti-tank guns then in use. The Army proposed to mount much more powerful 6-pounder guns on small vehicles operated by a single crewman. This meant that the gun would have to be fitted with an automatic feed system. The design of this mechanism was entrusted to the Molins company, which was already working on a belt-feed unit for the Hispano gun. They received a development order on 14 February 1942, and most of the work was done by Desmond Molins, assisted by a Frenchman, Felix Ruau.

The rounds were stored in groups of four or five. When one group was fired, an electrical mechanism moved the next group sideways into position over the breech, thus the heavy shells were fed into the gun without links. When fully loaded, the magazine held 22 rounds. It was found necessary to modify various parts of the gun to enable the recoil to operate the magazine, and the modified weapon was officially known as the Molins gun. In August 1942 the prototype was taken to the nearby Deptford Shooting Club firing range, and with the

This 'Magnificent Molins' was rebuilt in 1986 by apprentices at the Molins factory in Peterborough.

assistance of Ordnance Corps experts it was fired automatically for the first time. It was then despatched to Woolwich for exhaustive testing. Although the trials were successful, the project was terminated owing to the appearance of the PzKw VI (Tiger) tank, which was impervious to 6-pounder shells.

In early 1943 the Air Staff was discussing the replacement of the 40 mm anti-tank gun currently mounted on Hurricane IID aircraft. The Molins seemed to be a logical weapon for a larger aircraft, so a trials team under G. F. Wallace carried out ground firing tests. The gun was found to be trouble-free, the only possible problem being the ability of the feed to operate under the stress of manoeuvres. The head of the department, Captain Adams, sent a favourable report to Air Marshall Sir Ralph Sorley, the controller of Research and Development. J. E. Serby of the Ministry of Aircraft Production then wrote to R. E. Bishop of de Havilland Aircraft regarding the possibility of arming a Mosquito with the gun. He was to bear in mind that the recoil force would be 8,000 lb, and the weight of the gun and ammunition together could be 1,800 lb (981 kg). Bishop replied that the Mosquito could easily accommodate the gun – indeed in 1942 a feasibility study had been carried out for mounting a 3.7 in (93.98 mm) anti-aircraft gun!

A prototype installation was designed at Hatfield, and when the huge gun arrived for installation, Rex King of de Havilland asked when the horses would be coming! The gun was duly fitted in a written-off Mosquito FB.VI to confirm the effect of the muzzle blast on the wooden fuselage, and on 29 April 1943 five rounds were fired at the Hatfield butts. The blast shook the ground and left many ears ringing, and a cloud of dust showed that the solid shot had hit the aiming point. No damage had been caused to the airframe other than a sheared fixing bolt. Two days later the gun was installed in another Mosquito Mk VI to find the best method of attaching the gun and the auto-feed unit. The weapon was mounted 4 in (100 mm) to the right of the aircraft's centre line, with the muzzle protruding 24 in (610 mm) under the nose at a slightly downward angle and the recoil spring faired under the barrel. On 8 May a new FB.VI, *HJ732*, was wheeled into the experimental bay for conversion into the prototype Mk XVIII (soon to be dubbed the Tsetse, or Tsetse fly). The massive gun and all its associated electrical and cockpit controls were installed in four weeks, helped in no small measure by the easily adapted wooden structure.

The peaceful morning of Sunday 8 June 1943 was

Magazine loaded with 22 rounds, with the charger poised to load the first round.

shattered when all 22 rounds were fired from *HJ732* in a staccato thumping burst into the sand. After further firing tests had been carried out, the aircraft was air tested. Air Marshal Sorley arrived at Hatfield on the same day to confirm that the Mk XVIII would be flown to Boscombe Down next day. He also announced that 30 of the next Mk VI Mosquitos would be delivered as Mk XVIIIs, initially to be used by Coastal Command against U-boats in the western approaches of the Bay of Biscay.

When firing commenced at Boscombe Down it was found that the feed unit would not operate if the 'G' forces exceeded 2.5. Hundreds of rounds were fired during attempts to overcome the problem, but without success. On 22 June the machine was flown back to Hatfield where the charger arm and other parts were strengthened to overcome both positive and negative 'G' forces. Meanwhile,

Above *Automatic feed: when the first clip is fired, the chain drive brings the next clip into position in line with the feed arm.* (Molins)

Below *One of the Mosquito XVIIIs used for trials. The nose section has been cladded with metal and the recoil cylinder enclosed in a fairing.*

the second Mk XVIII was fitted with 900 lb (408 kg) of armour protection around the nose, to give the crew some protection against the heavy flak armament of the U-boats. While 30 conversion sets were produced at Hatfield, Desmond Molins supervised the modifications to the feed, and by the end of June *HJ732* was back at Boscombe Down. During August it was extensively tested at Boscombe and Exeter. After 400 rounds had been fired, the undersurface of the starboard flap was torn off by the muzzle blast, and the flaps were therefore strengthened.

It was decided to retain the two outer 0.303 Brownings, with double firing time from enlarged ammunition tanks, to discourage anti-aircraft gunners during the final stages of an attack. The empty 6-pounder cases could not be ejected overboard as they might hit the tailplane, so they were retained in the fuselage, where they could produce a loud clanging noise during manoeuvres. It was decided to use the Mk IIIa reflector gunsight rather than the larger Mk II because it gave a better peripheral view and had a dimming screen which minimized reflections off the surface of the sea. Alignment of the 6-pounder and the Brownings was achieved by using a central graticule for the former and a higher dot for the latter. As the rounds converged at 400 yd (366 m), the Brownings came to be used as an additional sighting aid for the big gun. As the pilots gained experience, moored targets were constantly hit during shallow diving runs, and rounds which fell short often ricocheted into the target.

By October, three Mk XVIIIs had been delivered to Boscombe with long-range tanks and armour, and were soon cleared for operations. Five crews from 618 Sqn were posted to Predannack to work up on the new aircraft, along with 30 ground staff and armourers who had been trained on the Molins gun, by now officially entitled the 'Airborne 6-pounder Class 'M' gun. Predannack was the home of the Beaufighters of 248 Sqn, which formed a composite unit with the Tsetse Mosquitos ranging over the Bay of Biscay. The first operation took place on 24 October 1943. Mosquitos *HX902* and *HX903* searched unsuccessfully for enemy shipping. On 7 November a U-boat was caught on the surface. After being hit by several shells aft of the control tower, it submerged amid a cloud of black smoke. In December No. 248 Sqn was re-equipped with the Mosquito FB.VI, and as more Mk XVIIIs arrived, the strike force began to wreak havoc on German shipping in the Bay. Information from the broken Ultra code enabled the Mosquitos to intercept the U-boats en route both to and from the Atlantic. The following combat report is typical:

Flying Mosquito *HP922* we broke cloud at the prescribed time and position. I immediately observed a submarine proceeding on the surface with an escorting minesweeper. Whilst making my approach an escorting JU 88 appeared in the sights. I pressed the gun button and the Junkers disintegrated. I then attacked the submarine which was seen to be hit.

About this time the German Admiral Donitz issued the following directive: 'Owing to the damage caused by enemy aircraft mounting heavy-calibre guns, surface passage to port will only take place during the hours of darkness.' After a gruelling patrol, a journey through minefields at night was not appreciated by the submariners.

In February 1944 No. 618 Sqn moved to Penreath, from where throughout the summer the Tsetse Mosquitos used their guns to good effect. On D Day (6 June) the squadron flew sorties from 04:45 to 22:15 hours protecting Allied shipping. On 7 June Mosquito *MM425* hit a submarine with 12 57 mm rounds; the U-boat dived amid a large patch of oil before one unfortunate crew member could not get inside – he was last seen swimming in the general direction of America. By the end of August enemy activity in the Bay of Biscay lessened, and the squadron moved to Banff in Scotland. Attacks took place against German shipping and coastal installations in Norway, often with up to six MK XVIIIs leading the formation. The advent of more versatile rocket-firing Mosquitos restricted the production of the

A Mosquito firing the Molins gun during trials.

Molins shells being loaded into the Autoloader Mk III at the Molins factory by Bob Hugget.

Mk XVIII to only 27, most of which served with 248 Sqn until the end of the war.

As with the early Hispano, it was found that AP projectiles were most effective, for they could penetrate a submarine's pressure hull and cause major damage to a ship's hull and superstructure. The AP shells weighed 7.1 lb (3.17 kg), being tipped with hardened steel.

Details of the Molins 6-pounder

Bore	57 mm (2.25 in)
Action	Recoil
Cyclic rate	60 rpm
Weight	1,800 lb (816 kg)
Muzzle velocity	2,600 ft/sec (791 m/sec)
Ammunition feed	Molins automatic feed
Magazine capacity	22 rounds plus one in breech
Sighting	Dual graticule Mk IIIa reflector sight
Length	12 ft 5 in (3.6 m)
Height	38 in (965 mm)

The FH Recoilless (RCL) Gun

After the war it was found that Germany had a number of large-calibre recoilless aircraft guns almost ready for use. The advantage of this type of weapon (as Commander Cleland Davis had propounded in 1914) was a low weight factor and no recoil problems. The Air Staff decided to explore this option in parallel with the Mauser revolving breech system, and they asked the Royal Armament Research and Development Establishment at Fort Halstead to develop such a weapon. It was decided to use the Rheinmetall Company's Gas Recoil Equalizing system as a basis for development, in which part of the propulsion gases were diverted along a pipe to the rear. The calibre was fixed at 4.5 in (114.3 mm), seven rounds being loaded into a revolving magazine with one ready to fire in the breech. The reciprocating action was powered by a piston actuated by propulsion gases bled from a position close to the breech. One hit from the huge shells would have destroyed any aircraft, but design never reached the stage where test firing could be carried out. The weapon had been earmarked for the Gloster Javelin initially, and preliminary design

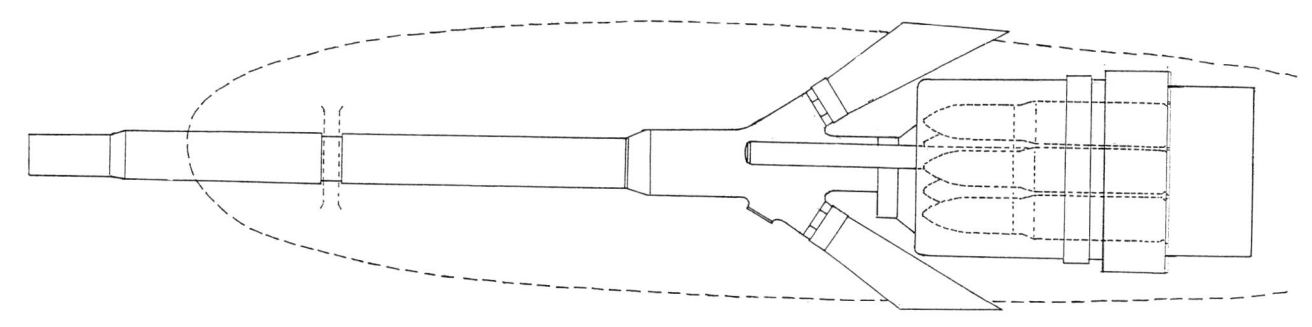

Preliminary design scheme layout of the 4.5 in cal. Halstead recoilless aircraft gun, the project was well advanced when it was cancelled in 1953.

work was carried out on a Hawker Hunter installation. The weight of the gun was kept down to 1,440 lb (653 kg), which was very low for a gun of this calibre, but with the prospect of many ingenious guided missiles in the future, and the promising ADEN gun, the project, using as yet unproven technology, was cancelled. With hindsight, considering the Rheinmetall system had not been conclusively proved to be successful, the RCL gun could have been as troublesome as its predecessor in the First World War, the Davis.

Details of the RH Recoilless

Bore	4.5 in (114.3 mm)
Action	Recoil
Cyclic rate	50 rpm
Weight	1,440 lb (653 kg)
Ammunition feed	Seven-round magazine (1 in breech)
Design principle	Propulsion gases equalized to balance recoil forces

The ADEN 30 mm

After the Second World War the world's air forces recognized the superiority of the German Mauser MG 213c 30 mm gun, and copies were soon made under various names.

IWKA Mauser at Oberndorf, one of the most famous arms companies in the world, had produced the MG 151/15 in 1936 and the MG/20 in 1940. These were high-velocity weapons of exceptional quality, featuring electric firing, and the 20 mm version was the most important Luftwaffe gun throughout the war. But by 1942 the Luftwaffe planning staff had decided that guns of a larger calibre than 20 mm were needed. A very large 30 mm gun, the MK 103, was already in use, and the much lighter MK 108 followed in 1943. To be hit by a 30 mm shell was a devastating blow to most aircraft. The most significant development, however, was a later project, master-minded by Diploma Engineer Linder of Mauser. Under Linder, the Mauser and Krieghoff companies were given a specification for a 30 mm gun with a very high rate of fire. By March 1943 a 20 mm prototype, the MG 213a, was ready for test firing. The gun performed well enough, but the ammunition feed could not cope with a cyclic rate faster than 900 rpm, and the project was shelved. In 1944 a Mauser engineer, Dr Maier, devised a new feed system using a cylindrical five-chambered breech. A test model proved that the idea was feasible, and the MG 213 project was restarted. Maier's system was incorporated into the MG 213c, which proved to be an outstanding success, 1,200 rpm being obtained in early test firings. The director of the project, O. H. Von Lossnitzer, made frantic efforts to bring the 213c into Service use, but such snags as excessive barrel wear held up production, and in spite of round-the-clock effort the gun was never used operationally.

On 3 May 1945 Free French forces captured the Oberndorf complex intact. French armament experts soon realized the significance of the 213c, and organized the transporting of machinery, drawings and partly finished 20 mm and 30 mm guns back to France, where their details were made available to the Allied authorities. Although the last thing anyone wanted at this time was more death-dealing weapons, the revolutionary design of the Mauser gun was too important to be ignored.

The American Pontiac division of General Motors developed a 20 mm version of the 213c known as the M-39, which had a cyclic rate of 1,500 rpm.

A, temperature-resistant-steel breech; B, gas-tight chamber; C, revolving cylinder (with five chambers, one being shown in the 12-o'clock firing position); D, electric igniter; E, twin feed-wheels; F, main recoil springs; G, cylinder-drive stud; H, reciprocating slide; J, short-stroke gas piston; K, electric/pneumatic cocking connection; L, ejection of cases (black arrow); M, ejection of open links (white arrow).

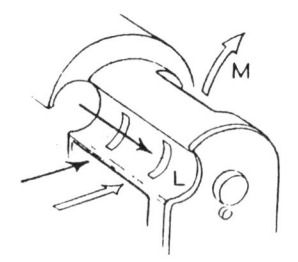

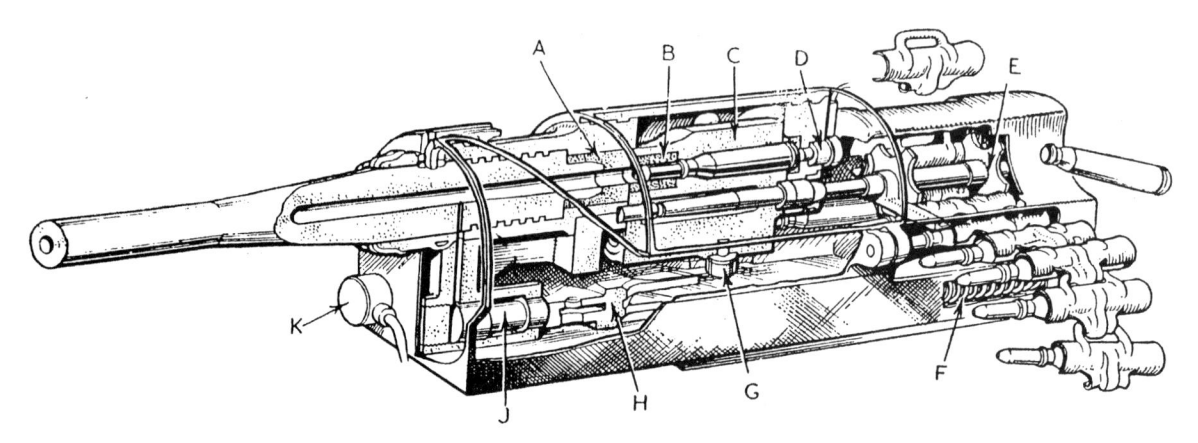

The Mauser 213C, the predecessor of the ADEN.

The French version was the DEFA, a 30 mm gun produced in many versions for the Armée de l'Air, and exported in large numbers. The British handed the 213c to the Royal Small Arms factory at Enfield and the Royal Research and Development Establishment. The resulting design, known as the ADEN (Armament Development ENfield), was almost identical to the German weapon, and was developed in both 20 mm and 30 mm calibres. At first the preference was for the lighter, faster-firing 20 mm version, but the 30 mm gun promised to deliver a much heavier punch at well over 1,000 rpm. The production of suitable ammunition caused some difficulty. It had been decided to use aluminium cases, but after much time and effort these were abandoned in favour of brass rimless 111 mm cases.

Six years after the war the 20 mm design had been abandoned, but the 30 mm was by no means certain of acceptance. A strong body of opinion favoured a complete change-over to rockets and guided missiles; indeed, it was even suggested that manned fighters would soon be obsolete. Fortunately, it was decided to continue development of the gun, and a pre-production batch was ordered from Enfield for air firing trials to be carried out by the Weapons Department at Farnborough. Two of the guns and their associated ammunition systems were installed in Beaufighter *RD388*, the fuselage was strengthened, and on 5 May 1954 F/Lt. Mitchell took off from Farnborough accompanied by W. Ainley of the Weapons Department as observer.

They headed for the Aberporth firing range, which tracked the aircraft on radar as it headed out from the Welsh coast. Mitchell fired a short burst followed by a longer one. The 'Beau' seemed to vibrate more than had been expected, but several more bursts were fired, and the shells churned up the water near the aiming point. When the aircraft landed back at Farnborough, huge holes were found in the wings and fuselage – spent links had been caught in the turbulence of the gun blast and had been hurled against the airframe. The chutes were shortened so that the links were discharged into the gun bay, the aircraft was repaired and the trials continued. When the guns were fired during high 'G' turns and dive pull-outs, Ainley had to curl himself up in a ball and brace his legs against a bulkhead. It was the last time this aircraft was to fly, for the rivet holes around the wing roots were pulled into ellipses, and the main spars were severely strained.

The Beaufighter was replaced by a Gloster Meteor F8 (*WK660*) and the trials continued. In January 1955 a trials team under Ainley, six guns and ammunition were despatched to Canada for low-temperature testing. With RCAF station Edmonton as the main base, F/Lt. Lowe carried out air-to-ground firing from the Central Experimental and Proving Establishment at Namao, Alberta. The firing took place round Cold Lake, in an area where trappers had been warned off. When high-altitude firing was carried out an area of 40 miles was cleared. Long bursts were fired at 42,000 ft, and again the guns had very few stoppages. Assessment after the guns had been fired took place in an open-ended hangar whose only means of heating was by captured German hot-air blowers.

Meanwhile, mounting and feed systems were designed for the ADEN-armed aircraft. The first was the Supermarine Swift, with two guns in the fuselage. However, the Air Staff requested heavier armament, so the Swift F4 was armed with four guns. To accommodate the extra ammunition, room was found in the swept-forward wing roots, but this caused a dangerous pitch-up under high g. This and other problems led to cancellation of the Swift programme. The naval Scimitar was fitted with four ADENs, and early marks of the Javelin had two guns in each wing. A singularly trouble-free installation was in the tiny Folland Gnat, in which a transporter system dumped the cases and links in the rear fuselage.

The Hawker Hunter day fighter was armed with four ADEN guns mounted in a removable package in the fuselage. When one of the first firing runs was made at 10,000 ft, the Rolls-Royce Avon engine suffered a flameout and stopped. The pilot was able to start it again, but after more flameouts it was obvious that something was wrong with the airflow round the engine intakes when the guns were fired, a problem which had also occurred on the Swift. Strangely, the AS Sapphire-powered Hunter F2 proved unaffected by the muzzle blast. Many bizarre remedies were tried, including the 'Lobster', which was a huge steel duct which gathered the muzzle gases, turned them through 180° into a large box under the fuselage and vented them into the airstream. It was thought this would prove that the actual muzzle blast was the problem. Luckily the ground test was fired remotely: when the switch was thrown the whole contraption was blown into the stop butts, followed by most of the nose section of the aircraft. The Lobster was not air tested!

Meanwhile, Rolls-Royce analysed the contents of the muzzle blast gases, and found that they were much more flammable than had at first been thought. When the gases were sucked into the engine they had the same effect on it as flooding a car engine with petrol. A device was therefore installed which cut off the fuel when the gun button was pressed. The engine burned the gun gases for this short period, the fuel being restored when the button was released. A great deal of research was also carried out on the high levels of gases found in the gun bays after firing, and as a result, 3 in ducts

Hawker Hunter F6 prototype with four ADEN guns. The bulged link containers were dubbed 'Sabrinas'; next to them were the case ejection holes. (British Aerospace)

The ADEN Mk 4 30 mm cannon.

were fitted to vent the gases into the slipstream via flush outlets. Two large collector tanks were enclosed in blister fairings under the fuselage, the centre section being detachable for the removal of the links, the spent cases being discarded into the slipstream clear of the fuselage. Fourteen squadron aircraft were flown into Farnborough to help perfect these modifications, and when the SBAC show intervened the testing continued at Blackbushe. Four years after the first flight the Hunter became fully operational.

The English Electric Lightning was the only British supersonic fighter to see RAF service. Various ADEN gun schemes were fitted: a two-gun pack

Removing an ADEN from a Jaguar at Coltishall using the built-in winch. The quick-release barrel has already been removed. Ammunition enters the gun bay from the duct near the electrical control box. (BAe Warton)

could be fitted under the fuselage as an alternative to guided weapons or rockets on the F.1, 1A and 2, and a ventral two-gun pack was designed for the export F.53 version.

With the development of computer-controlled head-up display systems, the guns became more accurate. The early versions of the Harrier carried two ADENs faired externally under the fuselage, each with 135 rounds. The Jaguar GR also has two internal guns to back up missiles and other offensive stores.

In the late '50s higher-velocity ammunition was introduced, increasing the range and penetration of the rounds. Various other modifications have also been introduced, culminating in the ADEN Mk 5. When the guns are fired at ground targets, solid ball practice rounds often ricochet into the air, and several aircraft have been hit by their own shells. (Reports from the Soviet Union suggest that 22 aircraft have been lost in this way.)

When the guns are removed from the aircraft, great care is taken to clean off any carbon deposits. This is done by immersing the barrels and moving parts in a bath containing a mixture of Stipolene 440 and paraffin. The electrical system and working parts are overhauled before the gun is reassembled, and the pod is then clamped into a harmonizing ring and aligned by means of an optical sighting system onto aiming marks on the armoury wall.

Gun functioning

Starting position

With the gun cocked and the first round in the uppermost chamber, with its cartridge cap in contact with the firing pin and the chamber in line with the bore. The second round is fully rammed into the adjacent chamber, and the third partially

A dismantled 30 mm ADEN: left – feed unit; centre – revolving breech; right – slide unit, chassis and return springs; front – quick-release barrel. (RAF Wittering)

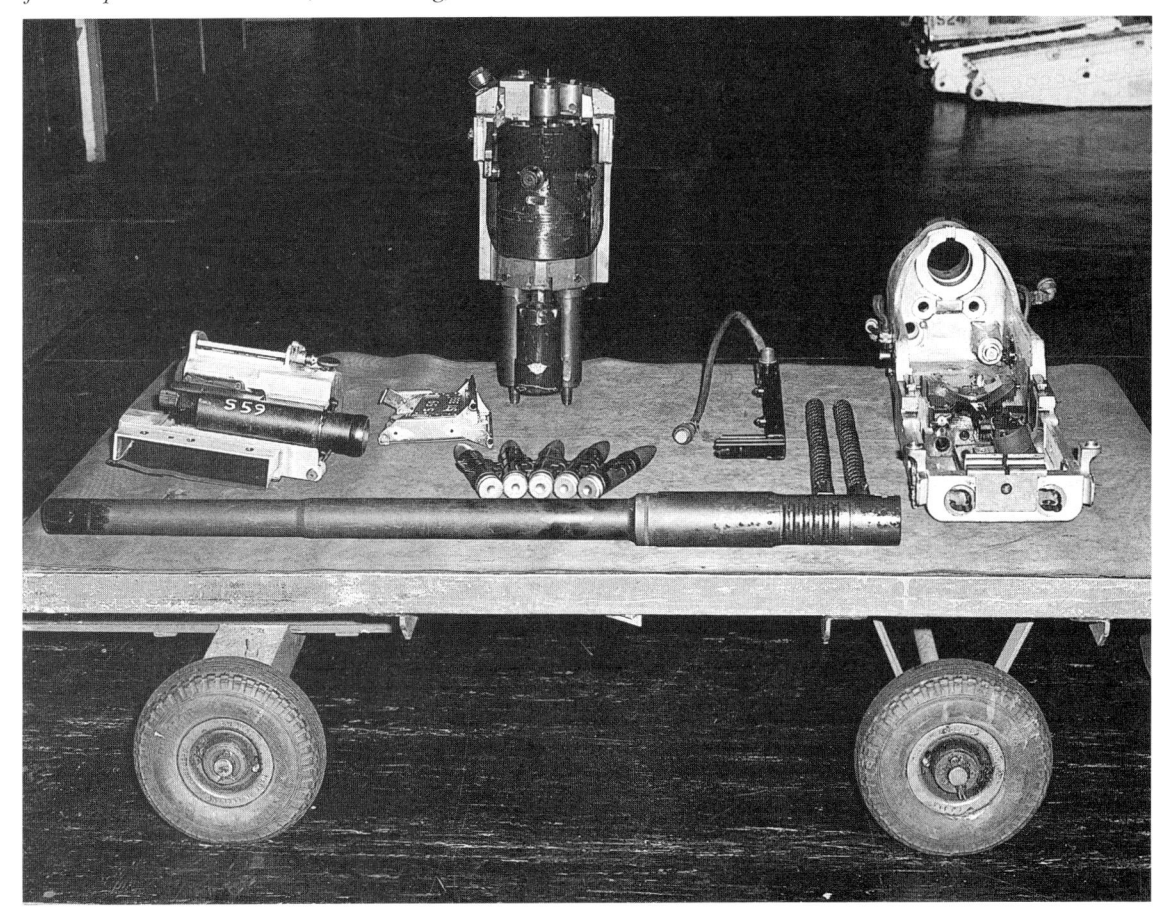

rammed into the third chamber. The slides are fully forward and in contact with the gas piston and the cradle switch contacts are closed.

Rearward stroke

As the first round is fired, some of the propellent gases are diverted into the gas cylinder and the gas piston forces the slides to the rear, compressing the return springs. The cradle switch contacts are broken as soon as the slide starts to move, and the firing circuit is interrupted. As the slides travel to the rear, the cylinder is rotated through 36° by the cam lever, which acts upon the bottom cylinder roller. The feed mechanism is also rotated by the cylinder, and the ammunition belt is drawn through the gun until the fourth round is positioned at the edge of the loading platform.

Forward stroke

As the slides move forward under the influence of the return springs, a deflector cam on the front slide contacts the incoming roller and causes the cylinder to rotate through a further 36°. The roller also forces the cam lever across the slide. At the same time, the feed mechanism draws the first round across the loading platform and in front of the rammer. The fourth round is now pushed from the belt and into the next vacant chamber by the rear rammer. The third round is fully rammed home by the front rammer. Towards the end of the stroke, the incoming breech cylinder roller enters a groove in the front slide, and a further rotation of the

Below and opposite *The functioning sequence of the 30 mm ADEN.*

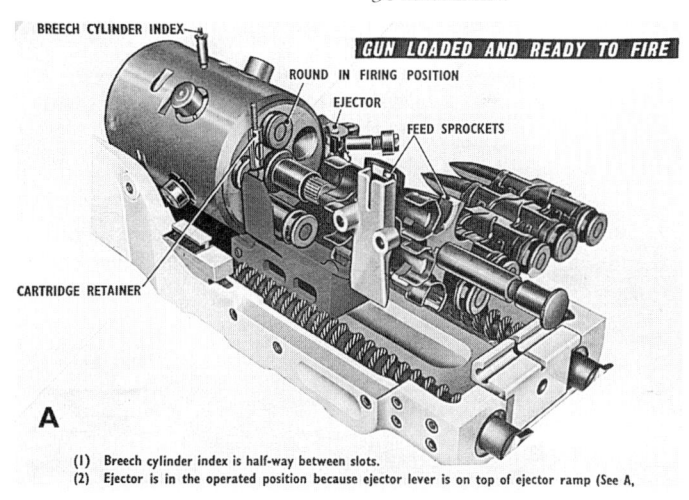

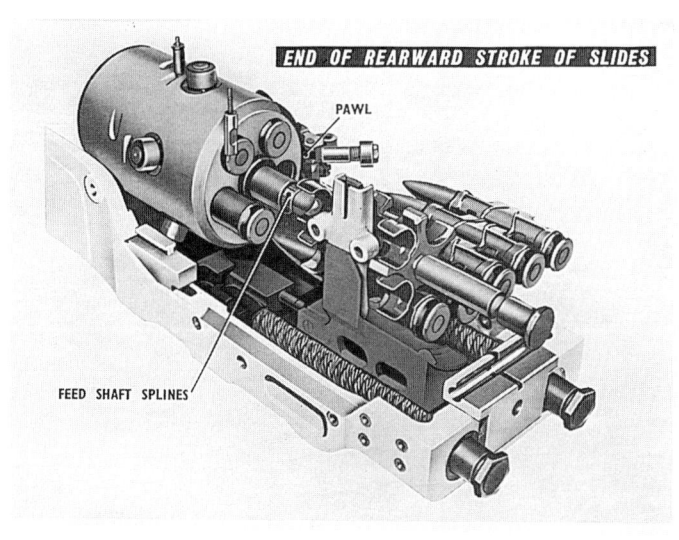

(1) Breech cylinder index is half-way between slots.
(2) Ejector is in the operated position because ejector lever is on top of ejector ramp (See A,

THE GUNS

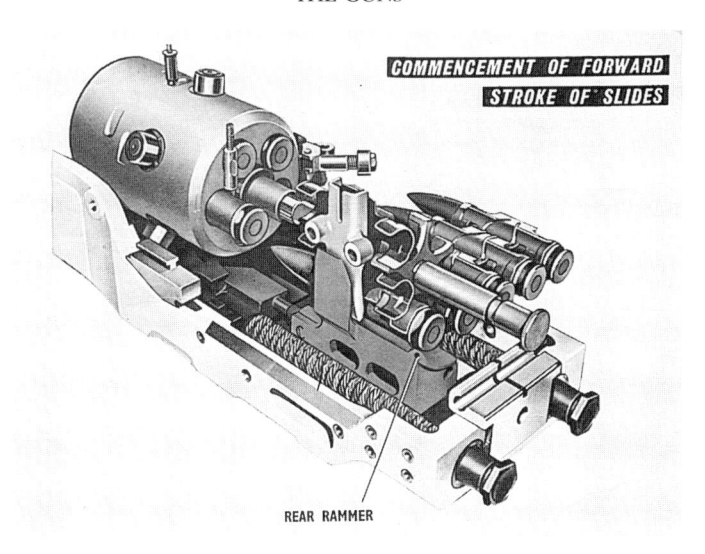

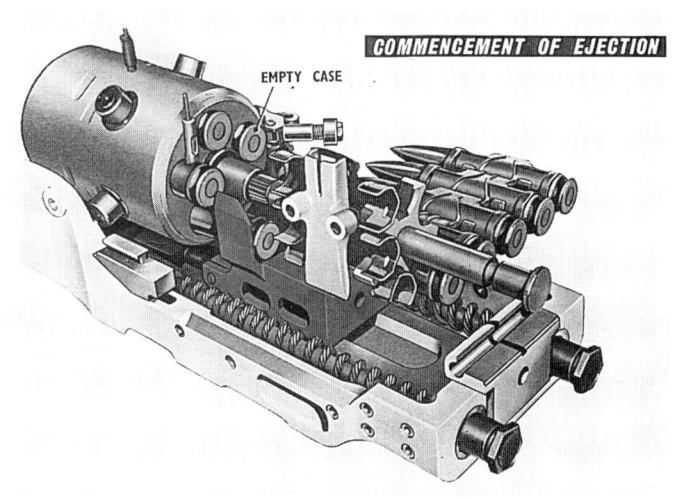

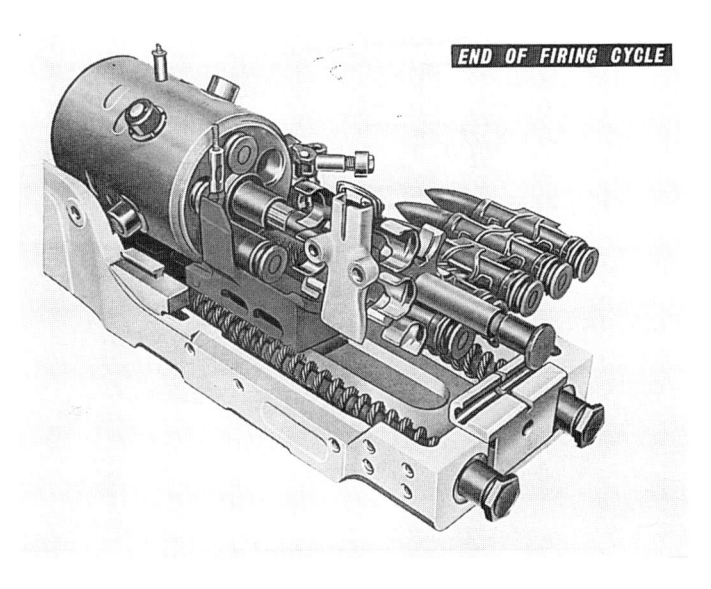

COMMENCEMENT OF FORWARD STROKE OF SLIDES

REAR RAMMER

COMMENCEMENT OF EJECTION

EMPTY CASE

END OF FIRING CYCLE

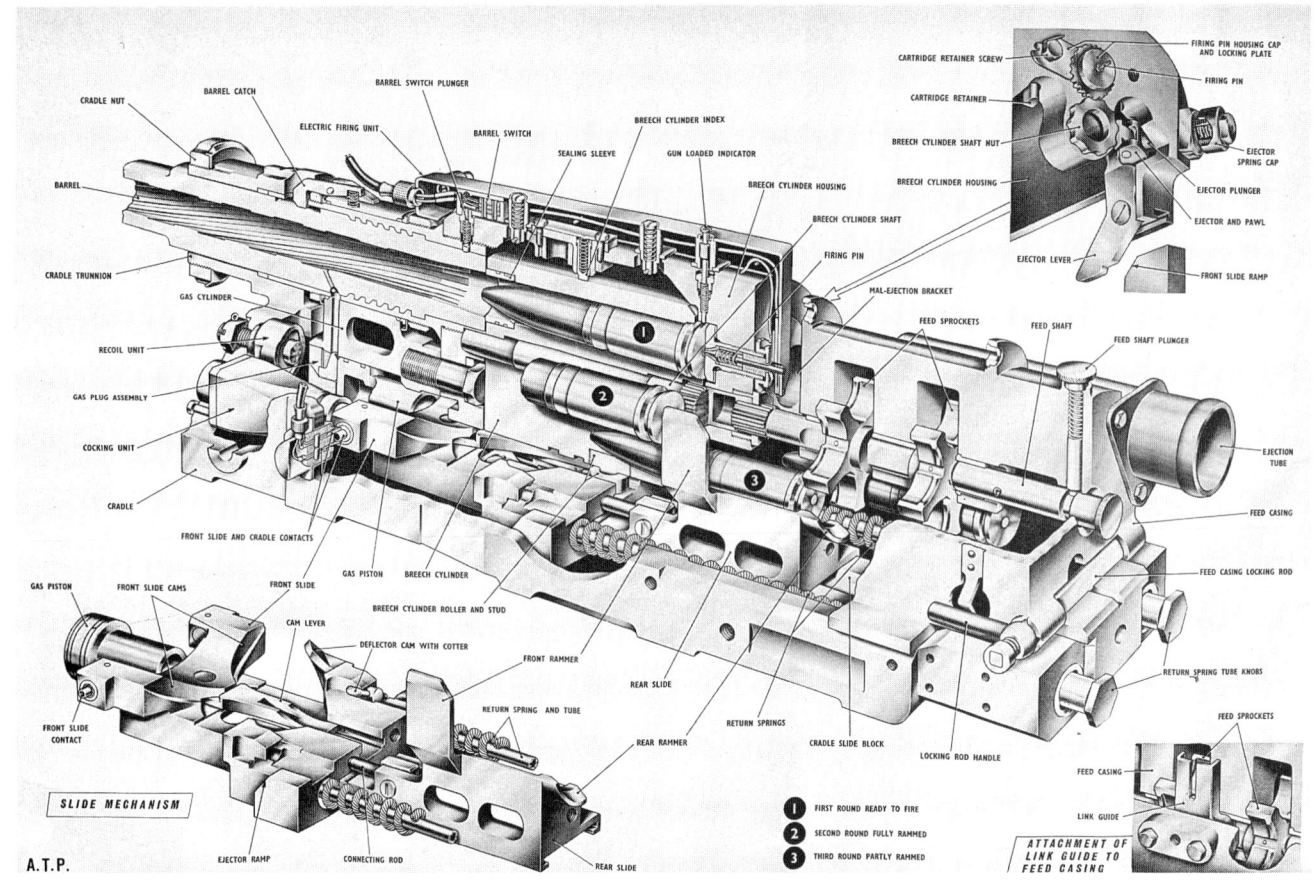

Rounds are fed into the revolving breech, the top round being fired on every stroke.

cylinder is prevented. The cam lever, which is guided by a pin running in a groove for part of the cycle, is now moved back to its original position. The rotation of the cylinder has caused the first cartridge to engage the ejector mechanism which is positioned at the side of the cylinder, and as the front slide moves forward the mechanism is triggered by a ramp to withdraw the cartridge case

A Harrier of No. 1 Squadron with gun pods fitted, just returned from the Falklands.

from the chamber. At the end of the stroke, the second round is positioned in line with the firing pin, the slides are fully forward and in contact with the gas piston and the cradle switch contacts are closed. Provided the firing trigger is still depressed, the firing circuit is again completed, the second round is fired, and the cycle repeated.

Gun cocking

Before the gun can be fired it must be cocked so that the first round is in line with the bore. This is achieved by a pneumatic cocking unit, consisting of a cylinder which operates the slides in the same manner as the gas piston. This unit is operated on the ground by an external air supply. The unit is operated three times, which completes the operational cycle, and the gun can be fired when the pilot activates the electrical system.

Ammunition

The following types of cartridge are used in the ADEN:

Mk 1Z AP	An armour-piercing projectile consisting of a tungsten core housed in a base and an aluminium nose assembly. The projectile weighs 10 oz (283.5 g) and is painted black.
Mk 2z practice	An inert projectile, ballistically similar to the Mk 32 HE, painted black at first, later light blue.
Mk 32 HE	An explosive projectile containing Torpex 5 and a No. 933 Mk 1 percussion fuse, this is painted buff.
Mk 5 HE	An explosive projectile containing Torpex 5 and a No. 944 Mk 1 percussion fuse. This projectile was painted bronze/green, and is now yellow (NATO).
Mk 6z HE	Similar to the Mk 32 HE, painted buff.

Fuses

Both the No. 933 and 944 Mk 1 fuses are direct-action percussion fuses, and are fitted with special spherical shutters to prevent arming until after firing. No. 933 fuse functions immediately on impact, while the 944 delays for a few milliseconds, enabling the projectile to enter the target before exploding.

Cartridge case design

When the gun was officially adopted in May 1949, experimental low-velocity cases made of steel and light alloy were tested, but these were not satisfactory, and all Service ammunition has been brass-cased. The initial cases were the type A (1945)* and the type B (1952), both low-velocity with a case length of 85.8 mm, differing only in rim diameter (31.9 mm and 33.3 mm). The need for high-velocity rounds resulted in extended length cases, for the type C (120.5 mm) and the D (111.2 mm). As a result of trials, a case based on the type D was adopted. Called the type J, it has a case length of

Initial drawing.

Firing circuit and case markings of an ADEN 30 mm round. Rounds are fired when the firing pin comes into contact with the electrically fired primer.

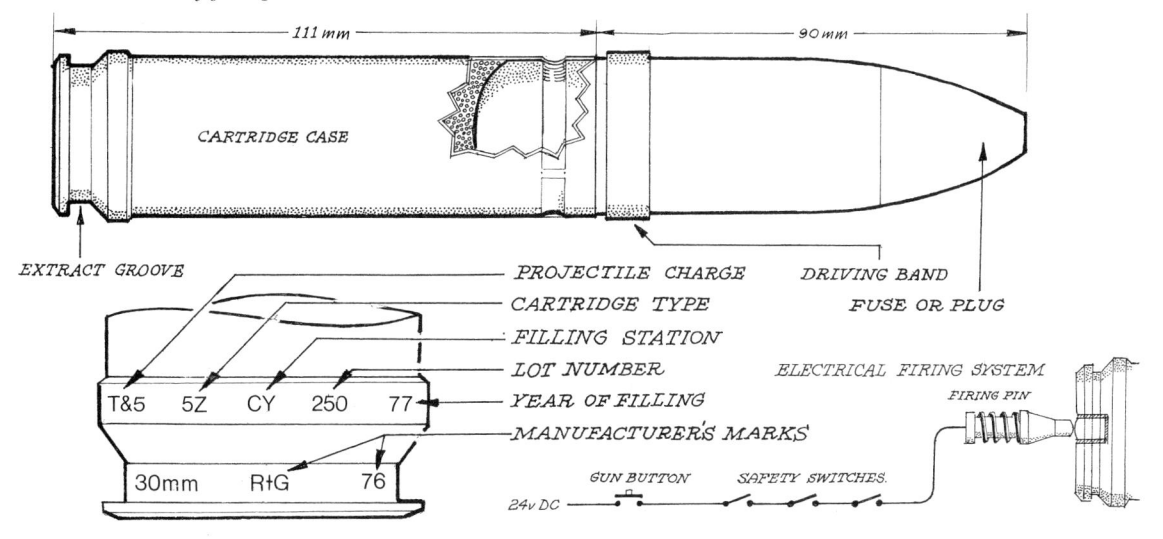

111.2 mm, a rim diameter of 33.3 mm and a belt diameter of 33.7 mm. This was approved for Service use, and is still in service today (1993). The overall length of the rounds has remained constant at 199.0 mm, the projectiles being adjusted in length to suit the breech mechanism.

Details of the ADEN 30 mm gun

Bore	30 mm (1.18 in)
Action	Gas-operated, electrically fired
Cyclic rate	1,200–1,400 rpm
Weight	192 lb (87 kg)
Muzzle velocity	2,625 ft/sec (800 m/sec)
Cooling	Air
Cocking	Pneumatic (ground only)
Length	64.5 in (1,638 mm)

The Mauser BK 27

In 1956 the Weapons Department at Farnborough was asked to carry out tests to find the ideal gun to arm a future fighter. After two-and-a-half years of comparative trials, the Department advised that the Swiss Oerlikon 304RK 30 mm offered the best all-round performance. However, 13 years later the Panavia Tornado aircraft programme committee decided that a new gun should be supplied by West Germany. No such weapon then existed, but it was agreed that, if a suitable single-barrel gun with high muzzle velocity could be produced in two years, it would be adopted.

The IWKA Mauser Werke of Oberndorf were given the contract. For some reason it was agreed that the ideal calibre would be 27 mm and, owing to the time factor, many features of the old 213C were

The 27 mm cal. Mauser Bordkanone produced at Oberndorf by IKWA Mauser.

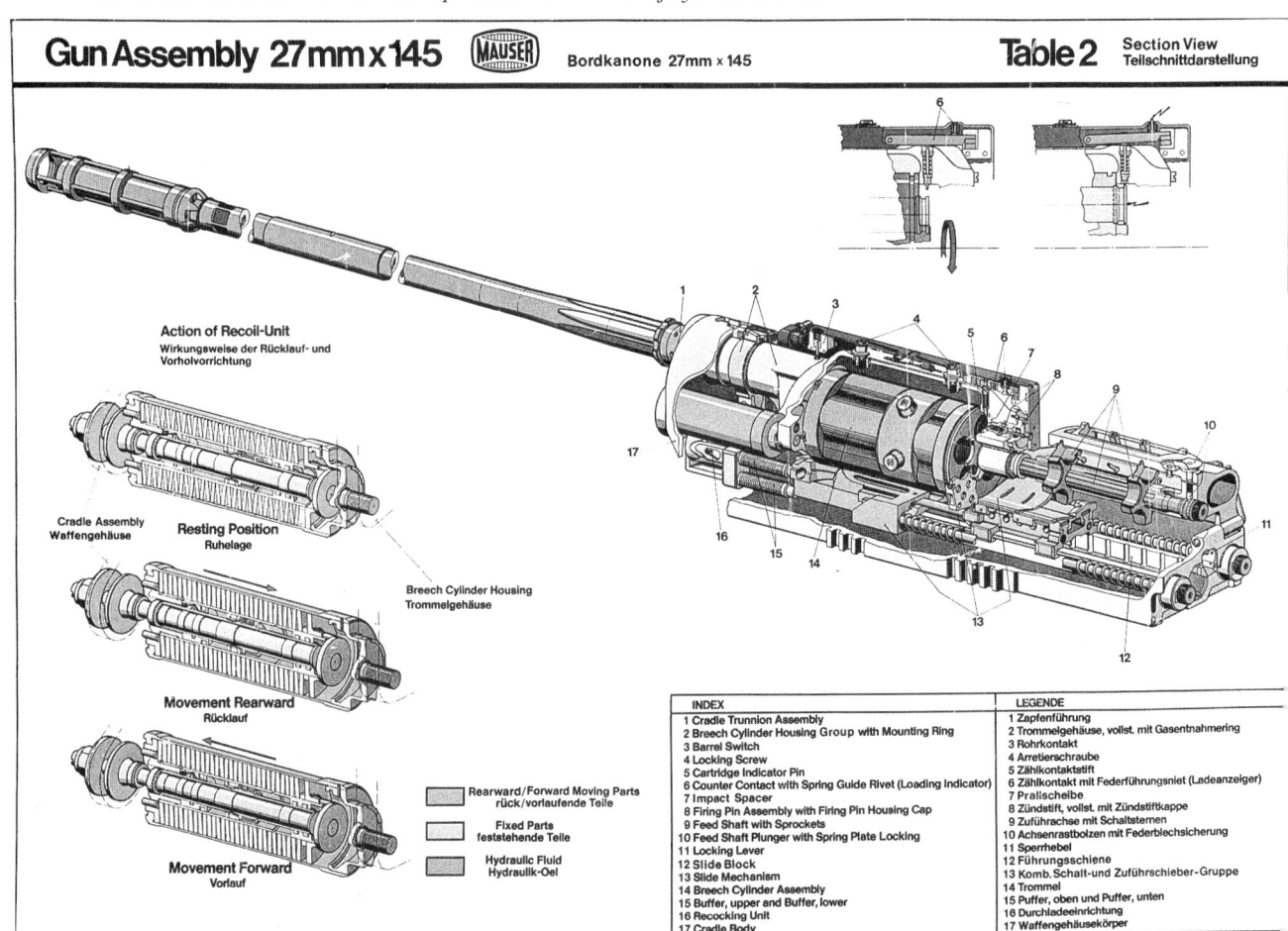

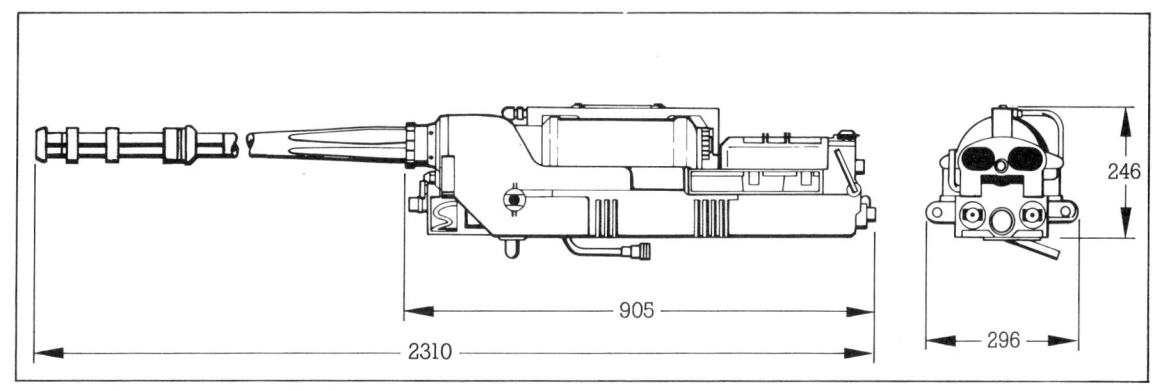

THE GUNS

Side and front views of the BK 27.

incorporated in the new design. The action is similar in most respects to the 213C system, in which firing gas pressure is used to feed rounds into a revolving breech chamber in four stages. An electrical firing system is retained and a high-velocity barrel is fitted which can be supported at the muzzle or in the centre. The cyclic rate can be regulated to suit different operational needs: 1,000 rpm for air-to-ground, or 1,700 for air-to-air use. Diehl and Dynamit Nobel have developed a suitable family of ammunition designed specifically for the weapon. The HE projectile is designed for air-to-air combat, having an electro-mechanical nose fuse which responds at extremely flat angles. It is supplied with

27 mm ammunition produced by Diehl & Dynamit Nobel for the BK 27.

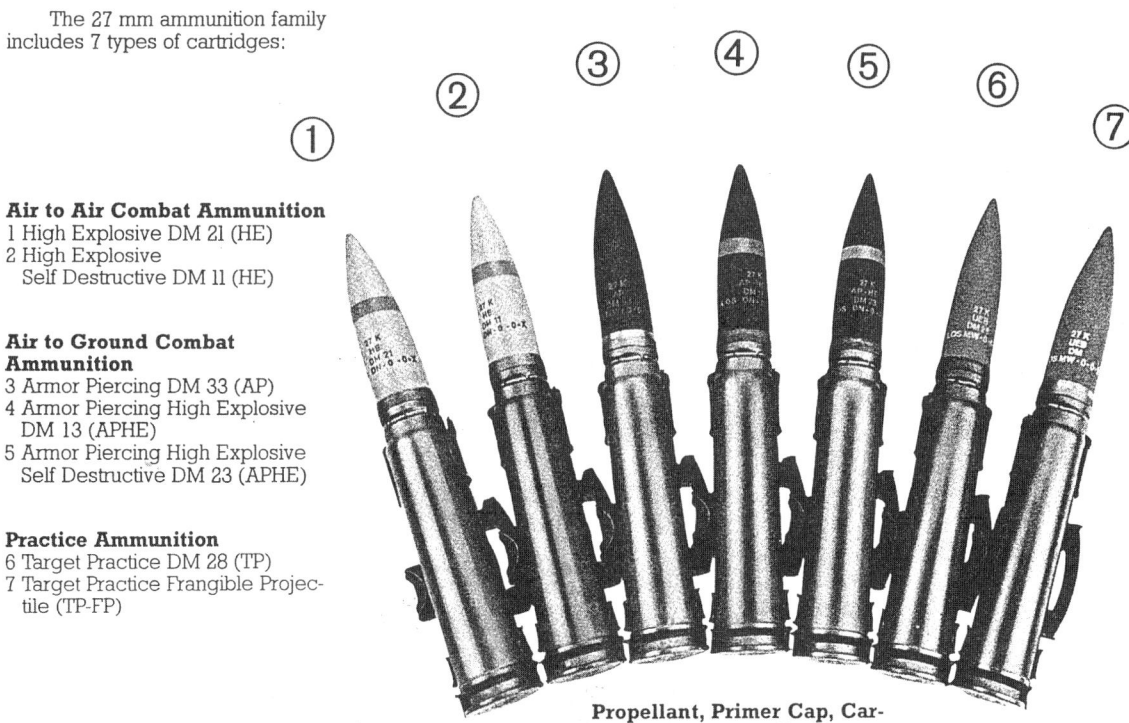

Ammunition family

The 27 mm ammunition family includes 7 types of cartridges:

Air to Air Combat Ammunition
1 High Explosive DM 21 (HE)
2 High Explosive Self Destructive DM 11 (HE)

Air to Ground Combat Ammunition
3 Armor Piercing DM 33 (AP)
4 Armor Piercing High Explosive DM 13 (APHE)
5 Armor Piercing High Explosive Self Destructive DM 23 (APHE)

Practice Ammunition
6 Target Practice DM 28 (TP)
7 Target Practice Frangible Projectile (TP-FP)

Propellant, Primer Cap, Cartridge Case as well as outer dimensions and weight are common to all 7 types.

or without self-destruct capability. The jacket contains a charge which provides an optimum fragmentation effect. An armour-piercing HE round is for use against armoured vehicles: incendiary composition is contained behind the nose cap, and a delayed-action fuse ensures fragmentation after penetration. The armour-piercing round contains a penetrator which is effective at flat angles, with zirconium for an incendiary effect. The practice projectile consists of an aluminium alloy with shear lines filled with iron powder pellets, to ensure that no effective fragments will cause a ricochet hazard.

The IDS version of the Tornado is fitted with two BK 27s in the fuselage, each supplied with 180 rounds, enough for 9 seconds' firing. The F3 interceptor has one gun on the right side only. The gun is also used on the Swedish JAS 39 Grypen mounted in the lower left side of the nose.

Though the BK 27 is a very powerful gun, the NATO standard round has been fixed at 25 mm. Another problem was the electrical firing system, which could be hazardous near high-powered radar transmitters, which could active the primer. However, this problem has recently been overcome.

Details of the Mauser BK 27

Bore	27 mm (1.06 in)
Action	Gas-operated revolver
Cyclic rate	Selectable, 1,000–1,700 rpm
Weight	220 lb (100 kg)
Muzzle velocity	3,363 ft/sec (1,025 m/sec)
Ammunition	HE, APHE, AP, practice
Length of round	9.57 in (243 mm)
Cooling	Air
Electrical power	22–30 volts DC, 2.7 amps

The ADEN 25

When the Harrier GR.5 was being developed it was decided to produce an updated ADEN gun using the new 25 mm calibre NATO 4173 ammunition. These rounds are longer than the 30 mm type, and the muzzle velocity and kinetic energy are increased.

The most significant modification is to the firing system. The original ADEN retained the German electrical firing system. After reports of some electrically fired ammunition detonating for no apparent reason, it was eventually found that high electromagnetic energy generated by powerful airfield radars could in some circumstances activate the electrical primers. This potentially dangerous situation had to be eliminated, consequently ordnance experts advised the Air Staff that future guns should not use electrically primed ammunition.

The designers of the new gun at the Royal Ordnance Factory devised a percussion system which could cope with the high rate of fire. The gun was similar in many respects to the original ADEN and, having the same dimensions as the older gun, can be retrofitted on the same mountings. The Harrier GR.5 installation has two detachable pods each containing a gun and 100 rounds. The complete installation weighs 948 lb (430 kg). The Aden 25s in a twin-gun installation give the Harrier and Jaguar a devasting firepower of 3,700 high-velocity projectiles per minute. The gun will also be used singly on the attack version of the Hawk.

The Aden 25 was the last gun to be designed at the Enfield Ordnance factory, which has now been relocated at the Royal Ordnance Factory at Nottingham, where the world-famous Enfield MOD Pattern Room has been rehoused.

Installationally interchangeable with the 30 mm, the ADEN 25 fires more powerful ammunition, using a percussion system. (MOD Pattern Room, Nottingham)

THE GUNS

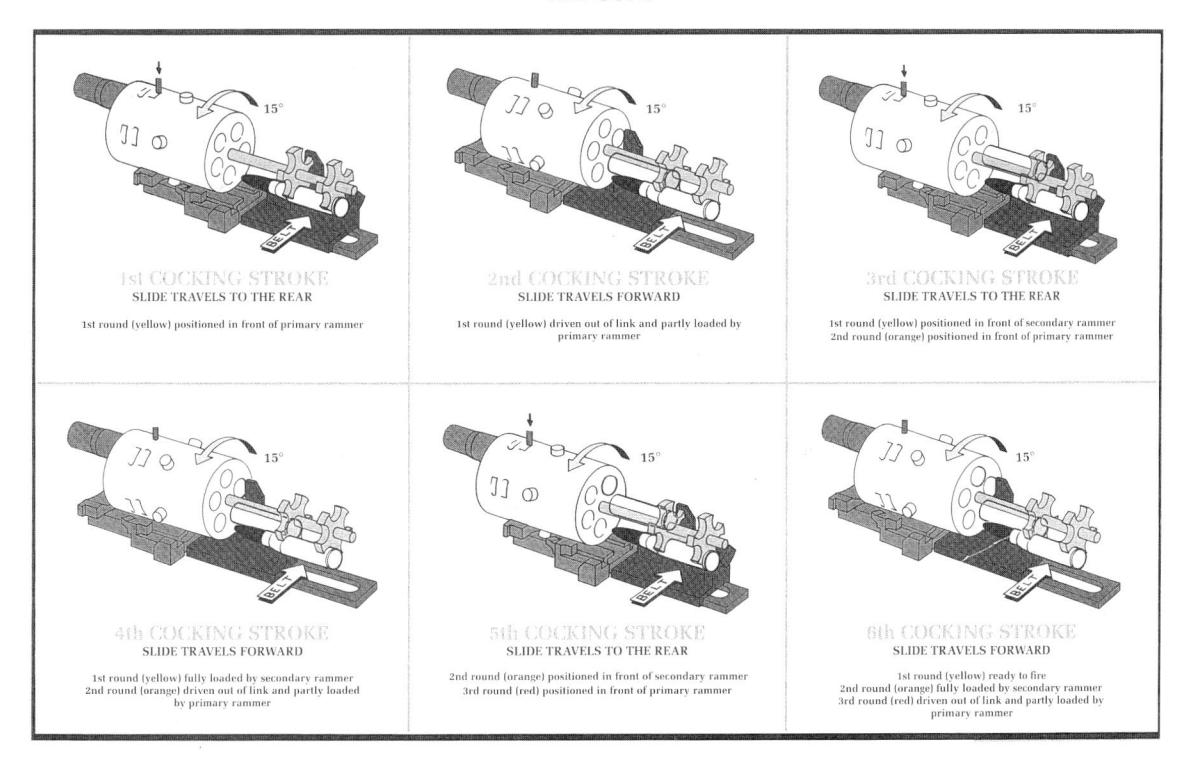

1st COCKING STROKE	2nd COCKING STROKE	3rd COCKING STROKE
SLIDE TRAVELS TO THE REAR	SLIDE TRAVELS FORWARD	SLIDE TRAVELS TO THE REAR
1st round (yellow) positioned in front of primary rammer	1st round (yellow) driven out of link and partly loaded by primary rammer	1st round (yellow) positioned in front of secondary rammer 2nd round (orange) positioned in front of primary rammer
4th COCKING STROKE	**5th COCKING STROKE**	**6th COCKING STROKE**
SLIDE TRAVELS FORWARD	SLIDE TRAVELS TO THE REAR	SLIDE TRAVELS FORWARD
1st round (yellow) fully loaded by secondary rammer 2nd round (orange) driven out of link and partly loaded by primary rammer	2nd round (orange) positioned in front of secondary rammer 3rd round (red) positioned in front of primary rammer	1st round (yellow) ready to fire 2nd round (orange) fully loaded by secondary rammer 3rd round (red) driven out of link and partly loaded by primary rammer

Above   *The functioning sequence of the ADEN 25.*

Below   *Sectioned drawing of the ADEN 25.* (MOD Pattern Room, Nottingham)

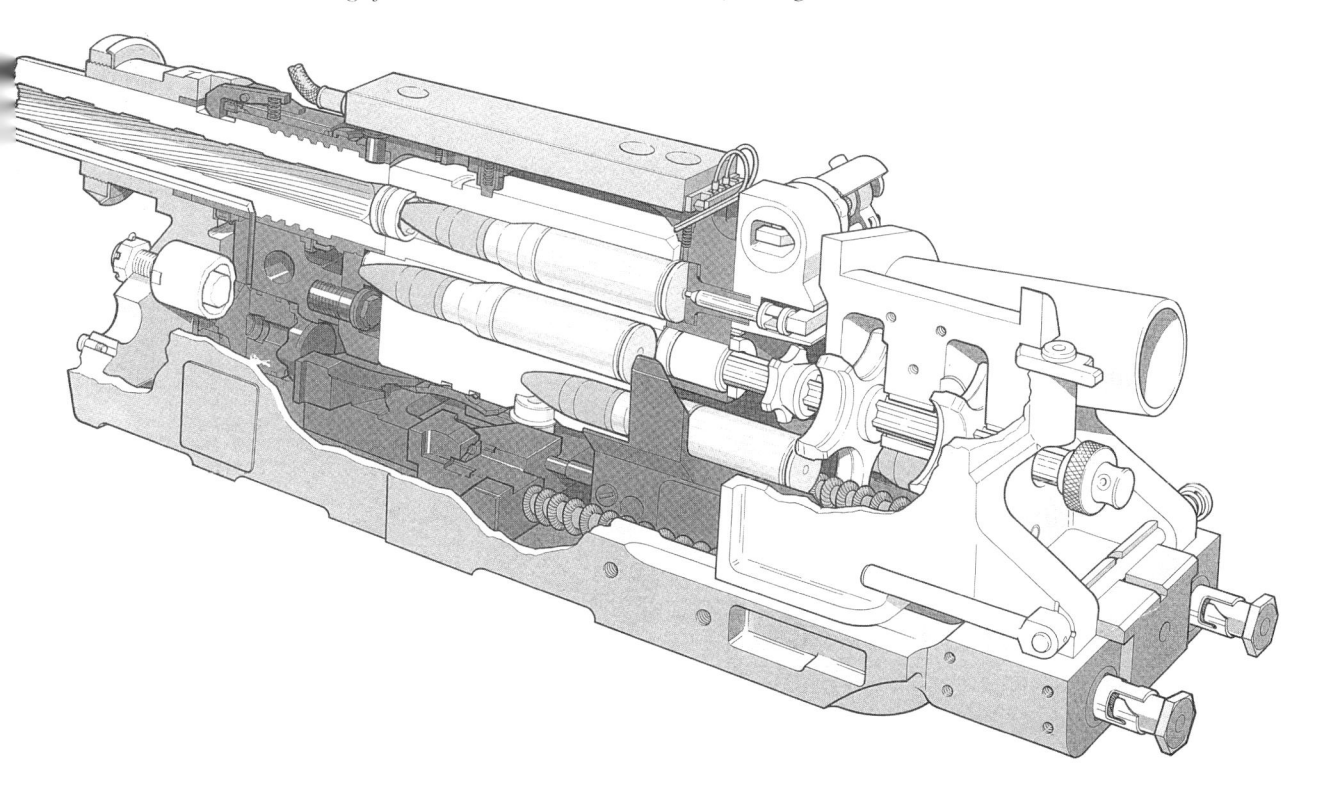

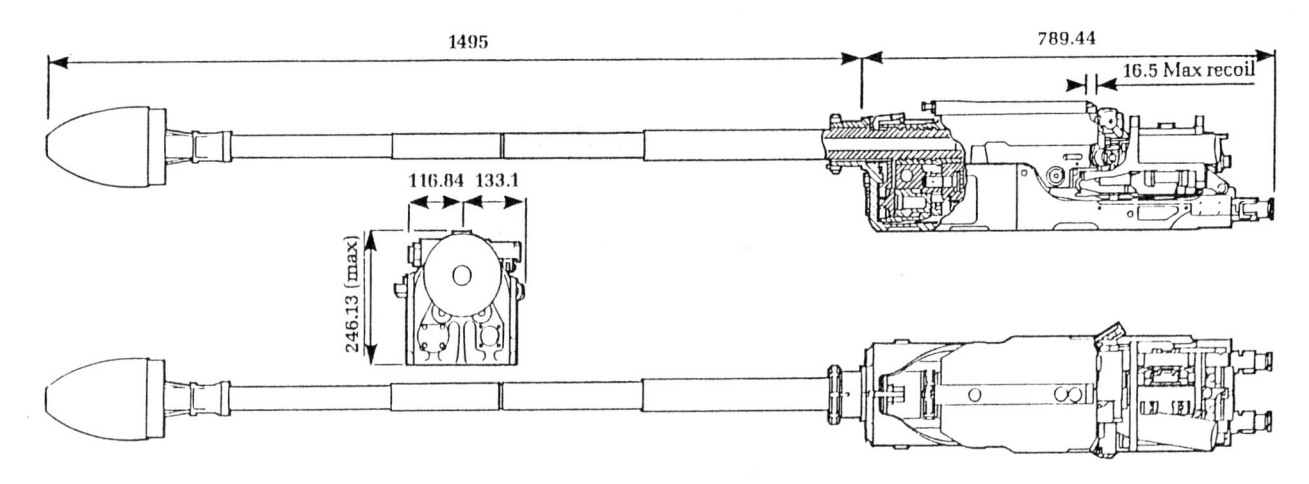

Above *General dimensions. The barrel cone is the blast diffuser.*

Below *One of four projectiles used in the ADEN 25, the multi-purpose round is designed to produce maximum penetration and fragmentation effect.*

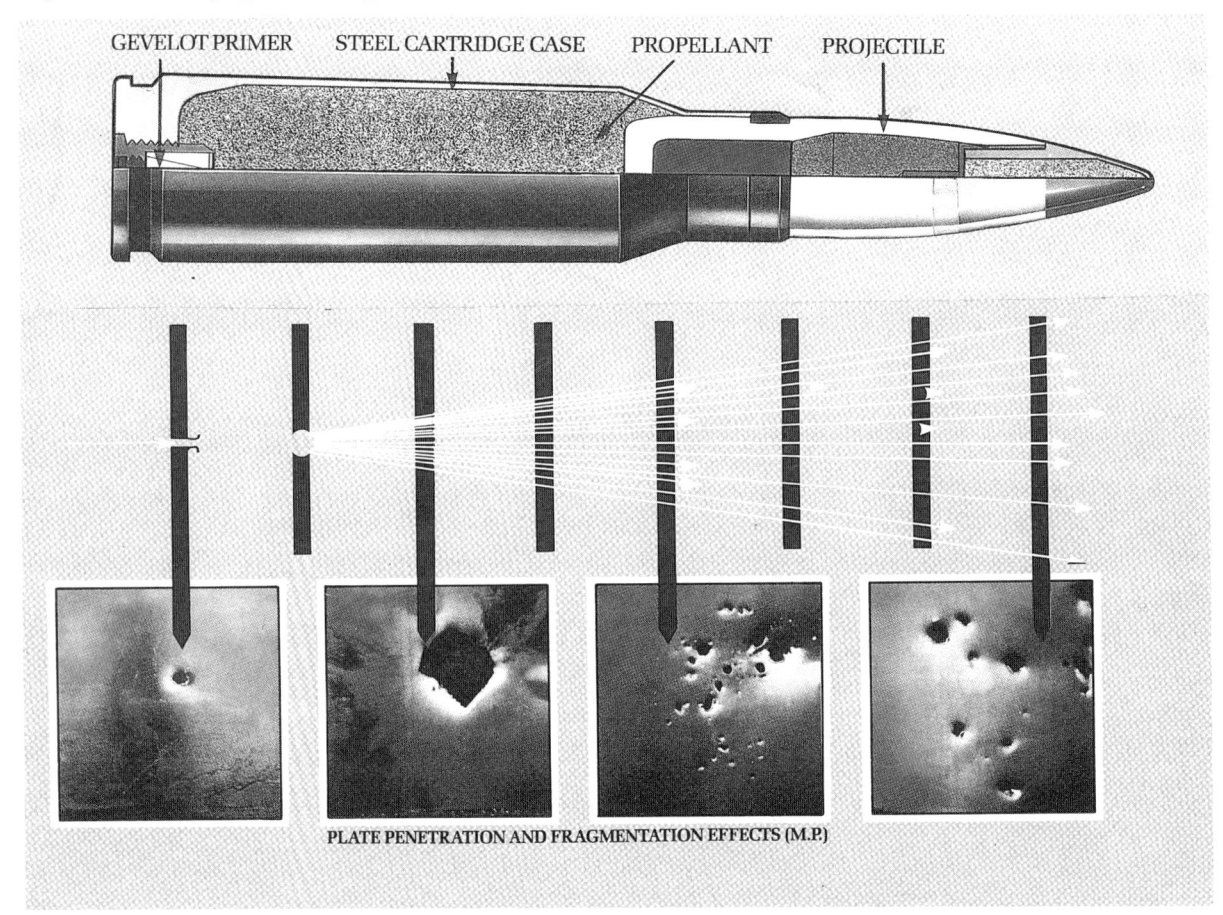

PLATE PENETRATION AND FRAGMENTATION EFFECTS (M.P.)

Details of the ADEN 25

Bore	25 mm (0.98 in)
Action	Gas-operated revolving breech
Cyclic rate	1,650–1,850 rpm
Weight	216 lb (98 kg)
Muzzle velocity	3,445 ft/sec (1,050 m/sec)
Ammunition	Nato 4173, HE, AP, TP
Recoil	9,790 lb (4,441 kg)
Rifling	Right-hand progressive parabolic twist, 16 grooves
Length	90 in (2,286 mm)

Below *The GAU-4 is a gas-operated version of the M61 Vulcan.* (General Electric)

Bottom *Action of the GAU-4.*

The General Electric GAU-4 20 mm Vulcan

The Royal Air Force Phantoms had no internal gun, but could be fitted with the General Electric SUU-23/A pod, housing the very potent GE Vulcan GAU-4 20 mm gun. This weapon is one of the family of 'Gatling-type' multi-barrelled guns developed from the original GE M61 and manufactured by the General Electric Company, Burlington, Vermont, USA. The GAU-4 is a gas-operated version; an electric inertia starter is used for the start sequence, after which it is self-powered.

The principle of operation can be understood by referring to the illustration, which shows the gun with one barrel removed. Four of the six barrels have small holes drilled in them which coincide with the holes in the piston chamber (A). As gas from the barrels is released into the piston

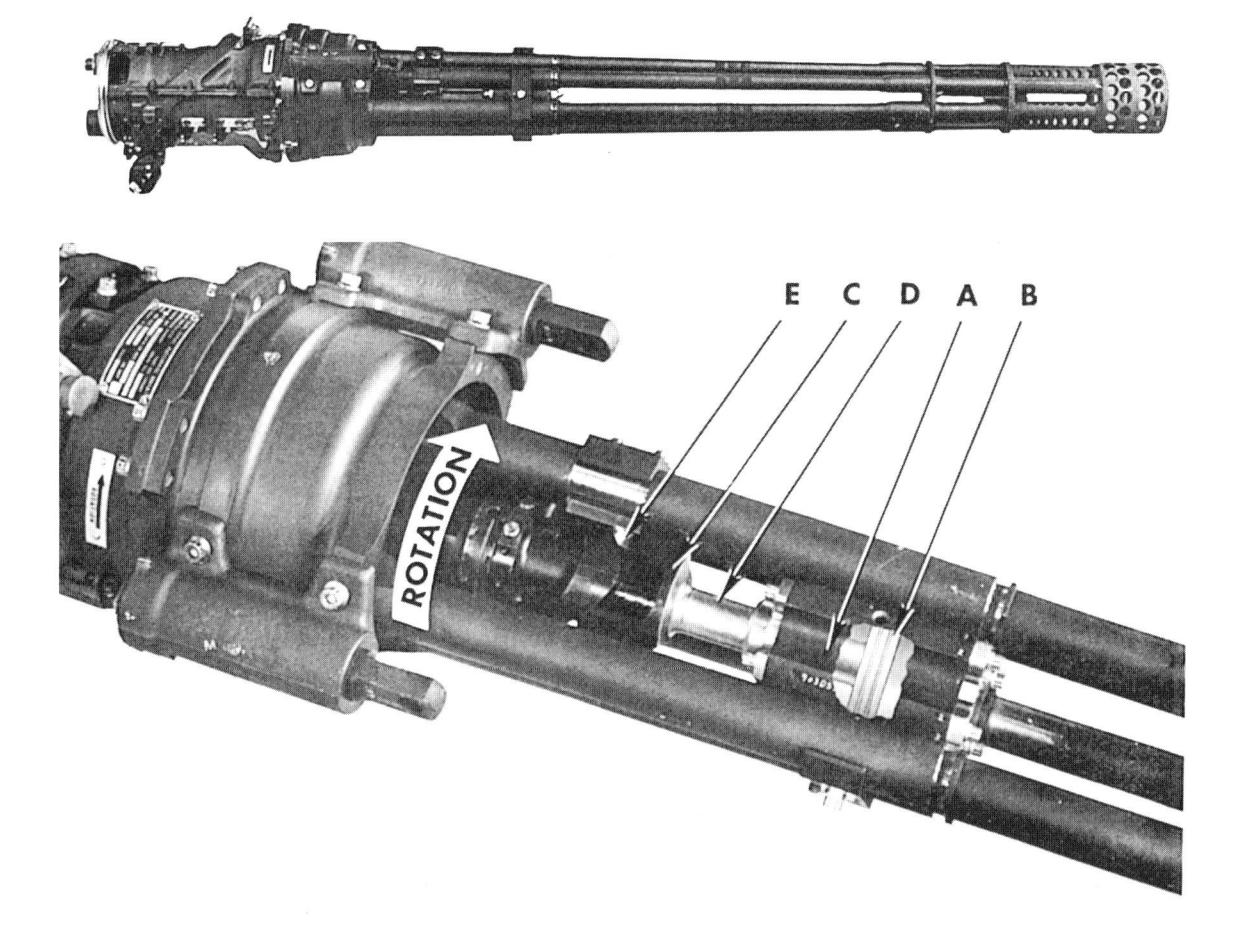

In RAF service the GAU-4 was mounted in the SUU-23A pod.

chamber, the pressure on the piston (B) pushes it forward, sliding the cam path (C) on the piston shaft. The shaft (D) can move fore and aft but cannot rotate. Cam follower (E) is fixed to two of the barrels, and follows the course of the cam path, rotating the barrel cluster. The first half of a revolution is complete when the piston shaft reaches its foremost position; the second cycle is the reverse of the first. Two of the barrels have holes positioned in front of the piston, so that when these barrels reach the firing position the gun gas pushes the piston rearwards, completing the cycle. Initial rotation is provided by an electric inertia starter, which accelerates the action to a firing rate of 5,400 rpm and then

A SUU-23A pod fitted to a Phantom.

Details of the GAU-4 20 mm Vulcan

Bore	20 mm (0.79 in)
Action	Gas-operated, electric start sequence
Cyclic rate	Variable to 6,000 rpm
Weight	276 lb (125 kg)
Muzzle velocity	3,380 ft/sec (1,030 m/sec)
Number of barrels	Six
Time to rated speed	0.2/0.4 sec
Stopping time	0.2/0.4 sec
Power required	Firing circuit: 250 VDC, .5 amps
	Clearing cam: 28 VDC, 298 amps
	Starting power: 10 amp 3-phase 400 Hz
Length	72 in (1,829 mm)

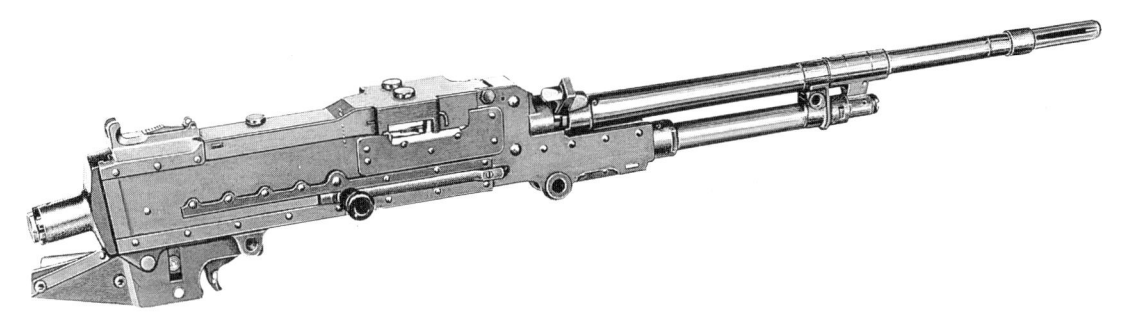

The Fabrique Nationale MAG60. (Fabrique Nationale).

disengages automatically. The cam and piston then drive the action at 6,000 rpm.

The GAU-4 fires electrically primed 20 mm ammunition such as M53A2 AP/INC, HE incendiary, and M55A2 ball. The six barrels are rigidly clamped together, producing a minimum dispersal pattern, though special muzzle clamps can be used which provide wider or elliptical patterns. The ammunition is stored in a 1,200-round drum which is fed by a conveyor system from the rear of the pod. The pod is hung by two suspension units, through which pass electric power, ammunition state signal cabling and firing circuit.

The Fabrique Nationale FN MAG 60/40

With the introduction of armed helicopters and light attack aircraft, the RAF has a requirement for a rifle-calibre machine-gun. The gun chosen to fill this role is the FN MAG 60/40 made by the Belgian Fabrique Nationale Company of Herstal. It is the aerial version of the FN58 GPMG, most parts being common to both guns.

The gun can be supplied in two versions. The free-mounted model has a rear spade handle and pistol grip, and is fed from a 250-round side-mounted box, spent cases being collected in a bag under the gun. This version is usually mounted on a pintle, with a sample 75 mm ring-and-bead sight. The fixed model has been used on experimental helicopter pods developed at Farnborough. It is also used on light attack aircraft in the fire suppression role. The fixed model is electrically fired, and can have either right- or left-hand feed. It is gas-operated and is of robust construction to withstand heavy use.

This weapon has been found to be relatively trouble-free, and has been used in small numbers by the RAF.

Details of the Fabrique Nationale FN MAG 60/40

Bore	7.62 mm (0.3 in)
Action	Gas-operated
Cyclic rate	1,000 rpm
Weight	10.2 kg (22.5 lb)
Muzzle velocity	2,756 ft/sec (840 m/sec)
Length with flash hider	1,255 mm (49.4 in)

FN MAG60/40 on a Sea King pintle mounting.

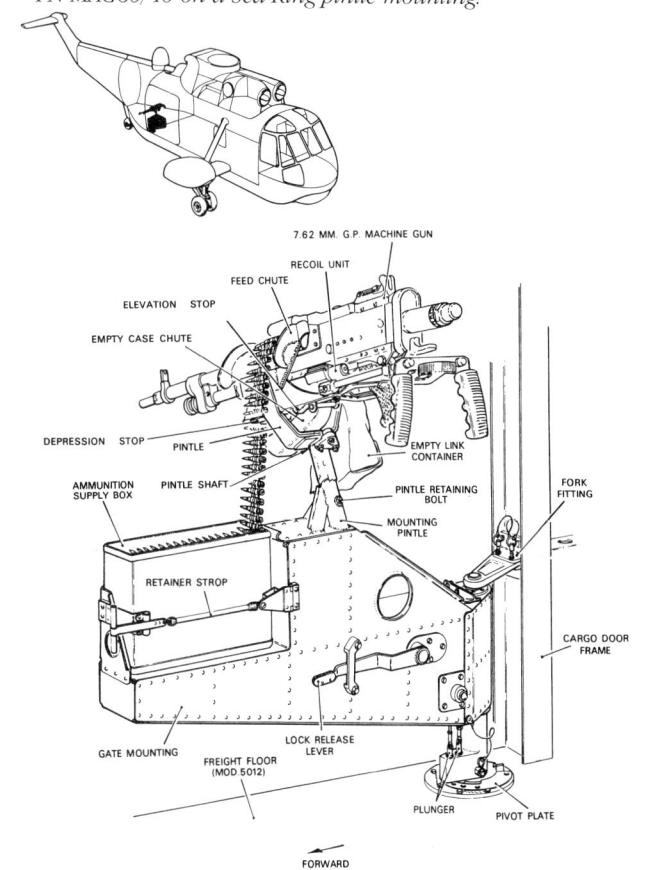

1—12 Fig 8 Machine Gun Installation

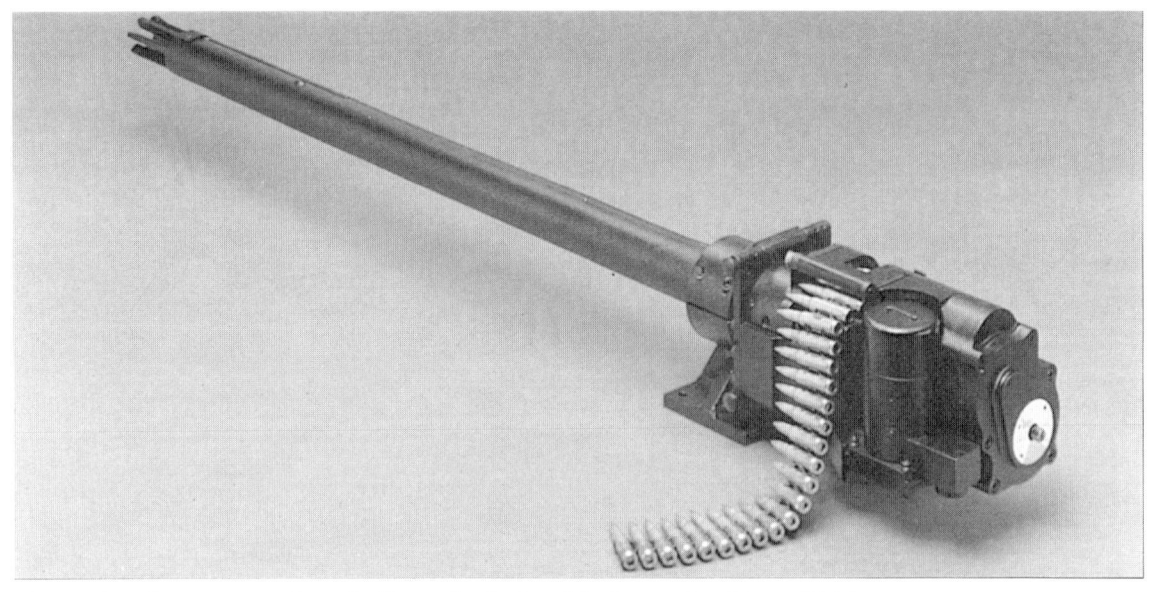

The Hughes Chain Gun, made under licence by the Royal Ordnance Small Arms Division.

The Hughes Chain Gun

The Hughes Chain Gun was developed by the Hughes Helicopter Company primarily for helicopters and aircraft. The Royal Ordnance Small Arms Division have manufactured the gun for trials in helicopters and armoured vehicles.

The action of the gun is unique in that it is powered by an endless chain and gearbox driven by an electric motor. All actions within the weapon are precisely timed and controlled, the chain drive permitting a gun cycle which operates from an open bolt. A conventional rotary bolt-locking action with fixed extractors locks the barrel extension and ensures that firing stresses are transmitted to the trunnions. A shoe on the endless chain engages in the bolt, carrying it forward to chamber the round, hold it closed, and then retracts it to eject the spent case. Empty cases are ejected forward, which prevents a build-up of cases on the floor of helicopters and cuts down gas fumes.

The makers claim the gun to be the most reliable ever built, owing to its uncomplicated design principle. It is available in three sizes, 7.62 mm, 20 mm and 30 mm, British interest being in the rifle-calibre weapon, which is being considered for helicopters and light fighters.

Details of the Hughes Chain Gun

Bore	7.62 mm, 20 mm, 30 mm
Action	Electric
Cyclic rate	520 rpm
Weight	Long barrel: 17.86 kg (39.38 lb); short barrel: 13.7 kg (30.2 lb) (7.62 mm)
Muzzle velocity	2,829 ft/sec (862 m/sec)
Ammunition feed	Disintegrating links
Rifling	Four grooves, one right-hand turn in 305 mm
Length	(Long) 1,250 mm, (short) 660.4 mm
Drive motor	24 v DC .3 hp, current 13 amps, .6, 400 rpm

Summary of guns used by the British Air Services

Gun	Bore in	Bore mm	Weight lb	Weight kg	Cyclic rate (rpm)	Muzzle velocity ft/sec	Muzzle velocity m/sec	Projectile weight	Action	Use
Lewis	.303	7.7	18.5	8.3	600–1,000	2,240	744	.4 oz	Gas	General service use, fixed and free-mounted, 1914–40.
Vickers 1½-pdr	1.48	37	265	120	Manual	1,300	395	1.5 lb	Recoil	Experimental only.
Vickers Mk I*, Mks II and III	.303	7.7	24.5	11.1	850–900	2,400	723	.4 oz	Recoil	General Service use, fixed pilot's 1916–34.
Elswick Ord. Co.	1.48	37	—	—	—	—	—	—	—	Proofed Eastchurch 12-7-13. No further details.
COW 1-pdr.	1.48	37	120	54	Semi auto	1,740	521	1.0 lb	Recoil	Experimental
Davis 2-pdr	1.57	40	45	20.4	Manual	1,200	366	2.0 lb	Opp. chg	Very limited op. use; mainly experimental.
Davis 6-pdr	2.24	57	165	78.4	Manual	1,000	305	6.0 lb	Opp. chg.	As above.
Vickers Crayford 'Rocket Gun'	1.59	40.4	47	21.3	Manual	1,050	320	1.2 lb	Recoil	Limited op. use; intruder ops. 1917.
Vickers 1-pdr	1.44	37	279	126	300	1,800	549	1.0 lb	Recoil	Intruder ops. 1917, esp. anti-Zeppelin.
Vickers 1 inch	1.0	25	110	50	150	1,542	470	.44 lb	Recoil	Experimental only.
Vickers/COW 1½-pdr	1.44	37	140	63.5	60	2,000	610	1.5 lb	Recoil	Experimental 1914–40.
Beardmore Farquhar	.303	7.7	16.2	7.3	450–550	2,427	740	.4 oz	Gas	Experimental only.
Adams-Willmott	.303	7.7	23	10.4	720	2,400	732	.4 oz	Gas	Experimental only.

BRITISH AIRCRAFT ARMAMENT

Gun	Bore in	Bore mm	Weight lb	Weight kg	Cyclic rate (rpm)	Muzzle velocity ft/sec	Muzzle velocity m/sec	Projectile weight	Action	Use
Vickers Class C	.5	12.7	52	23.6	340–600	2,550	777	.4 oz	Recoil	Experimental only.
Vickers Class F	.303	7.7	24	10.9	600	2,200	671	.4 oz	Recoil	Limited use and export.
BSA 0.50 in	.5	12.7	46	20.9	400	2,600	792	1.3 oz	Recoil	Experimental only.
Vickers Class K	.303	7.7	20.5	9.3	950–1,100	2,400	732	.4 oz	Gas	General Service use 1937–46.
Browning Mk II*	.303	7.7	21.8	9.9	1,150	2,660	811	.4 oz	Recoil	General Service use 1937–58.
Hispano Suiza Mk II	.78	20	109	49.4	650	2,880	878	4.4 oz	Gas	General service use 1941–70.
Vickers Class S	1.58	40	295	134	125	1,800	549	2.5 lb	Recoil	Used on Hurricane IID in Europe and Far East.
Browning .5 in	.5	12.7	64	29	750–1,200	2,750	838	1.17 oz	Recoil	Service use turrets, limited fighter use.
Rolls-Royce 40 mm	1.58	40	328	148	120	2,440	744	2.5 lb	Recoil	Experimental only.
Rolls-Royce .5 in	.5	12.7	49	22.2	1,000	2,340	713	1.4 oz	Recoil	Experimental only.
Molins 6-pdr	2.25	57	1,800	816	60	2,600	791	7.1 lb	Recoil	Armed Mosquito XVIII.
FH Recoilless	4.5	114	1,440	653	50				Recoil	Experimental only.
ADEN 30 mm	1.18	30	192	87	1,200–1,400	2,625	800	8 oz	Gas	Service fighter use 1956–93+.
Mauser BK 27	1.06	27	220	100	1,000–1,700	3,363	1,025		Gas	Tornado IDS and F3.
ADEN 25	.98	25	216	98	1,650–1,850	3,445	1,050	6.4 oz	Gas	Harrier, Jaguar, Hawk.
GE GAU-4 Vulcan	.79	20	276	125	6,000	3,380	1,030	4.6 oz	Gas	Phantom pod.
FN MAG 60/40	.3	7.62	22.5	10.2	1,000	2,756	840	.38 oz	Gas	Helicopter, pintle mounting.
Hughes Chain Gun	.3	7.62	30	13.7	520	2,829	862	.38 oz	Electric	Experimental only.

Appendices

1. Gun firing systems

When fixed pilots' guns were mounted out of reach of the pilot, various methods were used to fire them. As described in the Vickers gun section, rods, cranks and finally the hydraulic Constantinesco-Colley was used to fire the guns in the 1914–18 war. When metal cantilever aircraft were introduced in the late thirties the multiple guns of the fighters and gun turrets needed quite elaborate firing mechanisms. Three systems were used: electric or hydraulic for turret guns, and pneumatic for single-seat fighters.

Spitfire and Hurricane guns were fired in the following manner. A small Heywood or BTH air compressor was powered by a take-off on the aero-engine. Pressurized air was taken from the compressor to an oil reservoir through which the air was passed. It was then routed through an oil trap to an air container of 620 cu in capacity at 300 lb/sq in pressure. It was then taken through an air filter to a pressure-reducing valve. From the valve it was taken through 7/16 OD copper piping to the gun button on the pilot's control handle. When the gun button was pressed, air pressure taken to the

Control handle of the Hawker Hart showing the CC gear leavers. (Roy Bonser)

Spitfire (left) and Hurricane control handles. The gun buttons controlled the pneumatic pressure to the gun-firing actuators and fire-and-safe units. (Roy Bonser)

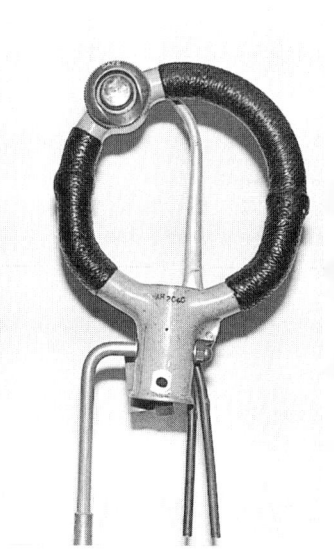

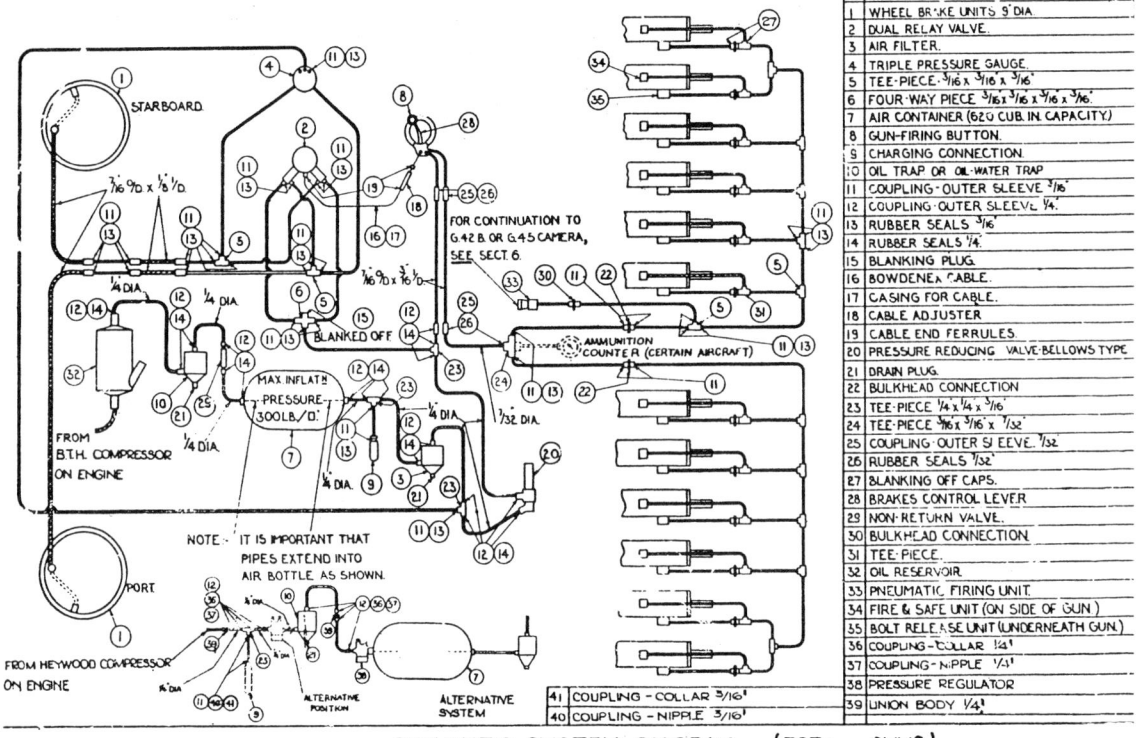

No	DESCRIPTION
1	WHEEL BRAKE UNITS 9" DIA.
2	DUAL RELAY VALVE.
3	AIR FILTER.
4	TRIPLE PRESSURE GAUGE.
5	TEE-PIECE - 3/16 x 3/16 x 3/16
6	FOUR-WAY PIECE 3/16 x 3/16 x 3/16 x 3/16.
7	AIR CONTAINER (620 CUB. IN. CAPACITY.)
8	GUN-FIRING BUTTON.
9	CHARGING CONNECTION.
10	OIL TRAP OR OIL-WATER TRAP
11	COUPLING - OUTER SLEEVE. 3/16"
12	COUPLING - OUTER SLEEVE 1/4".
13	RUBBER SEALS 3/16"
14	RUBBER SEALS 1/4"
15	BLANKING PLUG.
16	BOWDENEX CABLE.
17	CASING FOR CABLE.
18	CABLE ADJUSTER
19	CABLE END FERRULES.
20	PRESSURE REDUCING VALVE-BELLOWS TYPE
21	DRAIN PLUG.
22	BULKHEAD CONNECTION.
23	TEE-PIECE - 1/4 x 1/4 x 3/16"
24	TEE-PIECE - 3/16 x 3/16 x 1/32"
25	COUPLING - OUTER SLEEVE. 7/32"
26	RUBBER SEALS 7/32"
27	BLANKING OFF CAPS.
28	BRAKES CONTROL LEVER
29	NON-RETURN VALVE.
30	BULKHEAD CONNECTION.
31	TEE-PIECE.
32	OIL RESERVOIR.
33	PNEUMATIC FIRING UNIT.
34	FIRE & SAFE UNIT (ON SIDE OF GUN)
35	BOLT RELEASE UNIT (UNDERNEATH GUN.)
36	COUPLING - COLLAR 1/4"
37	COUPLING - NIPPLE 1/4"
38	PRESSURE REGULATOR
39	UNION BODY 1/4"
40	COUPLING - NIPPLE 3/16"
41	COUPLING - COLLAR 3/16"

PNEUMATIC SYSTEM DIAGRAM (·303 in. GUNS.)

Above *Layout of the pneumatic gun-firing system of the Hawker Hurricane.* (Dunlop)

Left *The Palmer hydraulic gun-firing system of a Frazer-Nash turret. Bowden cables from the gunner's triggers operated the valve.*

gun bays by piping was connected to a 'T' joint at every gun position. Flexible tubing from the 'T' went to a spring-loaded gun firing unit, and the fire and safe unit on the side of each gun. In some aircraft there was provision for an air-powered ammunition rounds counter. Pressure from the system was also used to supply the wheel brakes.

The system used on turret guns depended on the type of turret. Boulton Paul and Bristol used mainly electrical solenoid release units, which were usually trouble-free and could be accurately controlled by gunfire interrupters. Frazer-Nash turrets used hydraulic units, in which the gunner's triggers operated Bowden cables which controlled hydraulic valves, releasing oil to the gun firing units. The fire and safe units on turret guns were usually operated by hand.

The firing systems of aircraft such as the Beaufighter and Mosquito fighter differed slightly from single-seat fighters. A pneumatic system was used but, owing to the power needed to fire and recock the Hispano cannons, a more powerful compressor and storage system was needed. There was also a need to select which guns were to be fired; this was achieved by an electro-pneumatic system. A gun selector switch unit in the cockpit controlled electrically operated pneumatic valves which could be operated as needed.

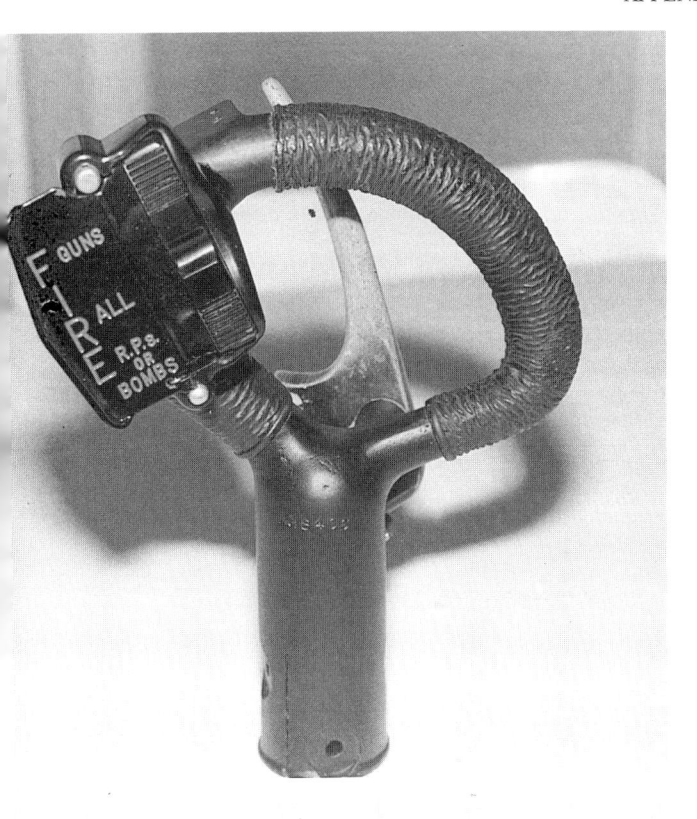

There were some problems with hydraulic and air systems: it was difficult to prevent oil fouling from the flexible glands and rotating joints of hydraulic mechanisms, and air-powered systems were sometimes prone to freezing at high altitudes. For these reasons, and the increasingly powerful electrical systems used on later aircraft, all-electric firing became standard.

The Palmer company produced the hydraulic firing systems of Frazer-Nash turrets, while pneumatic systems were developed by Dunlop, who also produced the elaborate pilots' control handles.

Left *A Vampire control handle. The pilot could select guns, rockets or both. The lever was for the brakes and the button for radio transmit.* (Roy Bonser)

Below *A Dunlop electrical firing unit for Browning guns (0.303 or 0.5 cal.). Current consumption was 5 amps.*

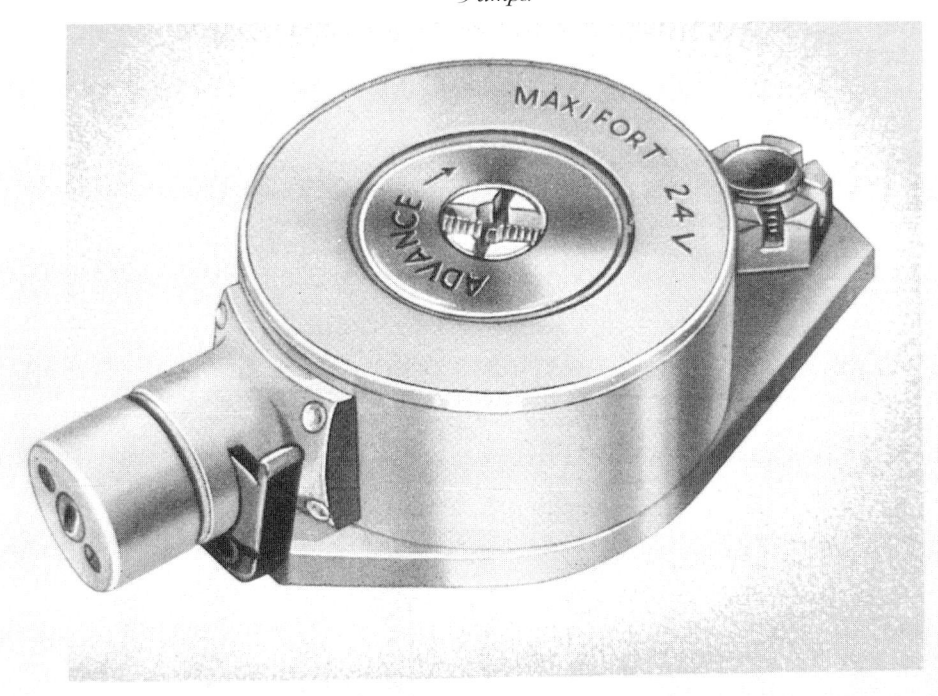

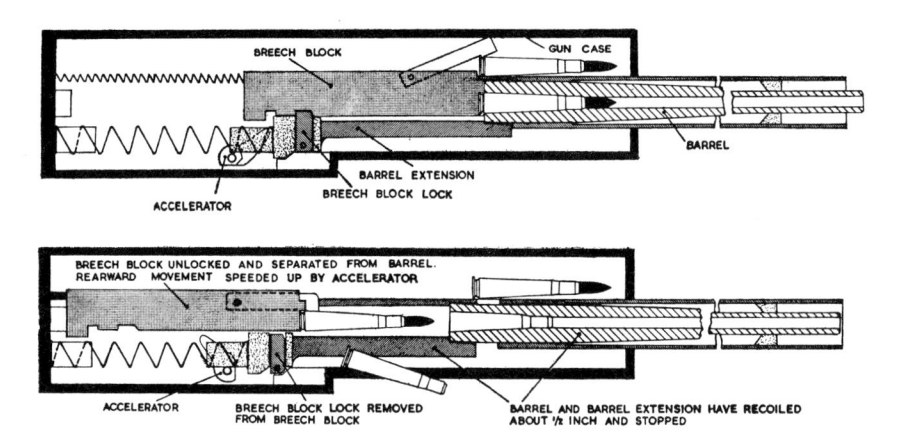

Recoil operation

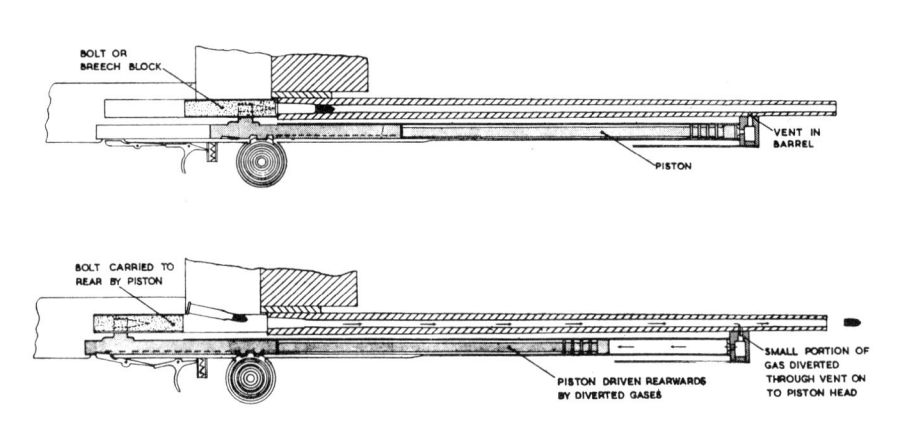

Gas operation

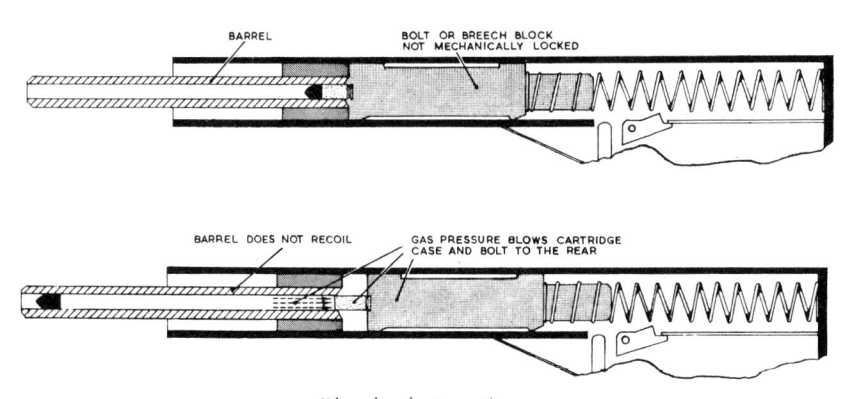

Blow-back operation.

Opposite *The three types of automatic gun action.*

When jet fighters were introduced, the pilot needed fingertip selection of his guns and rocket armament. This was provided by multi-switch units in his control handle. Dunlop produced a family of these units, and still provides the RAF with this equipment.

2. The action of automatic guns

Automatic guns used until the advent of modern electrically powered weapons operated using one of three basic systems: *recoil* – in which the force of the recoil is used to operate the reciprocating feed mechanism, the round being locked when fired; *gas* – where the mechanism is powered by a piston actuated by propellent gas pressure bled from the barrel when the round is fired, and used to force back a piston which operates the action; and *blow-back* – in which a heavy breech-block is fired unlocked, blown rearwards by the base of the round and returned by a strong spring, collecting a new round which is chambered by the face of the breech-block and fired by a firing pin. Recoil and gas-operated guns have been used in British aircraft since the early days of military aviation. Guns using the blow-back principle have never been accepted, the fact that the action is not locked at the moment of firing being considered unsafe.

3. Special ammunition used in the First World War

One of the more unpleasant aspects of aerial warfare is the fact that the loser in any engagement is likely to be burned alive, owing to the highly volatile nature of the fuel carried in aircraft. In the early days of aviation airmen often jumped to oblivion rather than stay in the aircraft and be incinerated. It followed that armament designers did their best to encourage this state of affairs, and designed explosive and incendiary bullets. There were many designs, some more successful than others. The rounds shown here are some of the more exotic fired from Lewis and Vickers guns.

The Pomeroy explosive bullet
The Pomeroy was used by the RFC from 1917. In the final version, the steel ball projected onto the highly sensitive nitro-glycerine warhead on impact.

The RTS explosive bullet
This cartridge was the ancestor of the 20 mm HE/I Hispano round. The warhead detonated at the target, followed by an incendiary flash. The incendiary mixture consisted of white phosphorus and tungsten powder, the warhead three grains of nitrocellulose in sawdust.

The Buckingham incendiary
The Buckingham was the first incendiary bullet used by the air services. On being fired, the solder over the flame holes melted, igniting the incendiary mixture of phosphorus and aluminium powder. This burned with a trail of white smoke which was found to be a useful trace.

Special ammunition (left to right) RTS explosive bullet, Buckingham incendiary bullet and Pomeroy explosive bullet.

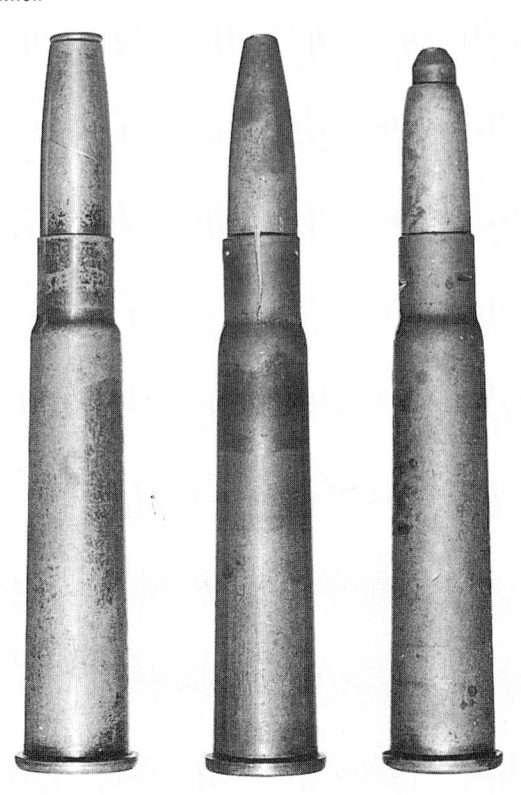

LEAD ANTIMONY STEEL BALL EXPLOSIVE WARHEAD

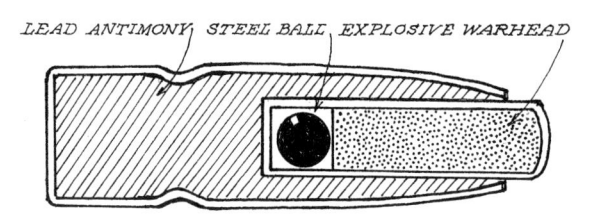

POMEROY MARK 2 EXPLOSIVE BULLET ·303 inch

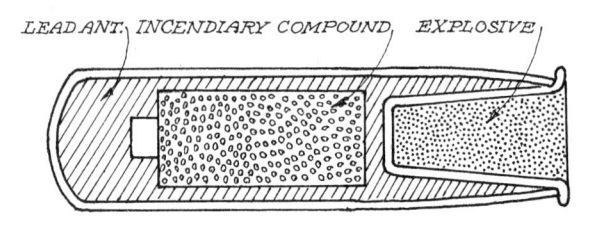

LEAD ANT. INCENDIARY COMPOUND EXPLOSIVE

R.T.S. MARK 2 EXPLOSIVE BULLET ·303 inch.

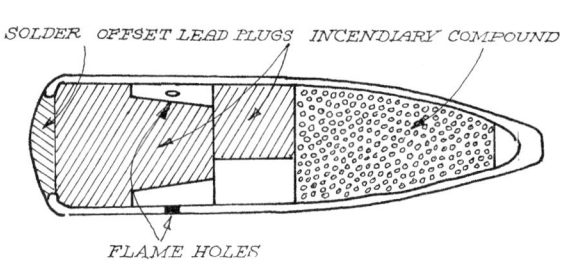

SOLDER OFFSET LEAD PLUGS INCENDIARY COMPOUND

FLAME HOLES

BUCKINGHAM INCENDIARY BULLET MARK VIIB ·303 in.

Above Composition of the Pomeroy, RTS and Buckingham.

Left The Short Gun Carrier used for air firing tests of the $1\frac{1}{2}$-pdr. It was reported that during air firing trials the aircraft nearly stalled and dropped 500 ft.

Part Two

GUNSIGHTING SYSTEMS

Introduction

When aircraft went to war in 1914, the first priority was to gather information on the disposition of enemy forces. On the few occasions when enemy aircraft were encountered, the more adventurous airmen on both sides exchanged fire, using a variety of single-shot weapons and, as they became available, machine-guns. It was soon discovered that, even when they got into a position to fire, their bullets seemed to have little effect.

The machine-guns of these first warplanes were identical to those used by the infantry, the sights being a vee or aperture backsight with a blade foresight. This system was quite effective when the gun was standing firmly on the ground, but of little use when it and the target were cavorting around the sky. There were three main problems. The first was to find some means of range-finding. With no fixed reference this was very difficult, and much ammunition was wasted firing at targets out of range. The second and most difficult was to estimate how far to aim in front of the target so that the bullets would hit it after both had moved after the gun had been fired: this became known as deflection shooting, or 'aiming off'. The third problem was the velocity imparted to the projectile by the host aircraft's forward speed.

These problems were all partially solved during the First World War, but some people have an inborn instinct with firearms that gives them an advantage over others. McCudden and Mannock were as adept with a Lee-Enfield rifle as they were with a Vickers gun, and Von Richthofen was a keen game hunter who could hit a running boar in the forests of Silesia as easily as knocking down lumbering BE.2s over the Western Front. The task of the gunsight designer is to give the average pilot or gunner a method of aiming his guns, eliminating as much guesswork as possible.

British companies and government establishments have been at the forefront of gunsight design, virtually every new development in aircraft weapon aiming up to 1970 having been initiated in the United Kingdom. The most widely used sighting system, the reflector sight, was in fact invented before the Wright brothers left the ground, Sir Howard Grubb patenting his invention in 1900. When it was finally adopted for aerial use it provided an illuminated aiming point in space to which the guns were aligned. This proved to be a big improvement over the ring and bead sight, which required two elements to be lined up with the target. However, the main problems of air firing – the correct estimation of deflection and range – remained.

A fighter pilot carrying out an interception on a bomber aircraft always had an advantage over the defending air gunners. Firstly he was presented with a relatively stable target, and in many instances, especially in the First World War, he could make his approach unseen by his intended victim. He was also provided with superior fire power, and was only vulnerable to defensive fire for the few seconds of his firing pass. Single-engined fighter pilots of WWII were protected by a bullet-proof windscreen and the aero-engine, and if the enemy gunner managed to hit his aircraft in a vulnerable spot he could take to his parachute. However, he had to fly his aircraft on a smooth linear approach, as any skidding or sideways movement would seriously affect the accuracy of his firing. Some aircraft were better than others in this respect.

An attack from the beam or side of his target required an accurate assessment of deflection. A bomber flying at 250 mph covered 122 yd (111 m) in a second, so at a range of 1,000 m he would have to align his gunsight eight aircraft lengths in front to obtain hits. He would also have to allow for the gravity drop of his projectiles. The average pilot could not cope with deflection shooting at ranges over 300–400 m, and most accurate firing was carried out from relatively short range. Even in a tail chase situation the average deflection needed was at least one aircraft length, as the pursuer was rarely directly behind his quarry.

An air gunner manning a turret had first to identify the approaching fighter, then commence tracking whilst the target was out of range. As the fighter closed in he had to keep his gunsight smoothly ahead of his attacker, and at the same time advise the pilot on the direction of evasive action. If the fighter was attacking from the port side he turned to port, giving the fighter pilot the maximum deflection allowance. During these manoeuvres the gunner fired whenever he could bring his gunsight onto the fighter, though, as the attack usually lasted only a few seconds, his most likely chance to fire a fatal burst was when the enemy broke away from his firing pass.

As with the fighter, the air gunner's most difficult attack to counter was from the side or beam quarter. In this situation the problem was not so much deflection as the forward speed imparted to his bullets by his aircraft's movement through the air, known as 'own speed allowance'. He was presented with a head-on view of the fighter which, after getting in position for the attack well out of range, turned and came in very fast. The gunner used his gunsight graticule ring to estimate the range, and made the 'own speed allowance' towards the tail of his aircraft until his target was under 200 m away when he adjusted his aim accordingly. It was the failure to appreciate this procedure which led to the inaccurate fire from the side hatch positions of USAF Fortress and Liberator aircraft during the American daylight missions in the Second World War.

The invention of the gyro sight solved most of the problems of air gunnery. It was found to improve the average fighter pilot's accuracy by 50 per cent, and air gunners' fire became almost magically accurate. This project, carried through by some of the top scientific brains in the country, proved to be one of the most important inventions of the war. As will be explained, this and other developments in the science of weapon aiming, using computerized display data, have given the modern pilot the ability to hit any target on the ground or in the air without any need for guesswork or estimation.

This survey is an attempt to trace the history of weapon aiming systems used by the British Air Services. Unfortunately, the work of some companies concerned with the design and production of gunsights could not be traced, but this book hopefully puts on record the drive and ingenuity of the major parties involved in this important aspect of aerial warfare.

The gunsights

As the first aeroplanes became reasonably reliable, much more money was devoted to adapting them for military use than to benefiting mankind. This attitude was partly due to the fact that the major European powers were being drawn into a situation where war was inevitable. Although the heads of state were busy entertaining each other, professing cordial friendship, politicians and military planners were frantically building up their forces in the name of national defence.

In Britain, whilst the Army and Navy were re-equipping to keep abreast of the Continental powers, intelligence reports were received about German army manoeuvres in which flying machines were being used to assist the artillery and for scouting duties. The British Admiralty were more concerned with the activities of Count Zeppelin, whose giant airships seemed to present a possible threat to the Grand Fleet. In one way or another, flying machines were seen by a few astute military planners as being a new factor that could change many of the accepted tactics of warfare.

Accordingly, the British Army and Navy formed specialized units to find the best means of operating the few practical machines available. The Army had used observation balloons in the South African campaign, and when the Royal Flying Corps was formed in April 1912 most of the work took place at HM Balloon Factory at a small camp near Farnborough, in Hampshire. The Admiralty announced the formation of the Royal Naval Air Wing of the RFC, which was to be based at shore stations and develop flying machines which could assist the fleet and find ways of countering hostile airships. Although this led to much duplication of effort, it also made for productive competition.

Aeroplanes could, it was hoped, reveal the secrets of the enemy's back areas, and what lay over the nautical horizon. It followed that the enemy could do the same, and one way to prevent this was to send up another aeroplane to intercept it. A meeting of the (later Royal) Aeronautical Society in 1911 was opened by Col. J. E. Capper of the Royal Engineers, who said:

> At present I would like my aeroplanes to give me information by reconnaissance. At the same time I would require them to be armed with some kind of light shooting weapon, as it might not prove unlikely that they might be required to fight an enemy aeroplane, either to secure information himself, or prevent the enemy from gaining any.

The first official air firing in Britain took place on 26 July 1912, when Geoffrey de Havilland took off in an FE.2 biplane with a heavy 1895 pattern Vickers Maxim gun precariously mounted on the front cockpit coaming. When he brought the machine in over the Farnborough range at 200 ft (61 m), the gunner in the front cockpit fired several bursts at sheets laid out on the ground, but the slipstream twisted the canvas ammunition belt, jamming it in the receiver. He reported that it had not been possible to lay careful aim, because as soon as he had located the target it had disappeared beneath the gun.

By 1913 air firing trials were being organized by both the military and naval wings. In most cases the guns were fired successfully, but it soon became clear that something had to be done to improve the sighting. The following report was received by the Admiralty from Cdr R. H. Clark-Hall, one of the leading naval air gunnery trials officers, after experiments at RNAS Eastchurch:

> Firing was first carried out at targets dropped in

the sea, and at the shadow of other aircraft flying as a target for the purpose, and finally at small target balloons of about 2 ft in diameter. The latter practice brought home in a practical manner the difficulty of accurate shooting, with no burst or flash to indicate where the shot was going, and the consequent need for tracer rounds.

No tracer bullets were available, and Clark-Hall invited suggestions on a suitable gunsight for air firing. It was agreed that the normal blade sights were too close to the barrel. These were designed to give a seated gunner only a few degrees' elevation for various ranges, whereas air firing introduced the need for free elevation and depression, plus the maximum field of view and the ability to align the gun and fire quickly.

It was suggested that a naval tubular optical sight could be mounted on pillars above the gun, giving an improved field of vision. A 12 in (305 mm) sight was mounted 6 in (152 mm) above the gun, and an optician adjusted the lenses. The gun was fired from a tripod into the station stop butts, and although the sighting tube vibrated it was found that the gun could be easily aimed at targets at various elevations. The gun was then mounted on the front cockpit of a Short S.38 aircraft, and Clark-Hall carried out test firing over the sea. He found that, although the sight was unwieldy and target location through the lenses not ideal, the arrangement was an improvement over the blade sight. After further trials a report stated:

An aerial gunsight should, in respect to free-mounted guns, be raised above the barrel to a position where the gunner has a good field of view, and be as robust as possible. Some means of range-finding should be provided to avoid wastage of ammunition on targets beyond the range of the gun. As air firing always involves the movement of the gun and a target aircraft, the problem of deflection shooting would have to be considered.

Army trials came to similar conclusions. Organized by the Small Arms Committee of the War Office in conjunction with the Royal Aircraft Factory at Farnborough, these featured the use of several kinds of weapons fired from a Henry Farman biplane at the Ash ranges near Aldershot. It was found that the Vickers Maxim was prone to stoppages without a second crewman to guide the belt into the receiver. The Danish Madsen and French Hotchkiss were lighter and more manageable but the new American Lewis seemed ideal.

Early gunsights

Vickers produced what was probably the first purpose-made aerial gunsight, for the gun emplacement in the nose of their Experimental Biplane No. 2. This consisted of an elaborate parallel motion sight which was automatically raised when the gun was elevated, and lowered when the gun was depressed. This was necessary as the gun was fitted to a conical mounting with a small slot through which the gunner could align the sight on the target.

When war was finally declared in August 1914 the Royal Flying Corps and the Royal Naval Air Service

Vickers EFB (Experimental Fighting Biplane) showing the elevated foresight of the Vickers Maxim gun and the minute gunner's sighting slot. (Bruce/Leslie)

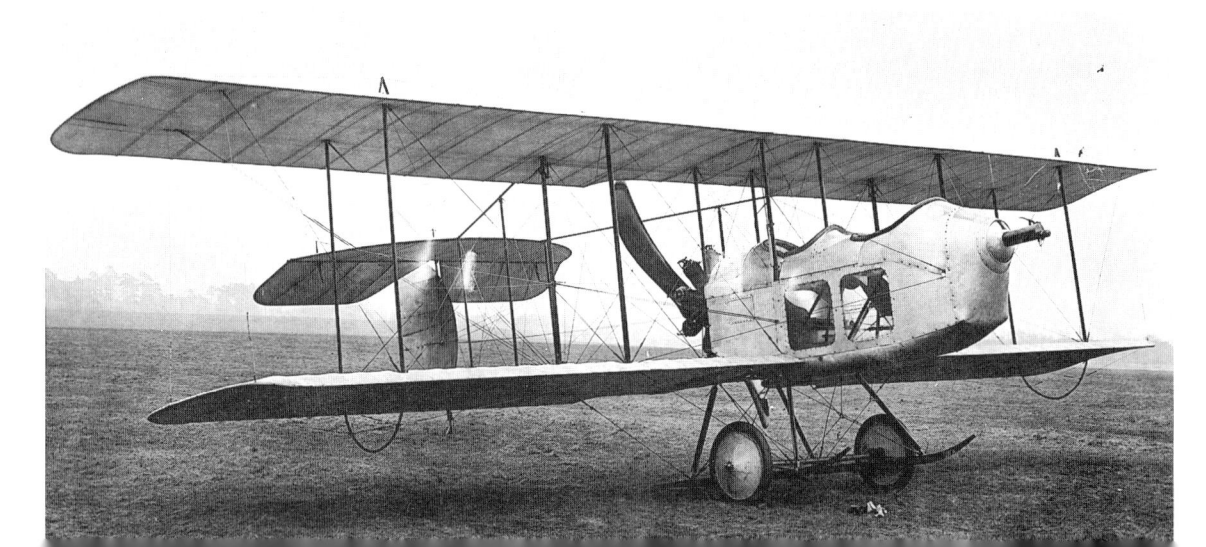

THE GUNSIGHTS

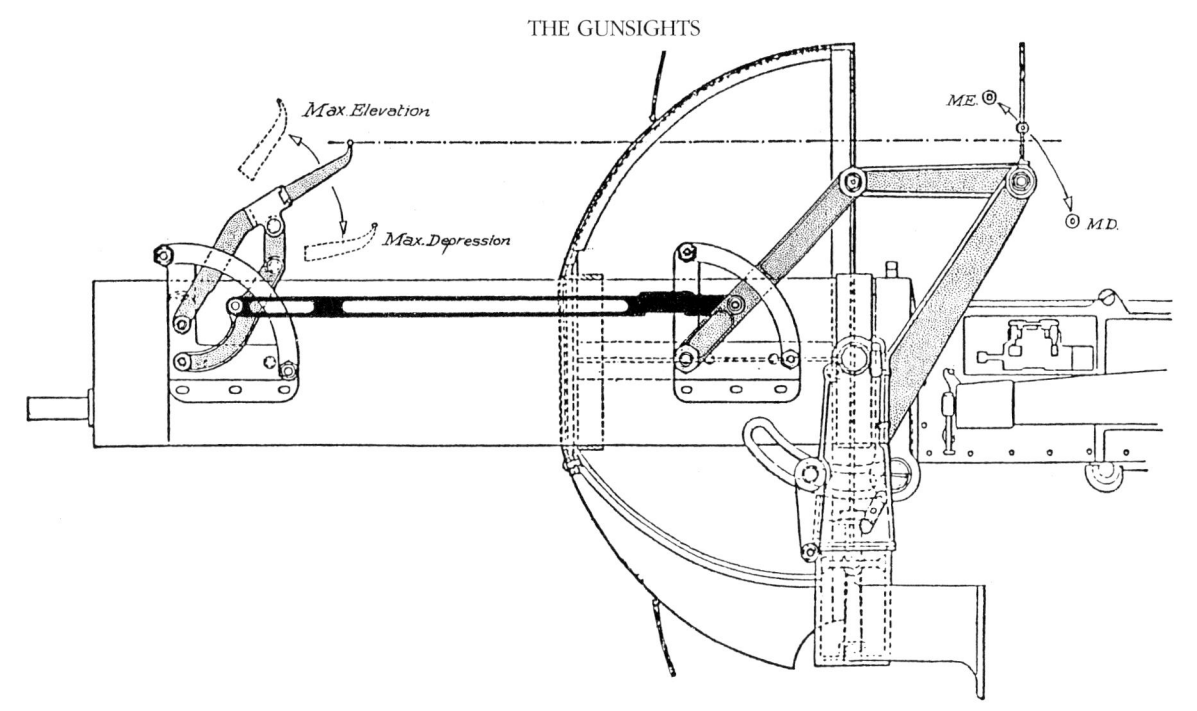

Vickers parallel-motion sight of the Vickers EFB.2. This gave the gunner an ideal line of sight at maximum elevation and depression.

took a motley collection of aircraft to France. They were used for scouting enemy positions and dropping hand grenades and other missiles on German troops. German and French aircraft were engaged on similar duties, and in the early months of the war there were few occasions when aeroplanes of opposite sides met. However, when the army commanders realized that their troop movements were being observed and that their own airmen were obtaining vital information, Capper's prophesy was proved to be correct. The most successful and numerous British aircraft were the very stable BE.2 biplanes, which proved to be ideal for observing and artillery spotting. When the RFC began to receive Lewis guns, they were used by observers in two-seat aircraft. The observer sat in the front cockpit of the BE.2, his field of fire limited by struts and rigging wires. At first this was no great problem. Deflection shooting had not been mastered and the loss rate was low.

In 1914 French aircraft designer Raymond Saulnier designed a synchronizing system for his company's Type L monoplane, enabling the pilot to fire through the arc of the propeller. This failed to be accepted, but when the company test pilot, Roland Garros, joined the Aviation Militaire, he persuaded the authorities to let him use it against the enemy. Using diving attacks from the rear, he shot down several German aircraft in quick succession in April 1915, but shortly afterwards was himself shot down during an attack on a railway station. The wreckage of his aircraft was inspected by Anthony Fokker, a Dutch aircraft designer working for the Germans. Realizing the advantage of the synchronizer, he instructed his engineers to design a similar system based on a device patented by a Swiss engineer, Franz Schneider, in 1913. He then designed a near copy of the Morane monoplane, and claimed to have invented the aeroplane and its armament system. In a few months the Fokker E series monoplane was accepted for service by the Imperial German Air Force. Armed with the German version of the Maxim gun, the MG.08/15, controlled by the cam-operated synchronizer. German pilots soon gained the ascendancy over Allied aircraft. Only 475 of the Fokker E series were produced in all, but from June 1915 they shot down hundreds of scout and reconnaissance aircraft. The Allies had ignored pre-war inventors who had offered synchronizing gear, but in early 1916 new types of fighting scouts were introduced which restored the balance.

The gate (frame) sight

The Fokker E monoplanes were among the first to use the frame gunsight, a smaller version of a naval sight used on the quick-firing guns of warships to counter fast torpedo boats. It consisted of an elongated rectangular wire frame which could be easily

A gate sight on a Fokker EIII, February 1916. Note its position at the end of the barrel housing.

aligned onto a target crossing the gunner's line of fire. Additional upright wires were fixed at intervals along the frame, giving the gunner graduations for estimating deflection. The aerial version was soon adopted by Allied and German airmen. It was found that if the upright wires were made adjustable, they could be used for estimating both deflection and the range of the target. When the two upright wires were set to correspond to the wingtips of an aircraft at maximum range, the pilot could avoid wasting ammunition.

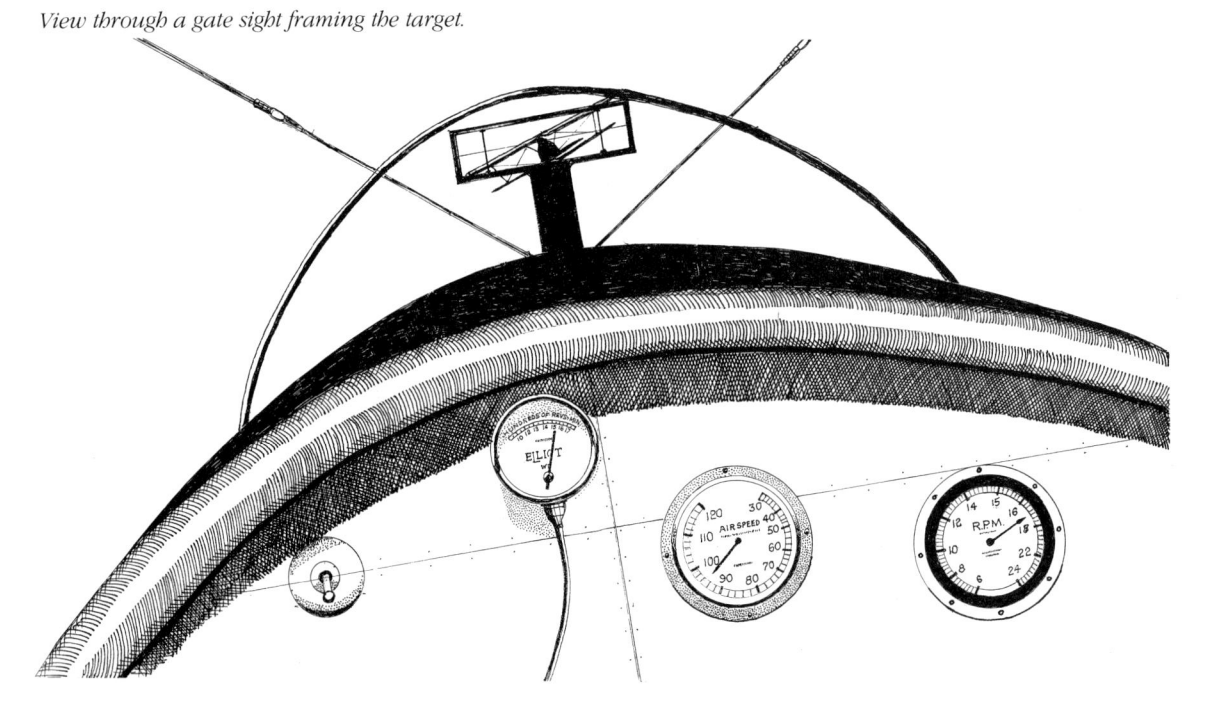

View through a gate sight framing the target.

THE GUNSIGHTS

The British model came to be known as the gate sight. It was fixed on the cowling just in front of the windscreen, and lined up with a small ring element further forward on the engine cowling. The RNAS version used on Sopwith Pups used a wider rectangle than that used on RFC aircraft. German pilots preferred to mount the sight at the end of the barrel housing, using it as a foresight which was aligned with a post or small ring. The pilot of a fixed-gun fighter had no great need for an elevated gunsight; his marksmanship depended on his ability to outfly his opponent, and when he was in a position to open fire his point of aim was relatively stable.

The ring and bead

The frame sight had shown that a rudimentary allowance could be made for 'aiming off' or deflection by placing the target on the edge of the frame before firing. However, it could be used in this way only if the target was moving across the line of sight horizontally. It was soon realized that, if the rectangular frame was changed to a circular shape, it would enable a pilot or gunner to engage an opponent flying at any angle across his line of flight, while some idea of range could be given by comparing the size of the ring to the enemy aircraft. The ring backsight was aligned with a small red bead mounted on a pylon at the muzzle end of the gun, and the system became known as the 'ring and bead'.

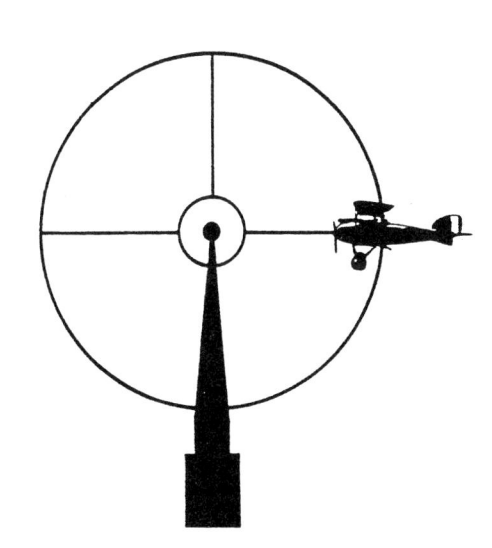

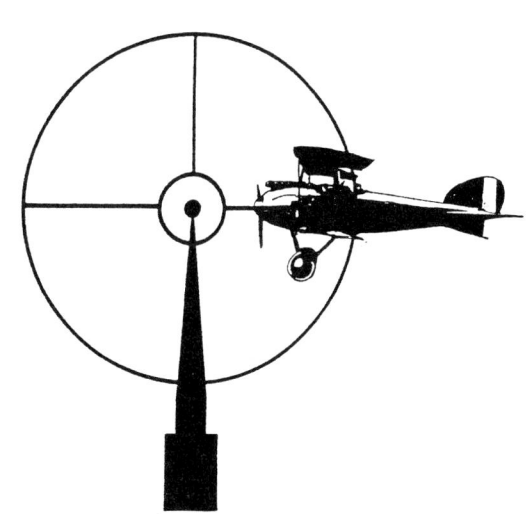

Above *A gate sight fitted to a Bristol Scout. There were various versions of the sight, the German type being more elongated.*

Below *View through a ring and bead sight, showing the method of allowing for deflection. The bead was centralized in the ring, the target placed on the outer ring.*

Hostile Aircraft crossing at a range of 200 yards.

Hostile Aircraft crossing at a range of 100 yards.

An observer manipulating a free-mounted gun had to be prepared to meet an attack from any angle, and his gun was if possible mounted in such a way as to cover the maximum field of fire. It was also essential that his sight gave him a quick sight line whenever a target came into view. By early 1916 the ring and bead had been adopted for free guns. There is no record of the first use of this sight; like so many other inventions, several people probably realized its logic at about the same time.

The dimensions of the ring and bead varied with its distance from the eye. Typical distances were 23 in (584 mm) for a 3 in ring, 38 in (965 mm) for a 5 in ring, and 36 in (914 mm) for the $4\frac{1}{2}$ in ring fitted to free-mounted Lewis guns. These measurements were calculated to give a full angle of deflection in an average combat situation. Typical parameters for a fixed gun were: own speed – 100 mph (161 km/h), target crossing speed – 100 mph (90° across the pilot's line of sight), range 200 yd (183 m). In this situation the bullet would take 0.254 sec to reach the target, which would have travelled 37.7 ft (11.5 m). The aiming procedure was to keep the red bead in the small ring in the centre of the ring element, and the target on the edge of the outer ring, flying towards the centre. In an ideal situation this would give the correct deflection angle, and the target would be hit. The parameters for a free gun were slightly different, and of course the actual angles, range and speed varied for every engagement, but ring elements gave fair allowance for an average encounter.

The ring was usually made of steel strip, with four radial wires supporting a central ring of 0.5 in or 1 in (13 or 25 mm) diameter, observer's rings often having only one supporting wire. The stem was hollow, fitted on a fixed post secured with a pin or screw. Observers removed the fragile rings after a sortie, handing the gun and ring to squadron armourers. The bead element was mounted on a shaped pylon; its distance from the ring was not critical, being used to align the ring with the centre-line of the gun or flight-line of the aircraft.

Pilots and observers who survived long enough soon found how to vary the allowance for range and bullet drop, and the ring and bead was the first effective gunsight used in air warfare. The first sights were made in squadron workshops, but these were soon replaced by factory-made units, easily detached from free guns when not in use. The ring and bead was eventually adopted by both sides, and remained in use until it was replaced by the reflector sight in the late 1930s.

The 'own speed' factor

As the ring and bead came into widespread use, observers reported that, even though they had a target coming in from the beam and had made the

'Own speed' allowance. Bullets fired at a target on the beam retain the forward impetus of the aircraft, so the vane sight gave a sight line allowing for this forward movement.

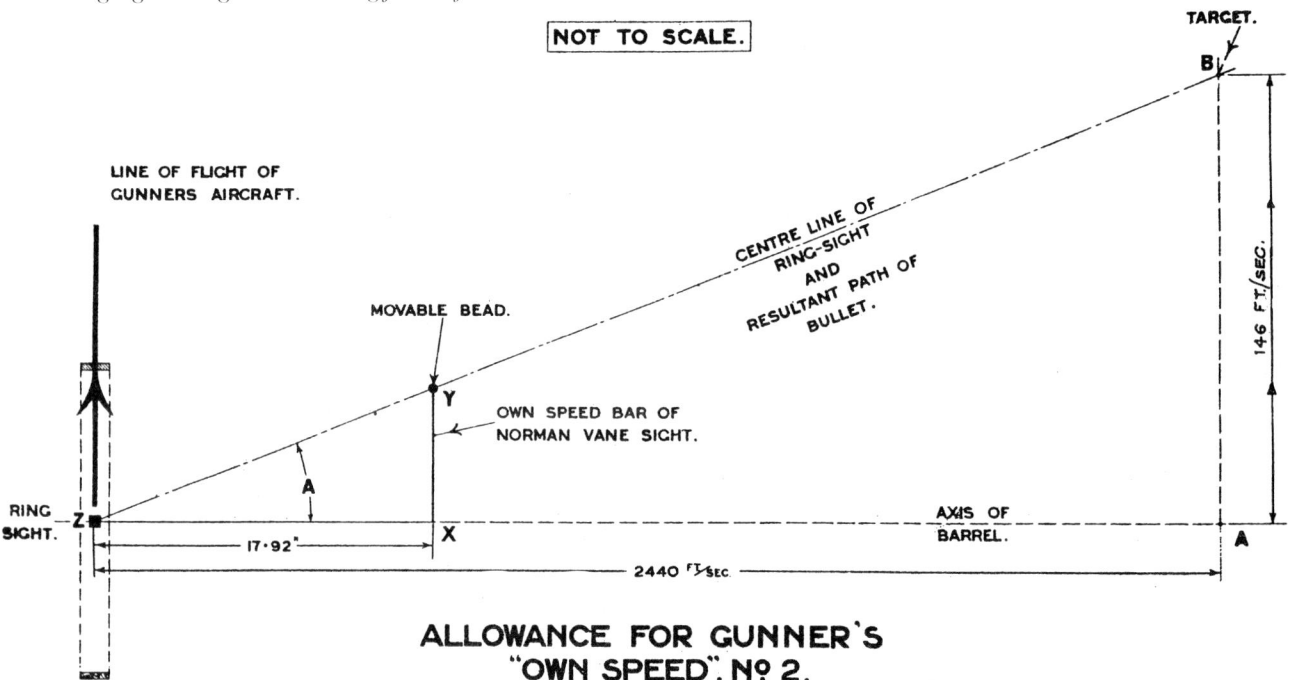

ALLOWANCE FOR GUNNER'S "OWN SPEED". N° 2.

THE GUNSIGHTS

NORMAN SIGHT.

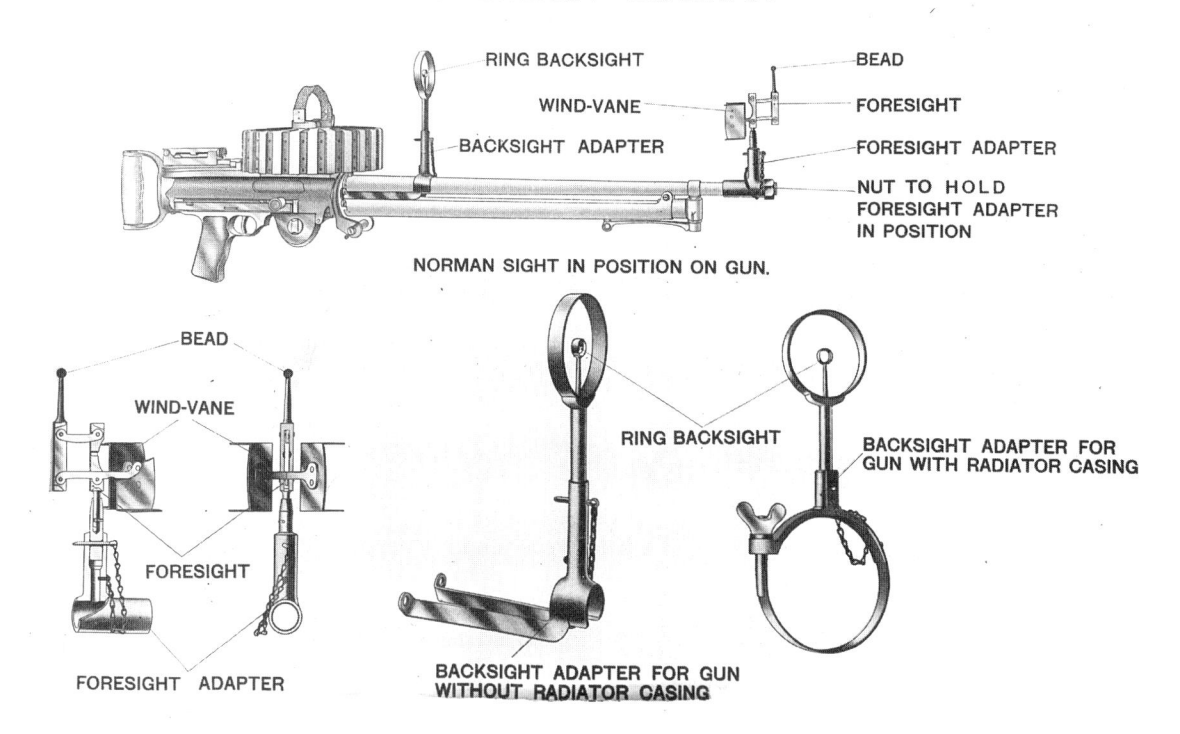

NORMAN SIGHT IN POSITION ON GUN.

necessary adjustment to their aim, a long burst seemed to have no effect. When experimental batches of Woolwich tracer ammunition became available in 1915 it was found that, when fired to the beam, bullets seemed to curve to the rear. This was due to the bullets losing the forward speed of the host aircraft, and became known as the 'own speed' factor.

The Norman vane

Lt. G. H. Norman, Workshop Officer of 18 Sqn RFC, listened to the 'own speed' factor being discussed in the Mess. He came up with an idea that was to be adopted by every major air force: the vane sight, which all but solved the 'own speed' factor. It consisted of a swivelling vane foresight, similar to a weathercock on a church steeple. The 'cock's tail' of the sight was a vane which was free to swivel in the slipstream. A small red bead was mounted at the cock's beak end. When the gunner aligned the bead with his ring backsight, the barrel always pointed to a position lagging behind the line of flight. The amount of lag was at its maximum when the target was at 90° (exactly to the beam); in any other position the sight automatically gave the correct amount of allowance.

Above *The general layout of the Norman vane sight. The two sighting elements were usually removed after an operation.*

Below *The vane sight gave the gunner an automatic allowance for the effect of the speed of his own aircraft on the bullets.*

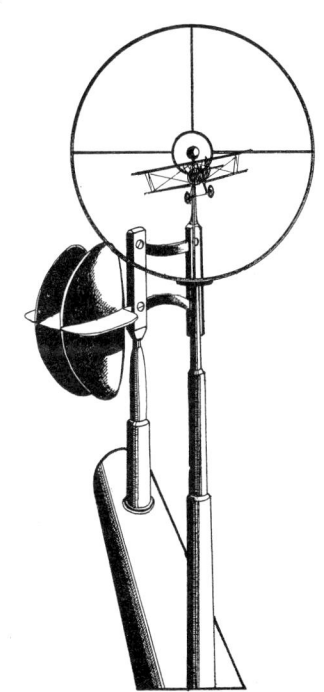

Above *A Norman vane sight with elliptical vanes, showing the swivelling pylon and foresight. This is a late version of the Mk III Lewis with a flash hider and spring clip securing the pin on the sight stem.* (MOD Pattern Room, Nottingham)

Below *The Norman vane Mk III with variable 'own speed' adjustment.*

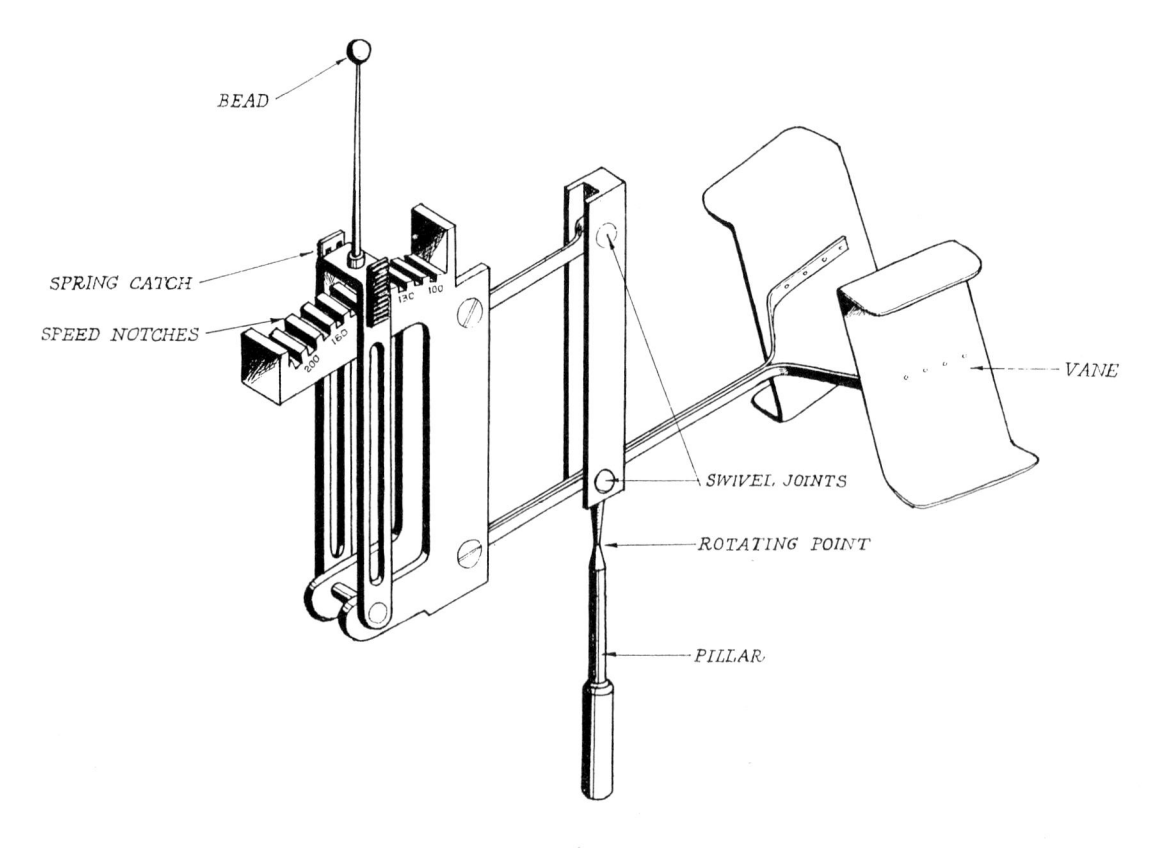

The first sight was made in the squadron workshops by Sgt. Steel and Cpl. Horsley, working under instructions from Lt. Norman. The fitters devised several improvements in the design, and many alterations were needed before the dimensions were found to be correct. The vane was used with a ring backsight 4½ in (110 mm) in diameter, the two elements being 18 in (450 mm) apart. Norman was sent back to supervise production in London. From early 1916 the sight became standard issue for free guns. The design of the vanes varied with the manufacturer, some being of sheet steel, others aluminium castings. The two elements of the sight were mounted on moving arms which, when the vane was reacting to the slipstream, rose into a horizontal position, allowing the elements to assume a position parallel to the line of flight. The effect of gravity on the bullet was not taken into consideration in the sight design: a Mark VII bullet would fall 14.4 in (360 mm) in 200 yd (183 m) and 67 in (1,700 mm) in 400 yd (366 m). Although this appears to be large, it was the centre of a cone of fire, and was reduced for any angle above or below horizontal. The bead element was a rod 6 in (150 mm) in length, the stem being threaded for 3 in (80 mm). It tapered from the base to a ball-shaped bead of .35 in diameter (9 mm) painted a brilliant red. As aircraft speeds increased, the connecting arms were lengthened, the speed allowance being stamped on the upper surface of one of the wind vane arms. A modified design was produced in 1932 in which the speed setting was adjustable, but this was not adopted for squadron service. When the US 8th Air Force became concerned at the inaccurate fire of waist gunners of B17 and B24 aircraft, a vane sight based on Norman's was designed by Capt. John Driscoll of the 389th Bomb Group USAF based at Grafton Underwood.

Tracer ammunition
When the first batches of RL and Buckingham tracer reached front-line squadrons in France it was thought that the answer had been found to many of the problems of air-to-air gunnery. How could a gunner or pilot fail to miss when the trajectory of his bullets was visible? Experience proved that when the spark-like bullets seemed to be hitting the target they were often missing it by a large margin. One problem was that the trace chemical often burned out prematurely, another that the trajectory differed from that of ball ammunition. Gunnery schools eventually stressed that, although tracer could be an aid, accurate shooting could be achieved only by the correct use of gunsights. Nevertheless, tracer was in widespread use in both fixed and free-mounted guns, sometimes as many as one round in three being requested by pilots and gunners. The reason was not entirely a quest for accurate shooting: it was an acknowledged fact that enemy pilots often broke off an attack when streams of tracer began to fly towards them. Tracer was not popular with armourers, the heavy barrel fouling meaning hours of extra work after operations.

Night sighting
When the Luftwaffe began night operations against

This night-sighting version of the Vane sight had a hollow bead with a pin hole and contained a 12 v bulb. It was used with an illuminated Neame backsight.

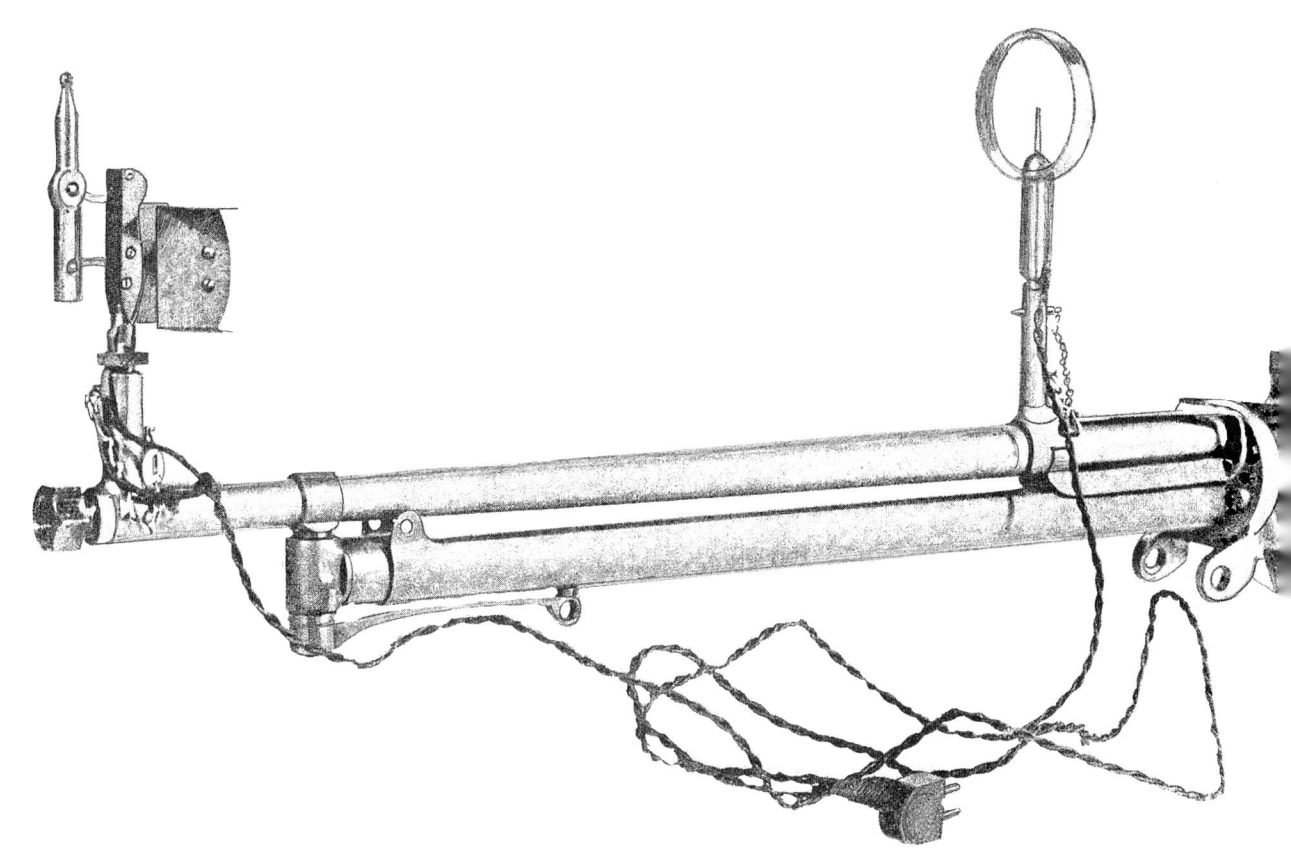

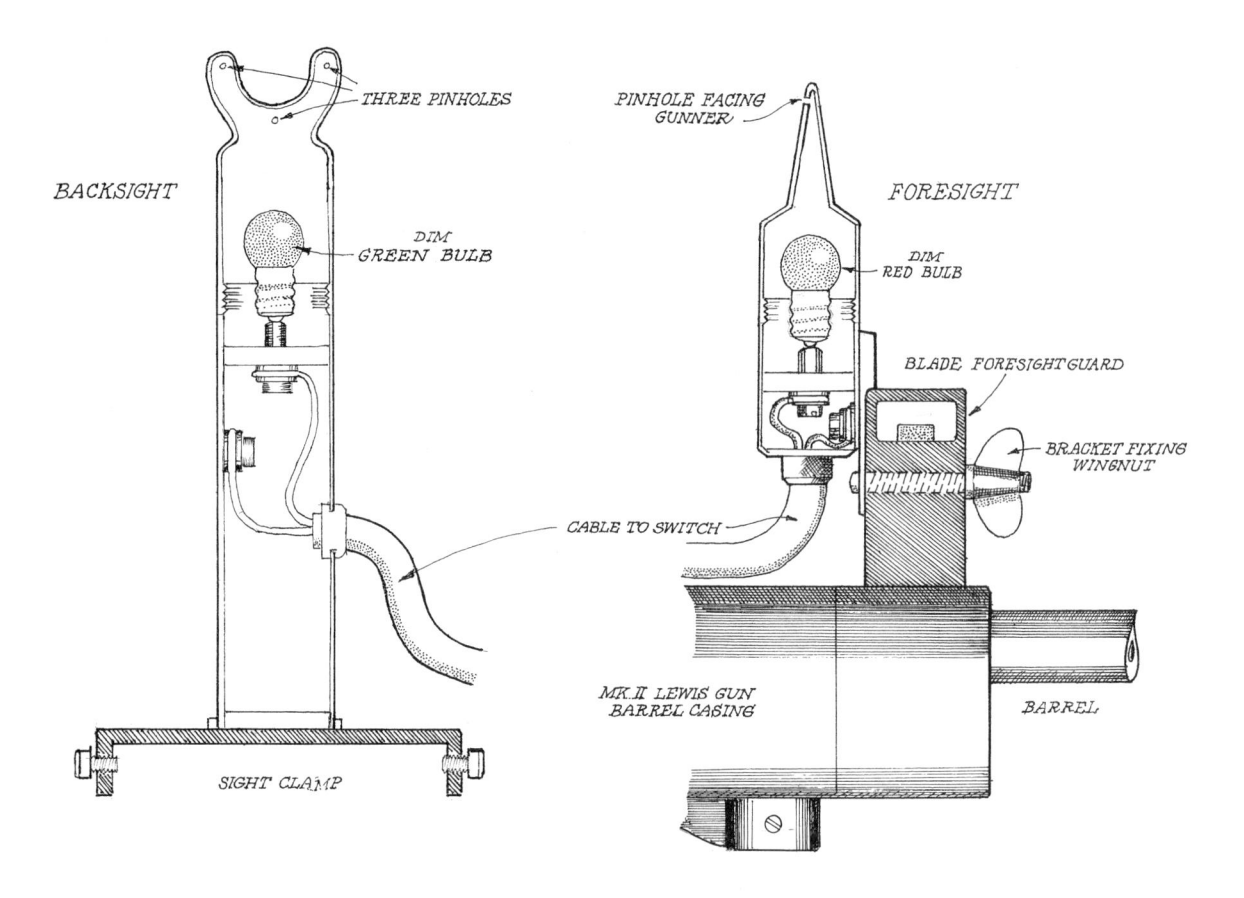

Above *The Hutton illuminated sight, found to be the most practical means of aiming guns at night.*

Left *Twin-Lewis mounting on a Home Defence BE.12. The right gun is fitted with a Hutton sight. A ring and bead sight can be seen on the right struts. This elaborate mounting could be lowered for upward firing and drum changing.* (Bruce/Leslie)

England in 1940, radar-equipped night fighters directed by radar-assisted ground controllers experienced great difficulty finding the raiders. Bearing this in mind, the problems faced in 1916–18 by the Home Defence Squadrons can be imagined. Apart from intercepting the hostiles, pilots were often involved in landing and take-off accidents which took a heavy toll of the unwary. Once airborne, they cruised round in an allotted area for hours in the hope of a sighting. Usually the only indications of enemy activity were the flashes from bombs or searchlights coning on an intruder. If, in spite of heavy odds, a pilot found his quarry, he coaxed his machine into a position where he could open fire. If the crew of a Zeppelin became aware of a fighter's

Above *A Hutton night sight aligned on a Zeppelin.*

Right *Batteries of the Hutton sights were usually taped to the barrel of free-mounted guns. This unusual mounting on an FE.2B night fighter enabled the pilot to choose his line of approach.* (Bruce/Leslie)

Below right *An Avro 504 night fighter of 77 Squadron RFC. A Hutton sight is fixed to the Foster-mounted Lewis, which could be drawn down for drum changing and upward firing. On the engine cowling is an illuminated Neame sight. The 47-round drum was used to save weight.*

approach, ballast was released and the huge gas bags lifted the craft quickly out of range. When the first interceptions were made it was found that firing was a very hit-and-miss affair: sighting was almost impossible, and the muzzle flash destroyed their night vision.

The Hutton sight

The first night sights were ring and bead elements treated with luminous paint, but these were found to be either too bright or too dim. The first practical night sight was invented by an armourer serving with 39 Squadron based at Hainault Farm. Sgt. Albert Hutton had been responsible for major improvements to gun mountings and ammunition feeds. He devised an illuminated sight which could be used on both fixed and free-mounted guns. It was simple, but, when tested at Martlesham Heath, trials' pilots preferred it to the more elaborate sights designed by scientists.

It consisted of a tubular foresight containing a red bulb. At the top of the domed tube was a 0.5 mm (0.02 in) hole through which showed a minute point of red light. The backsight was similar, but had a hallow vee-shaped tube at the top. At the base and two arms of the vee were holes which, when a green bulb was switched on inside the tube, showed as three green spots in the shape of a vee. The gunner simply aligned the red spot in the centre of the vee and on the target. The bulbs suffered severe vibration when the guns were fired, but they usually lasted long enough for what were often very short engagements. The bulbs were usually fed by batteries taped to the gun body.

The Neame sight

Unaware of Hutton's work, the Technical Directorate of the War Office had instructed their design department to devise a night sight. The project design leader was Engineer Lt. H. B. Neame. Several designs were produced, leading eventually to an illuminated backsight. It took the form of a hollow domed tube with a slot cut into the top, surrounded by a strip metal ring. A bulb inside the tube projected a dim light onto the inside of the ring, which was used to align a red point of light (similar to the Hutton foresight) onto the target. The sight was tested at Martlesham and accepted after minor modifications.

The diameter of the ring varied, the smallest being a 2 in (50 mm) No. 2 ring used on the Scarff compensating sight, the largest the 4.5 in (115 mm) No. 1, which corresponded with the wingspan of a Gotha bomber at 100 yd (91.4 m). For fixed guns the sight was fitted directly in front of the pilot's windscreen. As pilots gained experience it was found that the ideal attack was from below, when one's machine merged with the land mass and the target was silhouetted against the night sky. The first special night armament was a Lewis installed to fire upwards at an angle of 45°. W/O Scarff, one of the foremost RFC armament experts, produced a 45° Neame sight consisting of an 18 in (450 mm) bar carrying the two elements. The Neame was produced by the famous London gunsmiths Purdy and Co., the price being £1.75 per sight ex works without wire! They were used mainly on single-seat fighters, although some Bristol fighters were fitted with upward-firing guns, sighted by the pilot and fired by the gunner in the rear cockpit.

A typical engagement using a Neame sight took place on the night of 20 May 1918. Lt. Edward Turner of 61 Squadron, flying a Bristol Fighter reported:

It was my fourth night sortie with Sgt. Barwise. After an hour in the air we saw an aircraft which we soon recognized as a Gotha. After several attempts we managed to get into position below and behind, where the huge wings filled the illu-

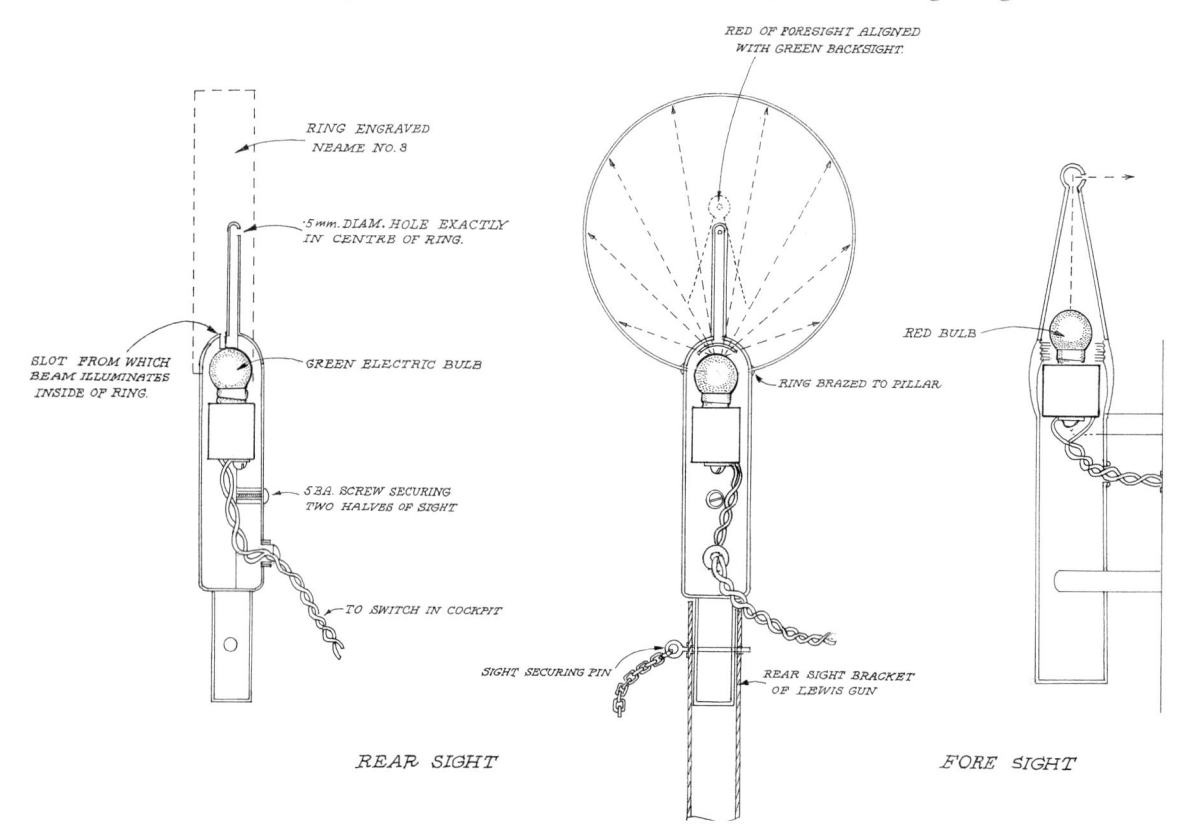

The Neame No. 3 illuminated sight.

REAR SIGHT

FORE SIGHT

View through a Neame sight aligned on a Gotha bomber.

minated ring of the Neame sight. On my signal Barwise fired, hitting the port engine with his first burst. After receiving the contents of two 97-round drums, the Gotha put its nose down and went into a flat turn. We followed until it dived into the ground with a tremendous explosion.

The Secretan sight

When night operations by German bombers increased in 1917, other ideas on night sighting were tried out. One of these was the subject of a patent taken out in 1917 by Major Secretan, RFC. This worked on the principle that, even on the darkest night, everything appears as various shades of grey. Secretan's sight consisted of a large ring backsight with two parallel cross wires, made of wood and coated with a black velvet substance. The foresight was a smaller ring of similar construction. Records show that the arrangement was tested at Martlesham and found to be reasonably successful, but showed no improvement over the existing night sights.

The Aldis sight

One of the drawbacks of the ring and bead was the need to align the two elements with the target. Whilst this was easy enough with a pivoted gun, a pilot with only a split second to fire in in a dogfight needed something which would give him an immediate point of aim. Naval guns were aimed by optical sights with collimated* lenses, and hair lines indicating a position in space harmonized with the bore of the barrel.

In 1915 Martlesham Heath tested an optical sight submitted by the Aldis Brothers of Sparkhill, Birmingham. It consisted of a 22 in (559 mm) metal tube containing lenses giving a magnification of × 3 and a cross-hair aiming mark. (The aiming mark of an optical sight should be referred to as the *reticule*, but Air Ministry publications refer to it as the *graticule*, which is how it will be known in this book.) The new sight was mounted on the cowling of a BE.2c with a Lewis gun on the upper wing, and firing tests were carried out at ground targets. The sight proved more accurate than the ring and bead, though on one occasion the pilot nearly flew into the ground whilst looking through it! The sight was then tested in air fighting manoeuvres. Magnification of the field of view often made it difficult to locate the target quickly. Another problem was that the front lens was prone to fouling by oil from the engine, and it was also no longer possible to estimate target range by comparing the wingspan with the ring diameter. Hugh Aldis was therefore asked to submit a sight without magnification, but with a circular graticule which could be used for range-

Adjusted to make parallel the line of sight.

Lt. C. R. Thompson RFC in his SE.5a, showing the Aldis sight mounting. (Bruce/Leslie)

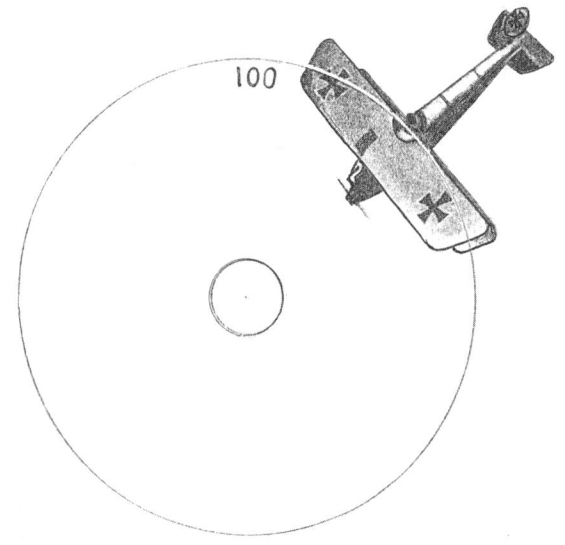

Pilot's view of typical deflection allowance with 100 mph ring.

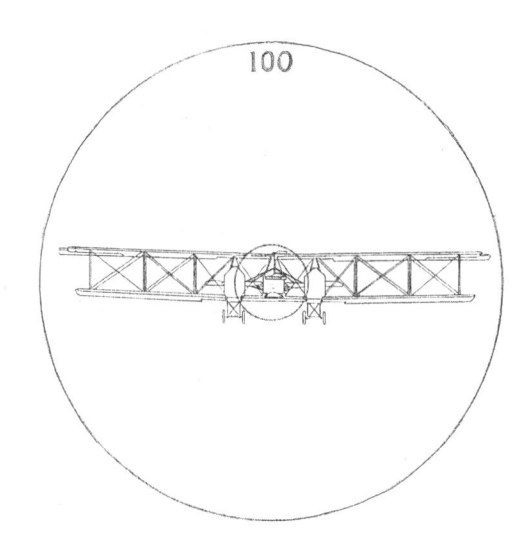

View through a 1.8 in Aldis. A Gotha fills the ring at 200 yd (183 m).

finding and, possibly, deflection. A spring-loaded oil flap was fitted, and leather-lined adjustable clamps mounted on short sturdy brackets were supplied.

The modified sight received the highest praise after testing under all conditions of temperature (inert gas had been sealed into the tube) and combat manoeuvres. Martlesham recommended that the Aldis should be the RFC's standard fixed gunsight. As air activity over the front intensified, casualties were reaching serious proportions. The ascendancy of the Fokker monoplane had been broken, but new German types were being encoun-

An Aldis mounted on an SE.5. It was more usual to fit it on the right, with a ring and bead on the left of the cowling. There is also no oil flap. Note the CC gear triggers on the control column.

THE GUNSIGHTS

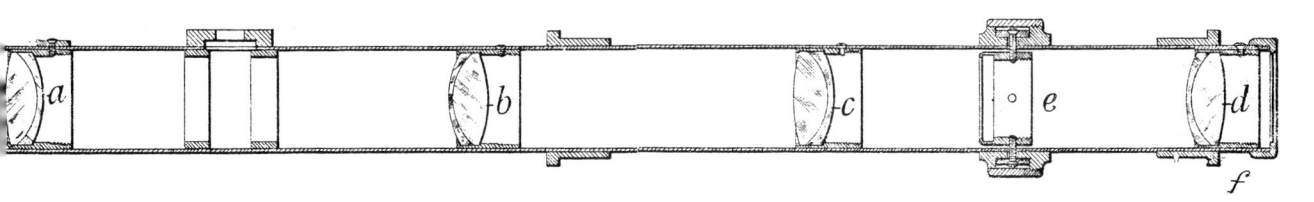

Sectional drawing of the Aldis sight. abcd: hermetically sealed double concave lenses; e: graticule unit (several patterns); f: viewing end, usually fitted with a rubber eyepiece.

tered, and if the new sight could give the pilots any advantage, it was decided to rush it into production. An initial order for 200 sights was issued, and the company was told to prepare for further substantial production.

The sight tube contained four hermetically sealed collimating lenses, with a graticule in the form of two concentric circles engraved on a plain glass screen. The outer circle gave the deflection needed for a target plane crossing at 100 mph, and a small circle gave the gun alignment point. On some later models the outer ring was modified to indicate the wingspan of a Gotha bomber at 200 yd (183 m), and with practice pilots found how to use the circle for other aircraft. The lens system gave unity (no) magnification, and ensured that the ring was always centred on the axis of the sight no matter where the eye was placed. The ideal eye distance from the rubber eyepiece was 5 in (127 mm), which gave a FOV (field of view) of 20°. The anti-oil flap was operated by a cable to the cockpit, where a ring was hooked to a small bracket; when unhooked, the shutter sprang open, giving a clear view. If the shutter was left open, oil fouling would occur, so it became standard practice to fit both Aldis and ring and bead sights. In the first installations the windscreen was cut away in the top right-hand corner, but factory-installed sights passed through a hole drilled in the screen.

Production Aldis sights were issued to selected front-line squadrons for operational trials in mid-1916. Pilots found the Aldis superior to the ring and bead, and news of a secret new gunsight soon spread to other squadrons. The Aldis was said to possess almost magical powers, and at a time of

SE5 gunsight brackets. The ring sight was clamped round the Vickers gun.

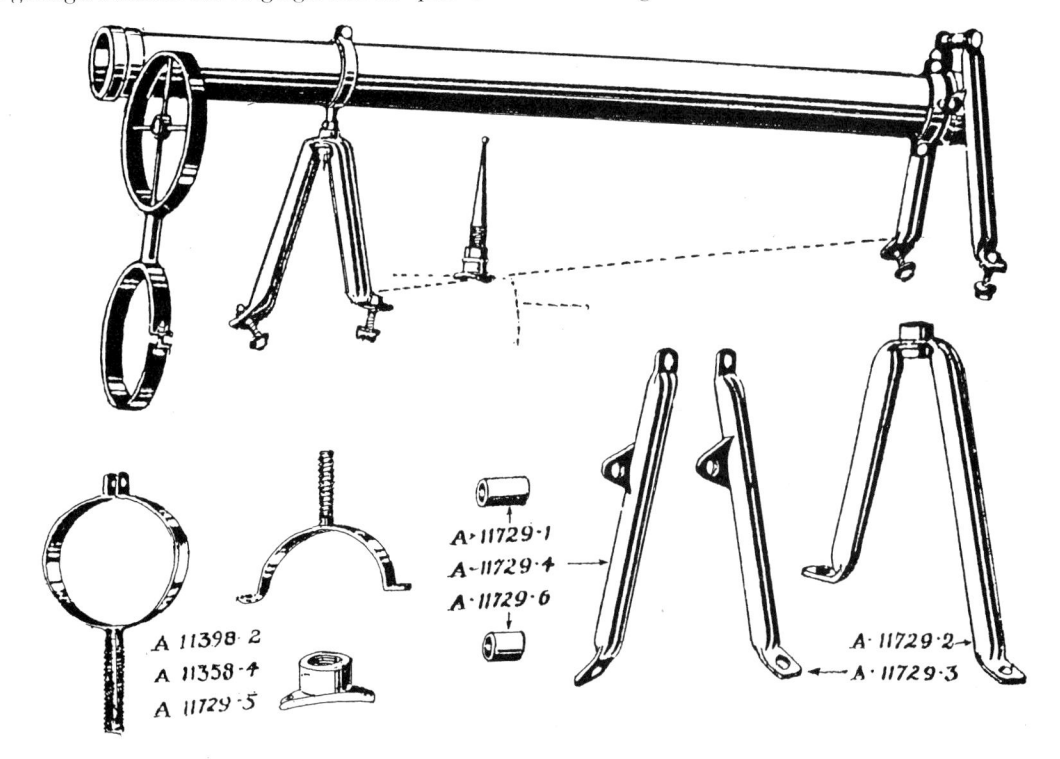

high casualties the authorities did nothing to dispel these rumours.

By 1917 the Aldis had become the standard British sight for fixed guns. It was usually mounted on the right side of the engine cowling, with the ring and bead on the left. A smaller version was used on large guns such as the Davis and COW, and a version for anti-aircraft guns had a prism which could be clipped into the field of view, giving a set deflection of 30°. In 1918 the French Aviation Militaire introduced a similar sight manufactured by Chretien, and the German Oigee company produced a very similar optical sight for use on both fixed and free guns. Aldis sights were much sought after by German pilots, who took them from the wreckage of crashed Allied aircraft.

The Scarff compensating sight

In late 1915 an elaborate free-gun sight was designed by Lt. Scarff, RNAS. It was based on the fact that the deflection required was at a maximum when the target was at 90° to the line of flight, and progressively decreased as the angle narrowed. The sight was in effect a gun mounting fitted to a purpose-made gun ring, on which a pillar supported speed cranks which automatically offset the sight line. The Lewis gun was fitted to a framework which was linked to a bar, on which was an aperture and bead sight. The gunner selected target range on a calibrated knob which elevated the gun and sight line. He then set his own and estimated target speed on knurled dials. As the gun was traversed towards 90° the speed cranks moved the sight bar progressively behind the gun line, giving 'own speed' allowance. The cranks also made allowance for deflection when this was needed. If all settings were accurate the correct sight line would be obtained, though in practice it was far too involved for a gunner in the heart-thumping situation of engaging an enemy fighter.

The Scarff compensating sight. Own and target's speed were set on speed cranks, the mechanism then offset the sight bar to give the correct allowances for deflection and own speed. Estimated range was then set on a dial. Although highly ingenious, it proved too lengthy a process, and was not adopted in any numbers.

A: Gun clamp
B: Sight bar pivot
C: Sight bar
D: Moving run ring
E: Fixed gun ring
F: Rotation cog
G: 'Own speed' setting
H: Target speed setting
K: Target range
L: Target speed setting knob
M: Speed crank support arms
N: Rotation lock
P: Peep backsight
Q: Amm. drum handle

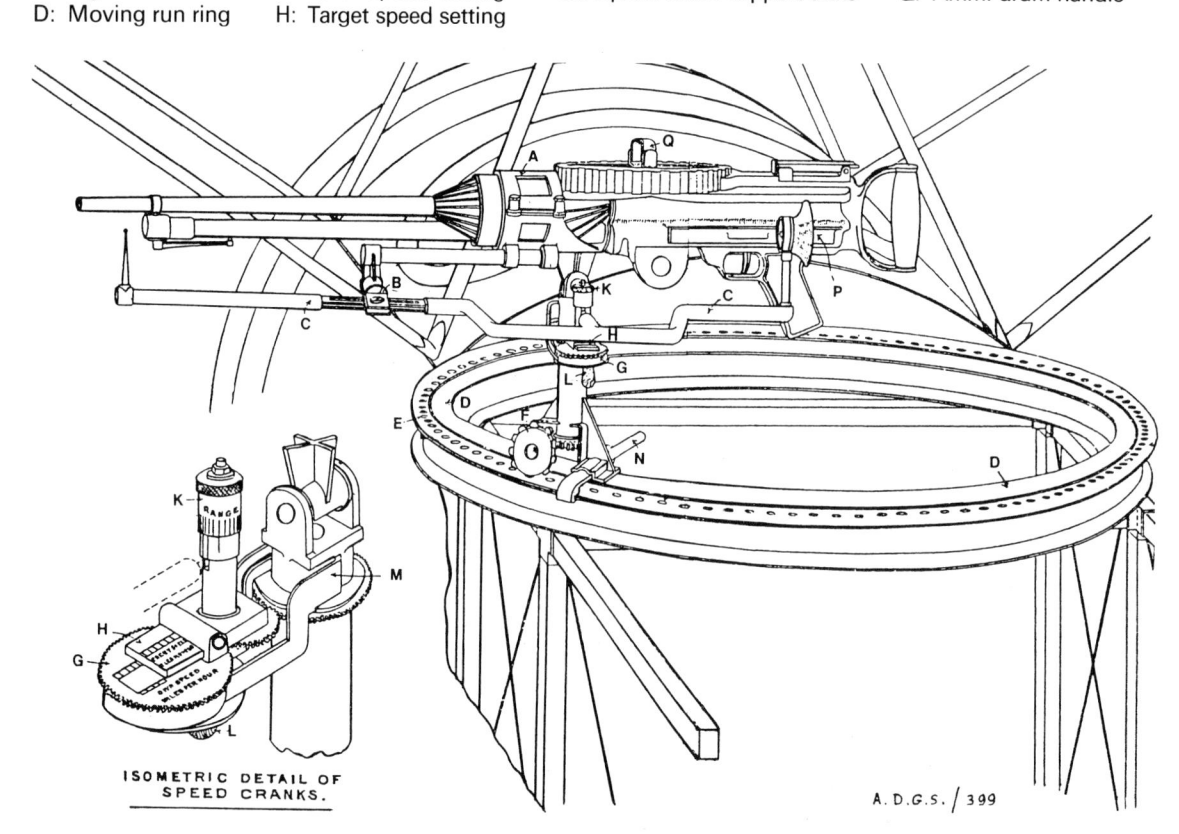

ISOMETRIC DETAIL OF SPEED CRANKS.

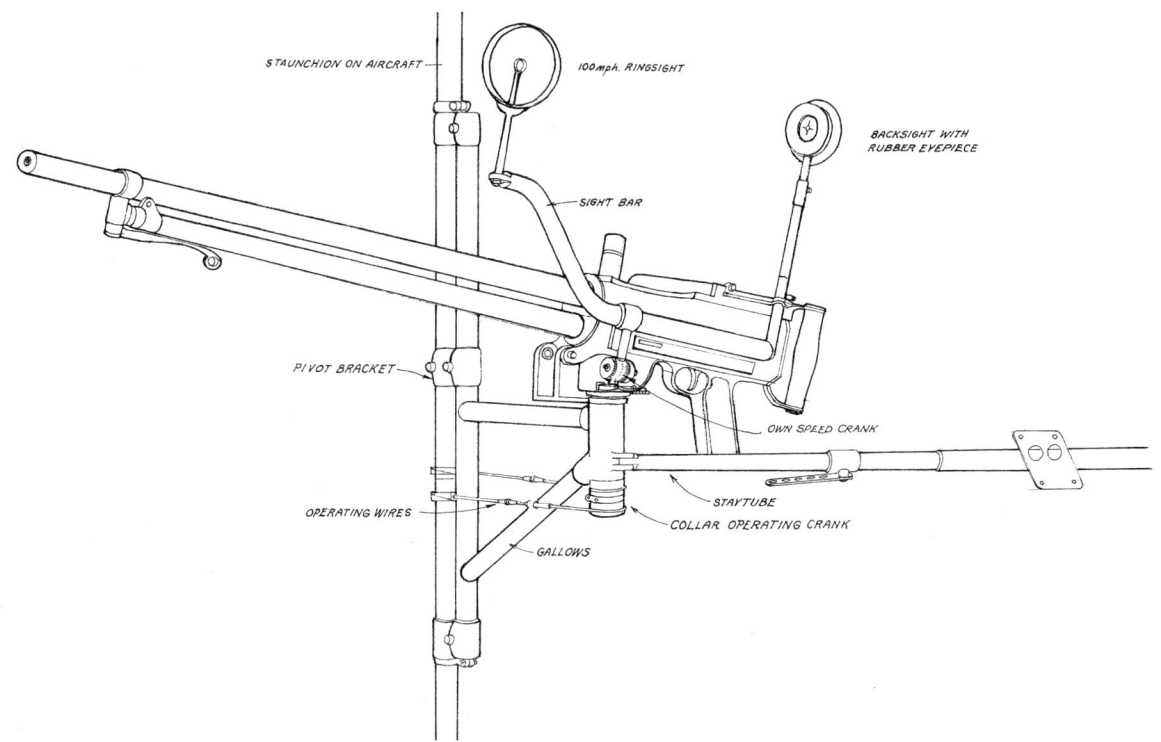

The Scarff 'gallows' mounting. Designed for large aircraft and flying-boats, this employed the same automatic sight line deflection as the ring mount version. It was used on some RNAS aircraft but was found to be no more effective than the Vane sight.

The mechanism was used on two free gun mountings. The first was the gun ring which was to be used in place of the normal Scarff ring; the second was known as the 'gallows' mounting and was designed for use on flying-boats or large aircraft. It comprised an upright tube on which was pivoted a second tube supporting a bracket holding a spiggot into which a gun was fitted. The sight was mounted on the compensating mechanism, the sight line being offset according to the setting on the dials. The foresight was a 3 in (76 mm) ring mounted high over the barrel; the backsight consisted of a peep-sight aperture behind a glass screen engraved with four small arrows pointing towards the centre, protected by a rubber eyepiece. For night sighting the peep-hole could be drilled out larger and used with an illuminated ring foresight.

The installation of both versions was very involved, needing careful setting up, but like all Scarff's inventions the system was well designed. The RNAS tested the ring mounting in a Sopwith 1½ Strutter, and the gallows in an F.5 flying-boat, but judged the ingenious mechanism to be no improvement over the more practical Norman Vane sight. A similar compensating sight was developed 20 years later by Messrs Barr & Stroud, and a compensating sight, the VE, used by Luftwaffe gunners, surprised many over-confident fighter pilots in the Second World War.

The DPG

In 1925 Martlesham evaluated a new version of the Aldis sight which was intended to upgrade it for faster aircraft. Lens sealing was improved and minor modifications made, but the main change was due to a requirement for long-range firing with heavy-calibre guns. Adjustable optics gave a magnification of × 2.5, achieved by a shaft rotated up to 270° by a flexible cable operated from a handle in the cockpit. A dimming screen could also be swung into place for use against white cloud or glare from the sun or water. This 'double purpose' gunsight was evaluated alongside a standard Aldis on a Sopwith Snipe. F/Lt. Pynches reported:

> The DPG is satisfactory in general principle for long-range work. The DPG is considered superior to the Aldis for short-range firing. For ranges of 400 to 600 yards, greater or less magnification than 2½ times would not serve any advantage. On a steady aircraft a line can be maintained on an objective with a steady magnification of 2½. During moderately quick movements, however, it is extremely difficult to keep a line on an objective, and unity magnification is preferable. It has a

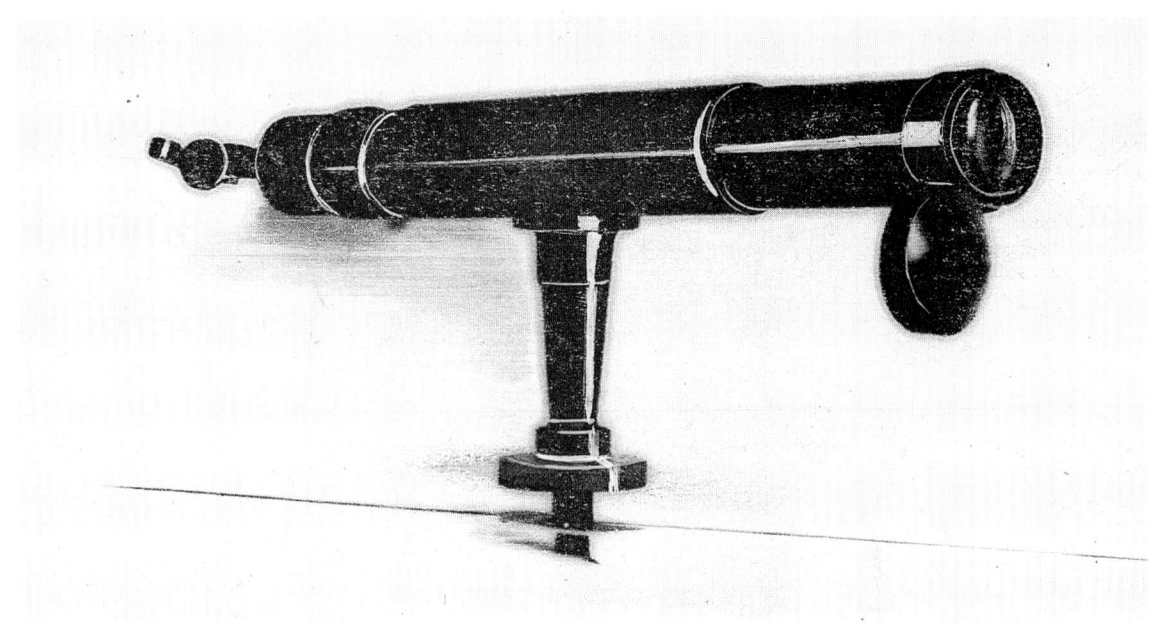

Double-purpose Aldis sight designed for long-range firing. It had an adjustable lens system giving magnification of × 1– × 2.5.

The reflector sight

View through the reflector glass of a reflector sight. It provided the gunner with an aiming mark which appeared to be at infinity, superimposed on the field of view.

The basic reflector system. The collimated lens system ensured that the graticule image was projected in parallel beams of light onto a semi-reflective glass screen.

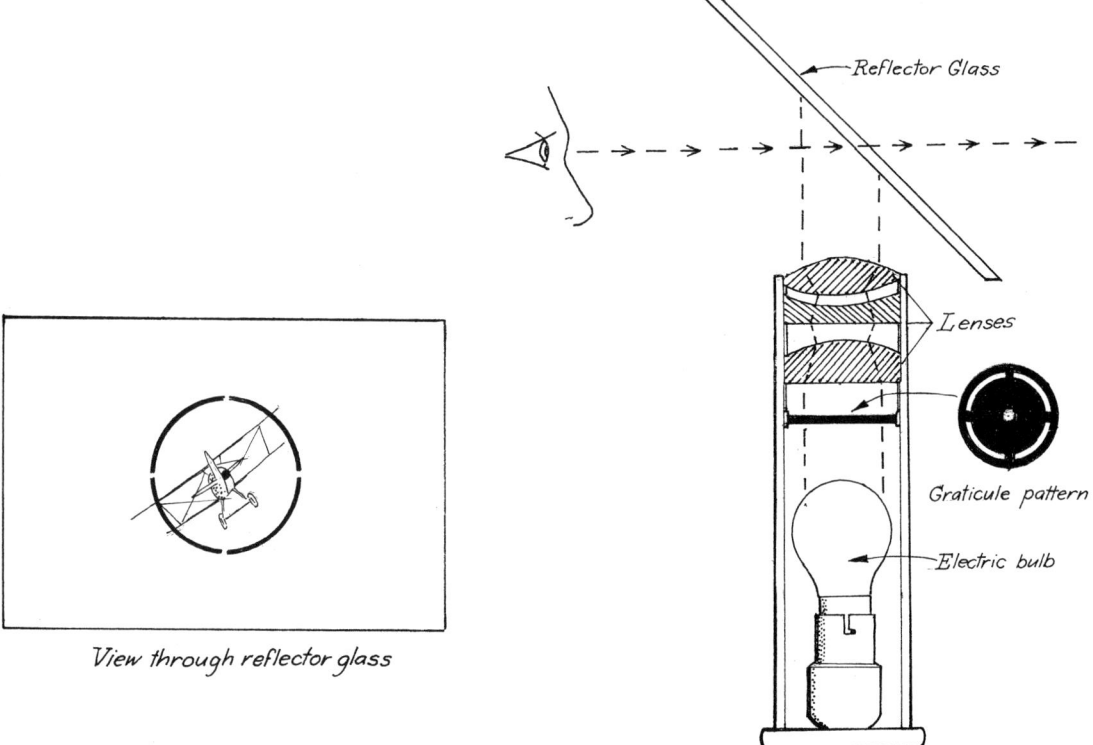

View through reflector glass

THE GUNSIGHTS

distinct advantage that on long-range attacks the DPG mists up considerably less than the Aldis, especially on long steep dives from high altitudes. The present shade of Sun screen is a definite advantage when sighting into the Sun; although it is considered that a slightly larger screen would be an advantage. (Report M/ARM/5S)

The reflector sight

This sight proved so effective from the day of its introduction that it has reigned supreme as the ideal aircraft gunsight. Ironically, it was invented before the Wright brothers left the ground. It was one of the many inventions of Sir Howard Grubb (1844–1931), a noted scientist and optical instrument designer. In 1900 he filed Patent No. 12108 for 'Improvements for sighting devices for guns'. In the following year the Royal Dublin Society published a paper by Sir Howard entitled 'A New Collimating Telescopic Gunsight for Large or Small Ordnance'. The basic concept was so comprehensive that even today no major improvement has been made. The principle of the reflector sight is as follows:

In the base of an upright tubular housing there is a light source. This is directed through an opaque glass plate on which is etched an aiming mark, or graticule. The image of the graticule is projected through the focal plane of a collimated lens, and reflected onto a glass screen mounted at 45° to the gunner's eye. This presents an aiming mark, usually a ring with a central dot on the reflector screen, giving the gunner a clear view of the target with the graticule pattern superimposed on it.

Sir Howard's first design used natural light reflected by mirrors, but after trials these were replaced by an electric bulb (such illumination was mentioned in the original patent). He approached various arms producers, including Vickers, but although they were impressed with the performance of the sight, no orders came. Vickers did commission Sir Howard to design a submarine periscope, and he was working on this in 1914. Vickers then decided to develop a reflector sight with Sir Howard as consultant. The first sight was produced in 1915 and, although a patent was applied for stating that the sight would be useful against aircraft, no mention was made of possible aircraft use.

The Oigee sight

The first record of a reflector sight, being used on aircraft was in Germany, where in 1918 the optical firm of Optische Antal Oigee of Berlin, working

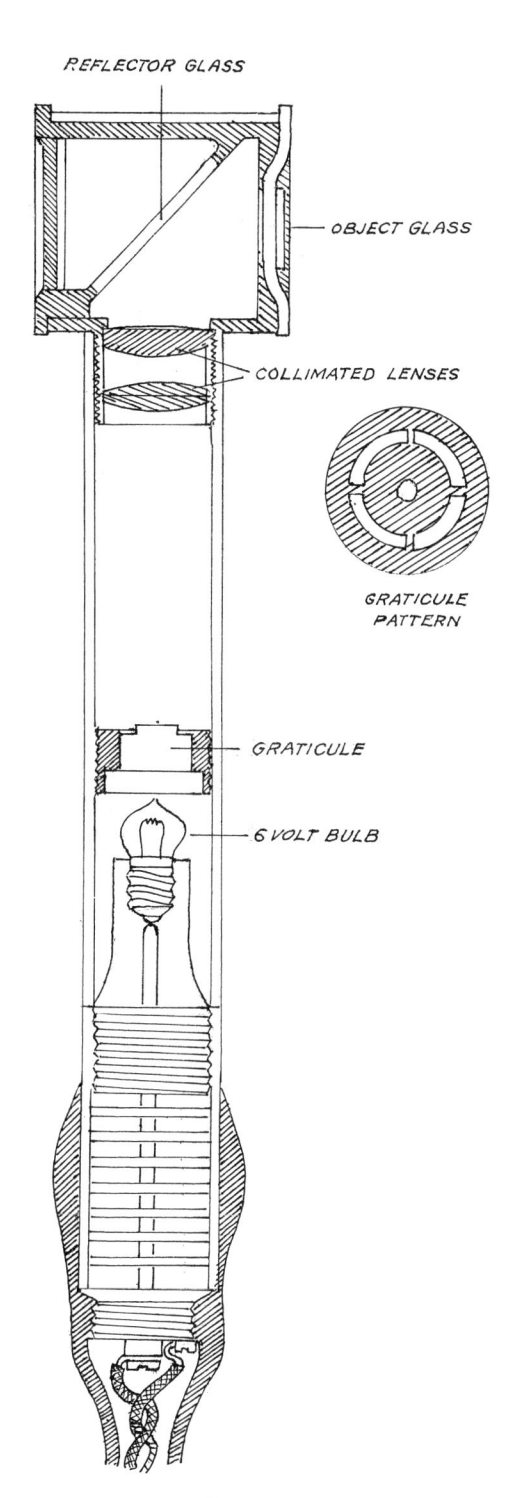

The Vickers Mallock reflector sight of 1915. Although it was successful in trials, the company did not proceed further with the design. (Vickers)

The German Oigee reflector sight of 1918 saw limited use on some Albatros and Fokker DR.1s. It is seen here mounted on a US DH.9 during trials at McCook Field in 1923. (Ross Whistler)

from the Grubb patent, produced two reflector sights for aircraft. One of these was fitted with a sun screen, and could be used by day and night. A smaller version was meant for night use only. A small electric bulb was used as a light source, the reflector being an elliptical glass screen. The sight could be harmonized with the gun by means of a screw and clamp.

Several of the larger version were issued to Jasta 12 Fighter Wing in 1918, when they were fitted to Fokker DR1 aircraft for operational trials. Others were tested on Albatros D. Va aircraft. In 1920 the US Military Attaché in Germany was given a demonstration of the sight. He was so impressed that he sent an example to the US Army Engineering Division at McCook Field, but no official interest was shown.

Barr & Stroud

Barr & Stroud was founded in 1888 by Professor Archibald Barr of Glasgow University and William Stroud of Leeds University, to develop a single observer range-finder for the Admiralty and War Office. From a small concern near Glasgow University, the company became a major supplier of optical defence equipment, and in the early twenties decided to develop reflector gunsights for the RAF.

Inter-war pilot sights

After Grubb's work and the designs of other manufacturers had been studied, an experimental pilot's sight was produced. It consisted of a horizontal lens tube mounted under a rectangular Triplex reflector. Light from a bulb was directed through a circular pattern stamped out of a metal disc, and a collimated lens train to a prism, which turned the image 90° up onto the reflector, which was fixed at 45° to the pilot's eye line. A twin-filament bulb was used to vary the brightness.

The prototype sight, known as the Barr & Stroud GD1, was dispatched to Martlesham Heath where it was tested on an Avro 504. The trials pilot reported that the graticule was not clear enough against white cloud, even with the smoked-glass sun screen in position, and the optics misted up after flying through cloud. The sight was completely redesigned in a new configuration, with the lamp at the base of an upright tube, the graticule being projected straight through the lenses onto the reflector screen, and heat from the lamp preventing misting. The lamp strength was increased, and a modified dimming screen was raised in position by a knurled

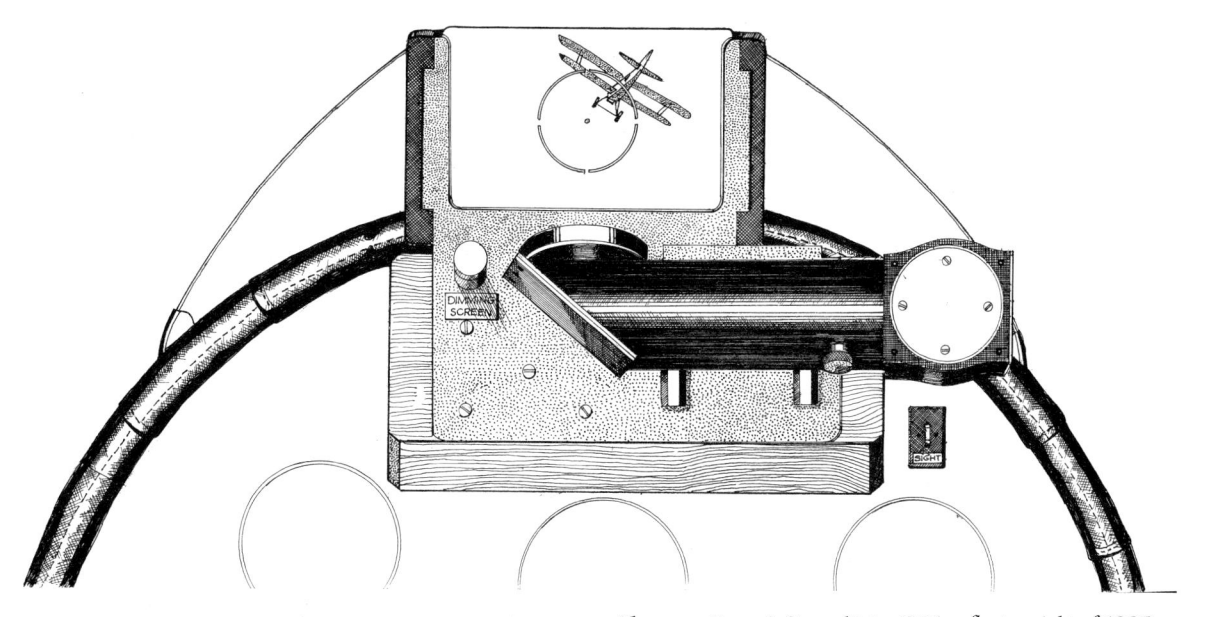

THE GUNSIGHTS

Above *Barr & Stroud type GD1 reflector sight of 1925.*
(Barr & Stroud)

Left *A&AEE Farnborough trials report on the GD1, July 1926.*

knob. This design, the GD2B, received favourable comments from the trials unit in 1927. Work then started on a sight for free-mounted guns (see the section on free gun reflector sights).

The ENI sight

In 1931 the German Oigee company produced another reflector sight, the ENI (Electrische Nivellier Instrument, or Electrical Levelling Instrument). Germany was forbidden to produce armaments at this

The GD2B reflector sight of 1927. The optical system was upright, heat from the bulb tending to prevent misting.

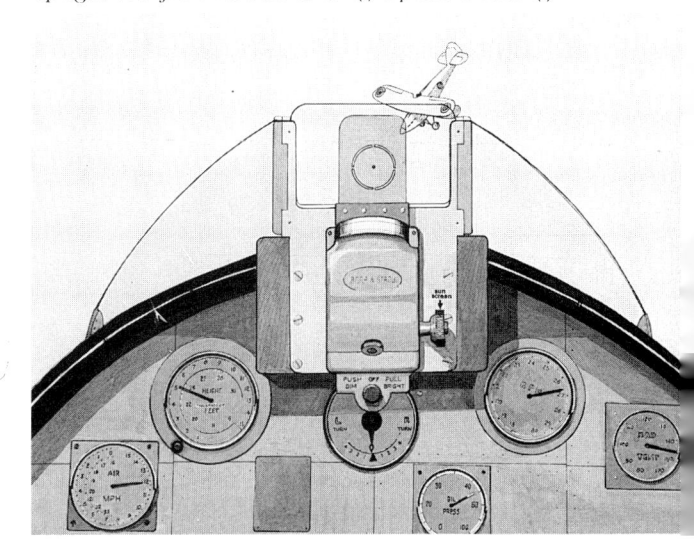

```
Aeroplane & Armament Experimental Establishment.
              Martlesham Heath.
          Interim Report M/Arm/124
                    on
        Trials of Reflector sight for
              Day and Night use.

Summary.
        The above sight has now been mounted on a D.H.9.a.
aircraft and tested in the air against a dark-cloud back-
ground.
        In the tests both Ne and Ar lamps supplied with
current by the special generator provided were used to
project a ring graticule on the grey screen.
Results.
        The ring graticule projected by both types of lamp
was only visible in day-time when a shade was placed over
the top of the screen. In no instance was it sufficiently
pronounced to be used as a day sight, even when opposed to
a very dark cloud back-ground on dull days.
Conclusions.
        It is considered that it will be necessary to
considerably increase the brightness and intensity of the
ring graticule before this type of sight can be used for
day work.
                                    Flight Lieut,
                          Experimental Armt Officer,

                          Chief Technical Officer.

Our Ref:- M/AT/1677.
A.M.Ref:- 814881/25.             Wing Commander,
Martlesham Heath.        Officer Commanding A & A.E.E.(H).
July 1st 1926.
```

The Oigee ENI reflector sight of 1931, tested on a Douglas O-25A at Wright Field. (Ross Whistler)

time, so the company used this acronym to promote sales. Again, Lt. Col. Jacob West, the US Air Attaché, sent an example to McCook Field, where it was tested by the 17th Pursuit Group on Boeing P-12F aircraft. As a result, the Armament Laboratory at Wright Field designed an American version, an L-shaped housing in which a central cross graticule was surrounded by eight small arrows to assist sighting. This was the forerunner of the American 'N' series used by the US forces for the next 20 years.

The GD5

All gunsight development was carried out in great secrecy. Comparing the sights of this period, Barr & Stroud probably had the edge over its contemporaries. A major problem was that there was little money available to produce the hardware. The Air Staff were faced with the impossible task of supplying new aircraft, guns and the new generation of ancillary equipment which was being designed. It was decided to concentrate on formulating specifications, hoping that by the time designers had completed new projects, the situation would have improved.

In 1934 a new Barr & Stroud sight was announced. The GD5 was a complete departure from previous designs. The sight consisted of a lens tube mounted in the centre of a 4.5 in (114 mm) bowl containing a parabolic mirror. The principle of the sight was to separate range-finding and the aiming graticule. The upright tube contained a collimating lens which projected a cross graticule. The parabolic mirror reflected a circular ranging ring which was adjustable to various diameters representing various wingspans. The system used a single bulb light source, and the images were projected up onto the pilot's windscreen. The ideal eye point was given as 16.5 in from the sight, giving an angular field of 7°40′ and allowing eye movement of 1 in (25.4 mm) either side of the centre of the windscreen.

The GD5 was specifically designed for use in the new Hawker fighter, the Demon. The prototype was tested at Martlesham on the prototype Demon by a pilot who made a rough landing and was struck in the face by the protruding sight. This was to be a common hazard with reflector sights. Thick rubber pads were later fitted to the rear of the housing, but the problem was never fully overcome.

Accuracy of the new sight left much to be desired, and it needed continual realignment. The main problem, was double imaging of the graticules, caused by the Triplex glass of the windscreen. It proved difficult to produce surfaces sufficiently parallel to avoid this fault. As the sight had been designed from the outset to use the windscreen as the reflector, it would have needed a complete redesign to adapt it to an integral reflector. Meanwhile, squadron Demons were fitted with the usual Aldis

The Barr & Stroud GD5 of 1935. The centre tube projected a cross, a ranging ring being projected from a parabolic mirror onto the windscreen, operated by the knurled ring at the base of the housing. (Barr & Stroud)

and ring sights.

Later in the year the company was asked to provide a reflector sight for the PV.3, which was to be the last of the famous line of Hawker biplane fighters designed by Sidney Camm. The Barr & Stroud drawing for the new sight, the GD12, specified its use in the Hawker Hawk. It was virtually identical to the GD5. It would appear that the company, realizing that a firm order would be unlikely, had submitted a design as requested, but also mentioned that a completely new pilot's sight was being developed which would be more robust and less complicated than the GD5 series.

Developments abroad

The major powers were not slow to adopt the reflector sight. The Armée de l'Air tested several, and in 1937 the Baille-Lemaire was adopted. Italian Fiat fighters used the San Giorgio reflector sight. The resurgent German Luftwaffe had more experi-

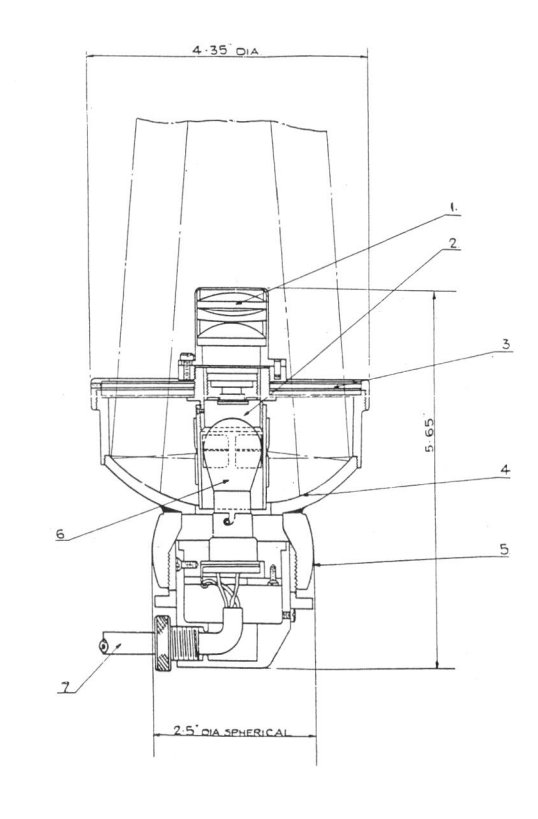

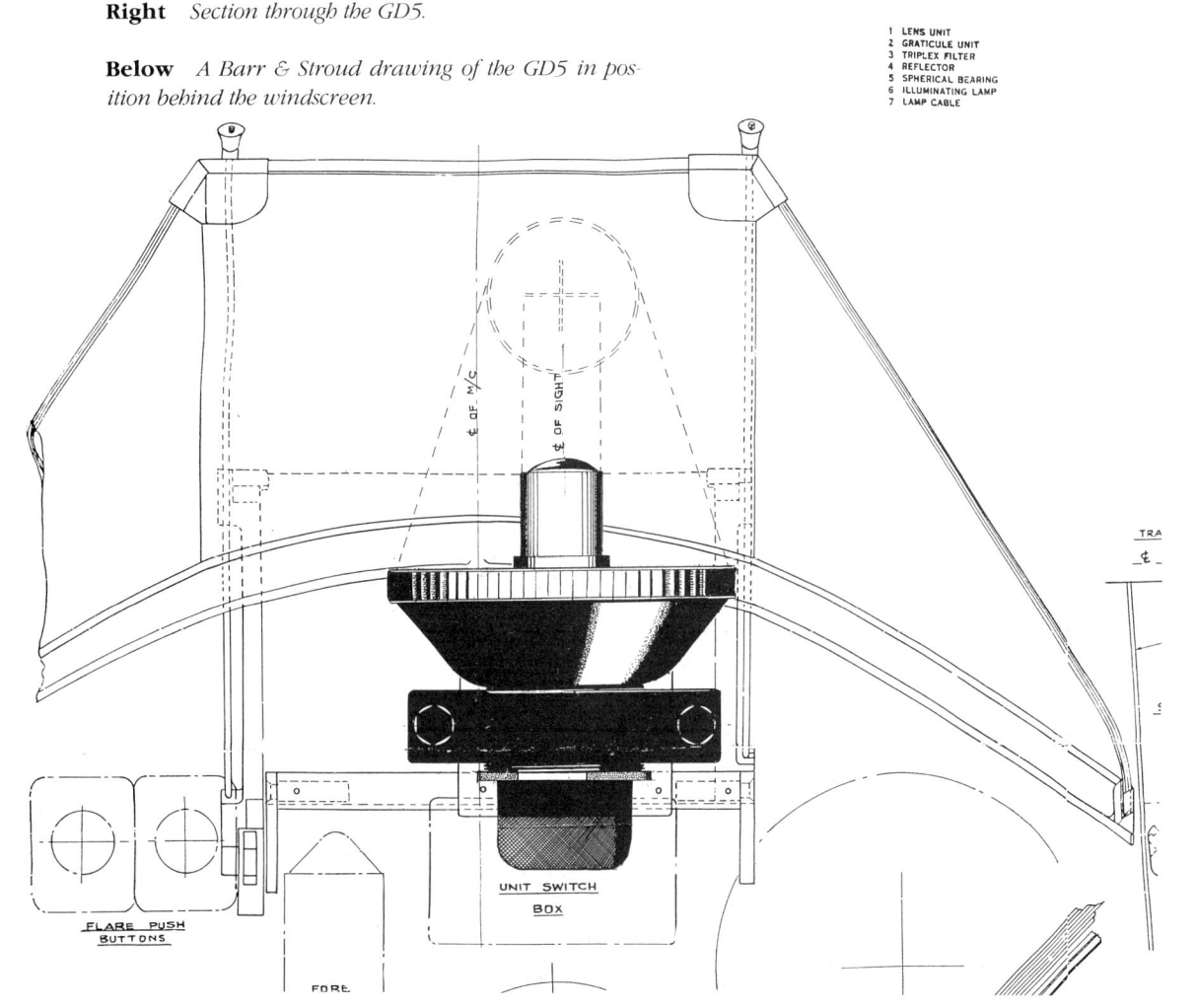

Right *Section through the GD5.*

Below *A Barr & Stroud drawing of the GD5 in position behind the windscreen.*

1 LENS UNIT
2 GRATICULE UNIT
3 TRIPLEX FILTER
4 REFLECTOR
5 SPHERICAL BEARING
6 ILLUMINATING LAMP
7 LAMP CABLE

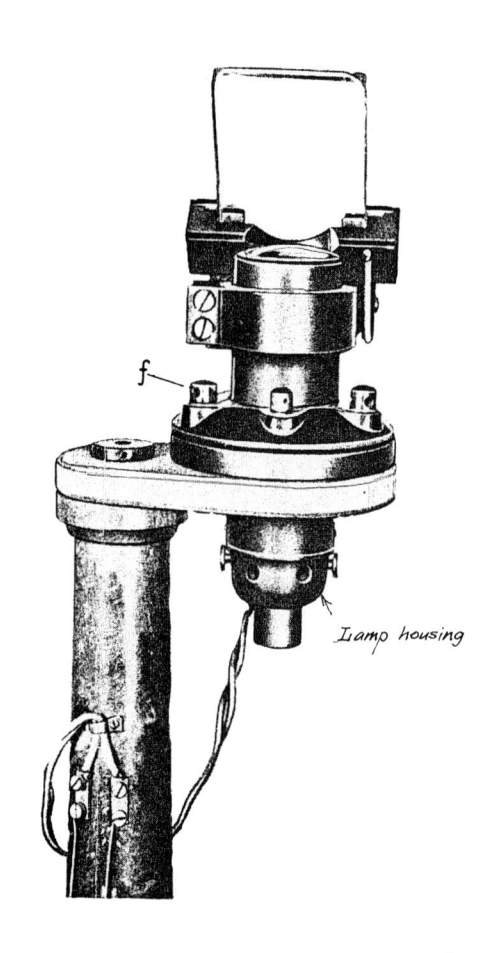

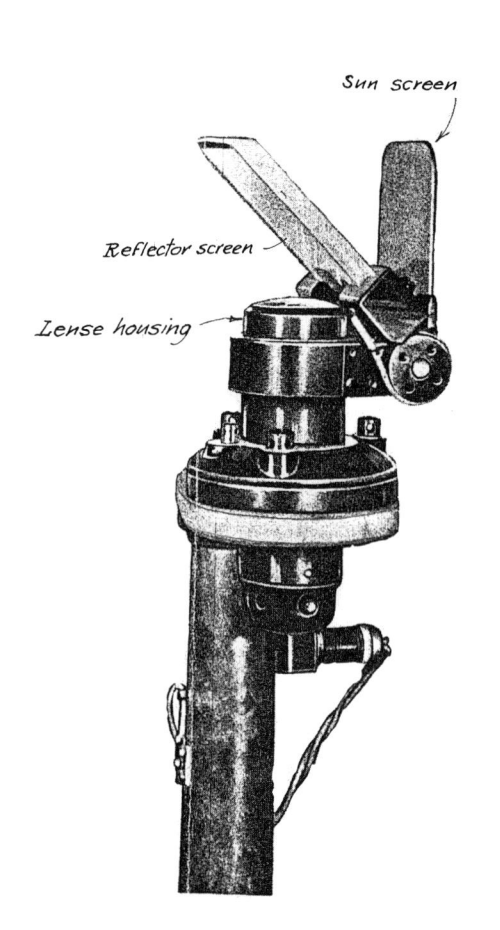

Above *The German Retiflexer Revi 1 was the first of a range of reflector sights used by the Luftwaffe.* (Bundesarchiv)

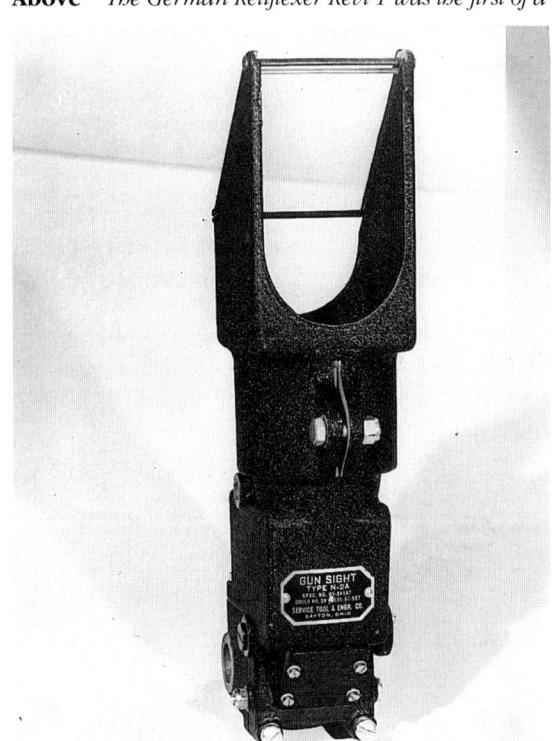

Left *The US N-2A was issued to US Army pursuit squadrons.*

Below *The first US Type N reflector sight designed by Ernest Baldridge. The sight head projected through the cowling in front of the windscreen, the graticule being taken through a prism. This was the type N-1.* (Ross Whistler)

ence of reflector sights than any other air force. Carl Zeiss produced the Reflexnivier Revi I which projected a cross-shaped graticule onto a reflector screen 4.33 in (110 mm) square. A dimming screen could be swivelled into position with a knurled knob. The design was progressively improved, and at the outbreak of the Second World War Revi sights were standard on Luftwaffe fighters.

The US Army Air Corps introduced the N-series sights developed by Ernest Baldridge, of the Armament Laboratory at Wright Field. The N-1 was tested on the XP-6H, and then by the 8th Pursuit Group in January 1933. These trials led to the adoption in 1936 of the N-2, which familiarized American pilots with reflector sights. The next model, the N-3, was used on American fighters during the early part of the war. The US Navy adopted the British GM2, with differences which will be described later.

The Soviet VVS was armed with aircraft weapons second to none, but it was not until 1940 that reflector sights were fitted in any numbers. The first design, the PBP-1, used a rectangular screen and bore a strong resemblence to the German Revi series.

Before the adoption of the new Barr & Stroud fixed gunsight described later, weapon aiming remained traditional. Hawkers fitted a long sight bar for the ring and bead on which the pilot could

An N-2C mounted on a Bell P-39 Airacobra. The handles are gun chargers.

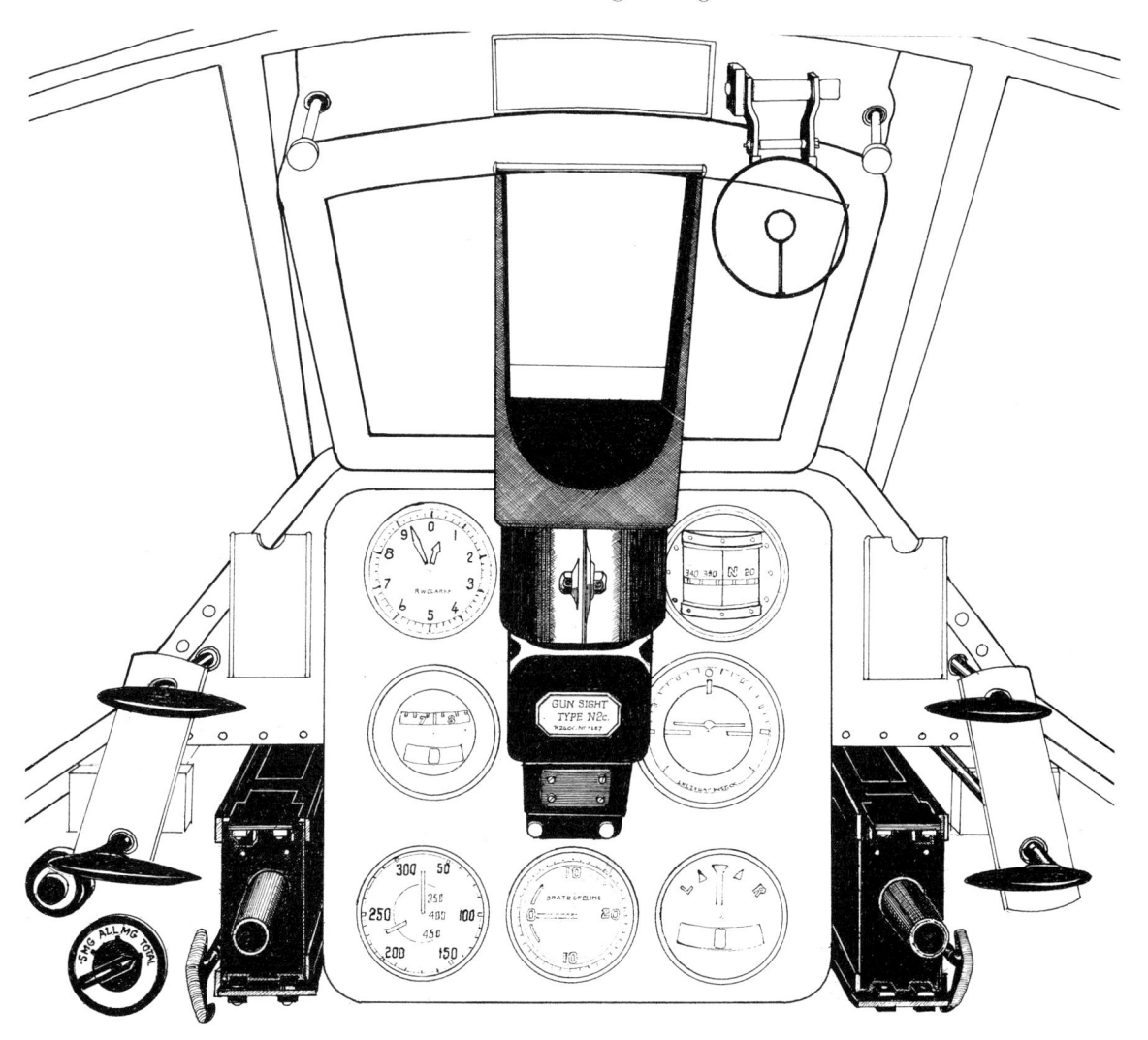

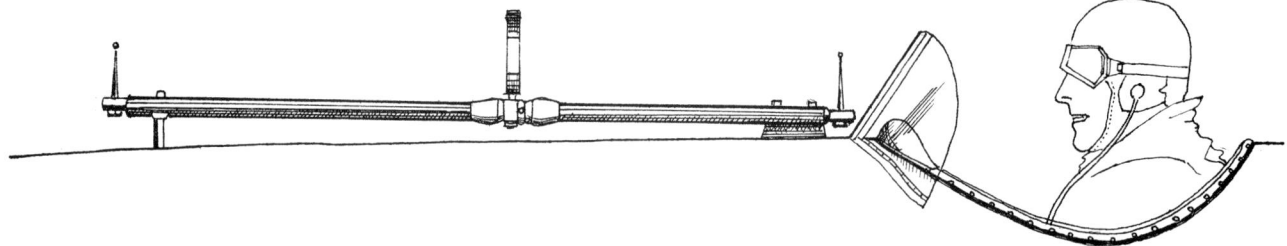

Top and above *The ring and bead sight bar mount of a Hawker Hardy.*

choose the position of the two elements, some preferring the ring in the centre with a bead at either end. Although the Aldis was more accurate, some pilots preferred the ring and bead, especially during fast manoeuvres when 'g' forces made it difficult to align the small eyepiece. Another reason was that, when engaged in low-level air-to-ground firing, the pilot needed full peripheral vision to avoid flying into the ground. The Aldis also restricted the field of view at a time when targets of opportunity could appear. Another advantage of the 'iron sight' was that when engaging ground targets, the point of aim was in front of the impact point, and it was not easy to do this with the Aldis.

Sight recording cameras

Camera guns have been used as a gunnery training aid since the 1914–18 war, when in 1916 a sight recording camera was designed at the RFC machine-gun school at Hythe in Kent. This camera

Below *The Thornton Pickard camera gun used from 1917 until the mid-thirties. The weight and balance were identical to the Lewis gun.* (MOD Pattern Room, Nottingham)

HYTHE CAMERA GUN · MARK III · H

(Sheet N°2)

Above The Thornton Pickard was known in the Service as the Hythe Camera gun Mk III. Single exposures were taken on 120 mm film. The film holder, graticule unit and shutter are shown here.

Right The Williamson G22 gun camera of 1934, shown here on a Hawker high-speed gun ring. This recorded single exposures and was mounted on a replica gun.

Below The Williamson G22 gun camera for over-wing mounting. A special telephoto lens was used, and the action was triggered from the pilot's control handle via a Bowden cable.

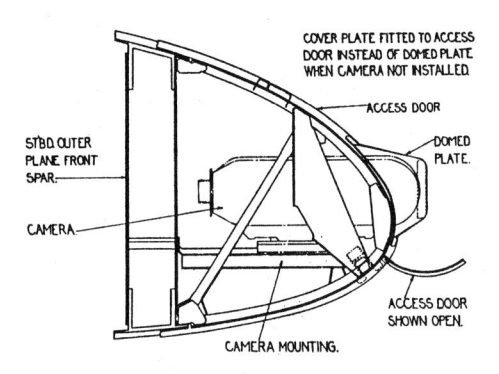

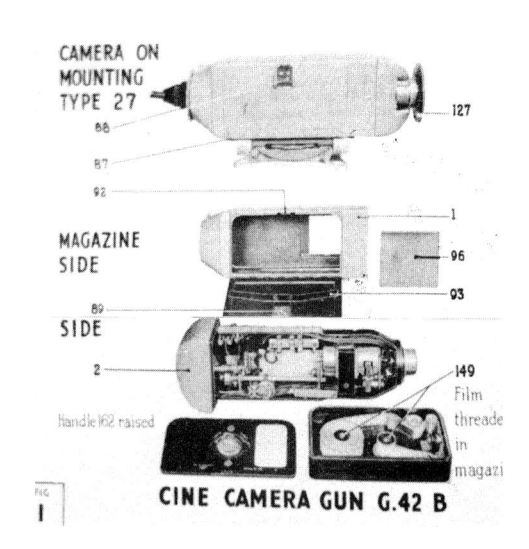

*The Williamson G42 cine camera gun gave a much improved assessment of air gunnery, being used in both fixed and free mounted positions. The drawing (**above**) shows the method of fixing in the wing leading edge; the diagram (**right**) shows the chassis and film magazine. The G42 was fixed to a replica gun for observers use.*

gun, made by the Thornton Pickard Co. of Altringham, was similar in shape and weight to the Lewis gun. It proved to be very successful in assessing the standard of gunnery in the air. Known as the Hythe Gun Camera Mk III, it took still exposures on a 120 mm film roll. Provision was made for a multi-ring graticule and the time of exposure. The trigger, cocking action and balance were all identical to the Lewis, and a drum magazine was clipped into place. It was used mainly on Scarff ring mounts, but was also fitted onto the top wings of fighter aircraft, controlled by a Bowden cable.

The Hythe was used for gunnery training until the Williamson camera gun was adopted in 1934. The Williamson G22 fixed gun camera was fitted onto the upper surface of the wing, and each still exposure was 'fired' from the spade grip of the pilot's control column. For observers' use the camera was mounted on a replica Lewis gun with the usual spade and pistol grip. The still photographs taken gave a good enough record of performance, but the next model to be adopted was to provide a cine-film record, giving an improved assessment of a pilot's aim. This was the Williamson G42B, which could be mounted in the leading edge of the thicker wing sections of aircraft then coming into service.

Free gun reflector sights

The first sight designed by Barr & Stroud for free-mounted guns was the Type J1 of 1926. This was fixed on a bracket clamped half-way up the barrel of a Lewis gun. A circular reflector screen was mounted over a cone-shaped lens housing, with a sun screen which could be raised into position at the rear. Although the sight was well elevated above the barrel, the need to clear the ammunition drum put the gunners' eye position too distant for easy sighting. Its position hard on the barrel also ensured a short bulb life. However, the new sight was found to be accurate when it was tested at Farnborough, where several improvements were suggested before it could be considered for Service use.

The Barr & Stroud type J1 of 1926. This was the first free gun sight produced by the company. The ring mounting was also a company product. (Barr & Stroud)

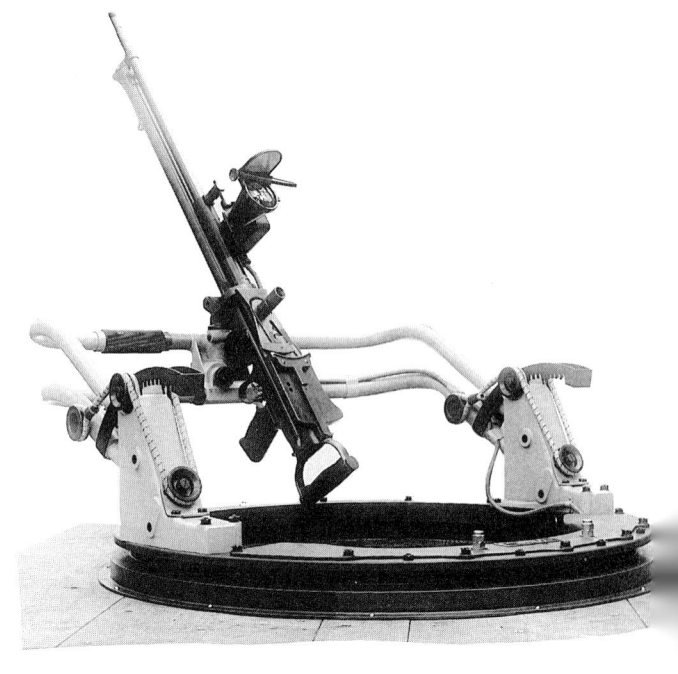

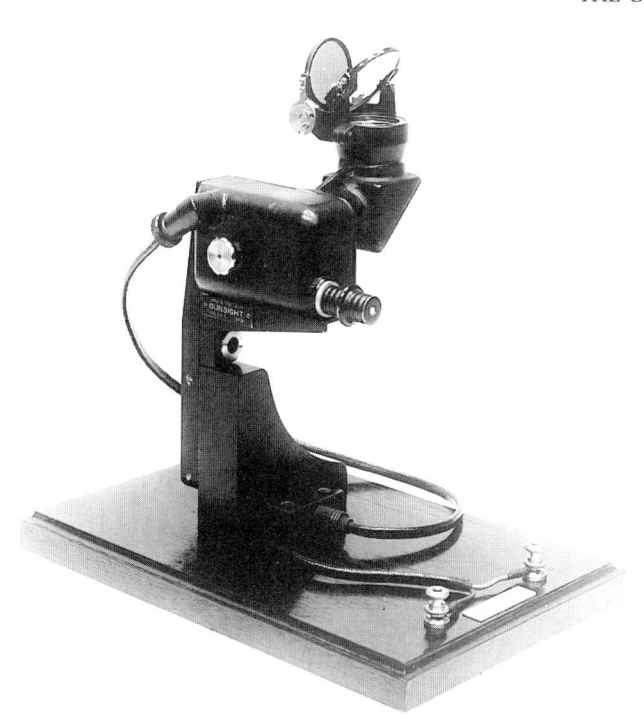

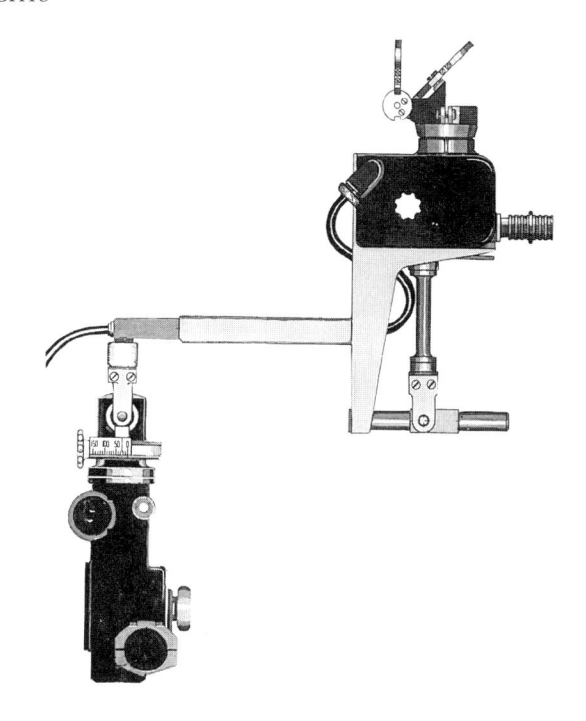

Above *The prototype Barr & Stroud GH6 of 1930, on which the sight head was offset.* (Barr & Stroud)

Above right *This deflection compensator was designed for the GH6 using the same principle as the Scarff compensator mounting.*

Right *A GH6 fitted to a Lewis gun.* (Barr & Stroud)

After the J1 was evaluated at Farnborough it was decided to change the format. The new design consisted of a rectangular lamp housing which was offset from the gun centre line. A bulb projected a ring graticule from the side of the housing onto a prism, which directed the image through a lens system onto a small circular reflector screen behind which was a swing-in smoked-glass sun screen. To be known as the GH6, the sight was fixed on the side of the gun clear of the drum, with the sight head protruding from the side to a position in line with the gunner's eye. The GH6 was found to be effective and was accepted for small-scale trials while an automatic 'own speed' and deflection system was devised on the lines of the Scarff compensating mount. This consisted of a pillar on which were speed settings, the sight being fixed to an arm fitted at right angles from the top of the pillar which varied the line of sight according to the setting. This required deft manipulation of the adjusting knobs, not easily accomplished with heavily gloved hands. The compensating mounting received the title Type GH2 No. 14, but was never issued for Service use.

The Mk I free gun reflector sight

The GH6 was used by gunners of a flight of Handley Page Heyfords taking part in the 1934 exercises. During a night attack on a 'blue-land' city, searchlights picked out the 'raiders' and the gunners reported that the beams had caused dazzle on the reflector screens (there were also reports of sunlight having the same effect). After various experiments, it was found that a small hood over the reflector prevented most of the dazzle and also protected the screen from damage. This feature was incorporated in a new sight, the Mk I free gun sight, which had similar optics to those of the GH6 but was much more compact and practical. The lamp housing and rheostat was a separate unit, quickly detached for bulb changing. Like the GH6, a prism projected the graticule onto a hooded screen, offset from the sight body above the centre line of the gun barrel. The double-filament lamp could be dimmed from full brightness to extra low to suit conditions. The diameter of the bright orange graticule represented the deflection for a target crossing at 50 mph (80 km/h). The graticule could also be used as a range-finding aid: if the span of a twin-engined fighter filled the graticule, the range would be 300 yd (274 m); if it filled the radius it would be 600 yd (549 m). Barr & Stroud then decided to develop a turret sight using the same features, but without the prismatic system.

The GJ3

As bomber speeds increased in the 1930s, windscreens were fitted to protect gunners. The next step was to enclose the gunner in a glazed cupola which could rotate. Some manufacturers then decided to develop enclosed, power-operated turrets, with the guns remote from the gunner's eye line. It followed that the sight also had to be remote from the guns, but linked to the gun movement. Barr & Stroud's answer was the GJ3, which, when

Left *The GJ1 of 1936, which was accepted for limited service in the RAF where it was known as the Free Gun Reflector Sight Mk IA.* (Barr & Stroud)

Below *Prototype of the Barr & Stroud GJ3, which was developed into the highly successful Mk III series.* (Barr & Stroud)

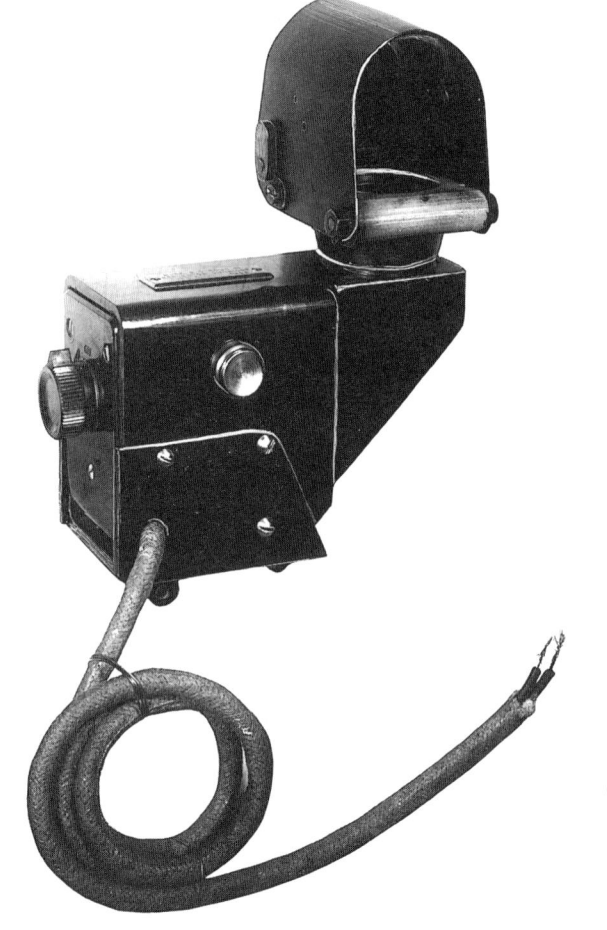

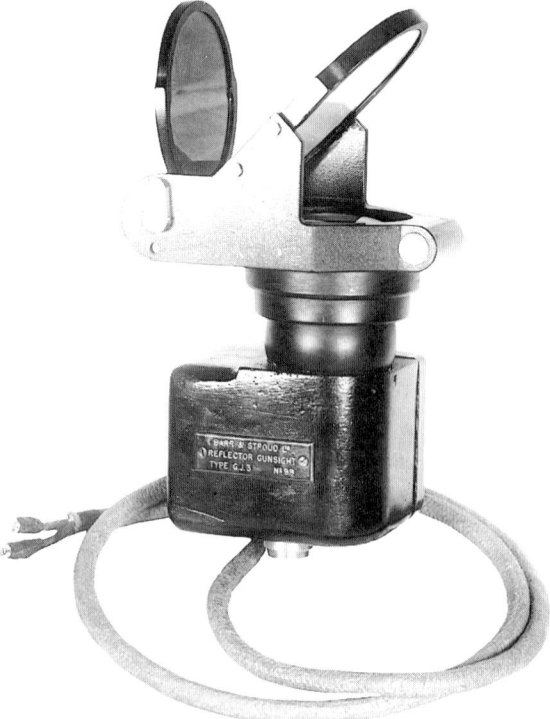

THE GUNSIGHTS

fully developed, was to be produced in larger numbers than any other British sight. It consisted of a detachable lamp unit and an upright sight head, the glass reflector screen being fitted to a housing surmounting the lens tube. The reflector housing supported a swing-in-type sun screen operated by a small knob at the side.

The prototype sight was tested at the weapons department at Farnborough, where the optical system and electrics were proved to be satisfactory. It was then sent to Martlesham Heath, where it was installed in a Parnall Hendy Heck fitted with two Browning guns, and air firing trials were carried out at various heights and at ground targets. The trials reports were favourable, although it was

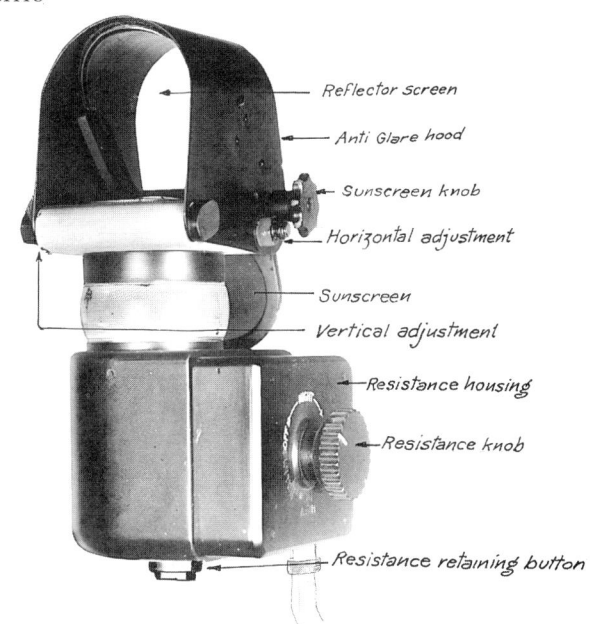

Right *The production GJ3, adopted by the RAF as the standard turret and free gun sight, designated the Mk III and Mk IIIA.*

Below *The Mk IIIA is seen here on the twin Brownings of a Westland Lysander.*

400 YARDS   600 YARDS

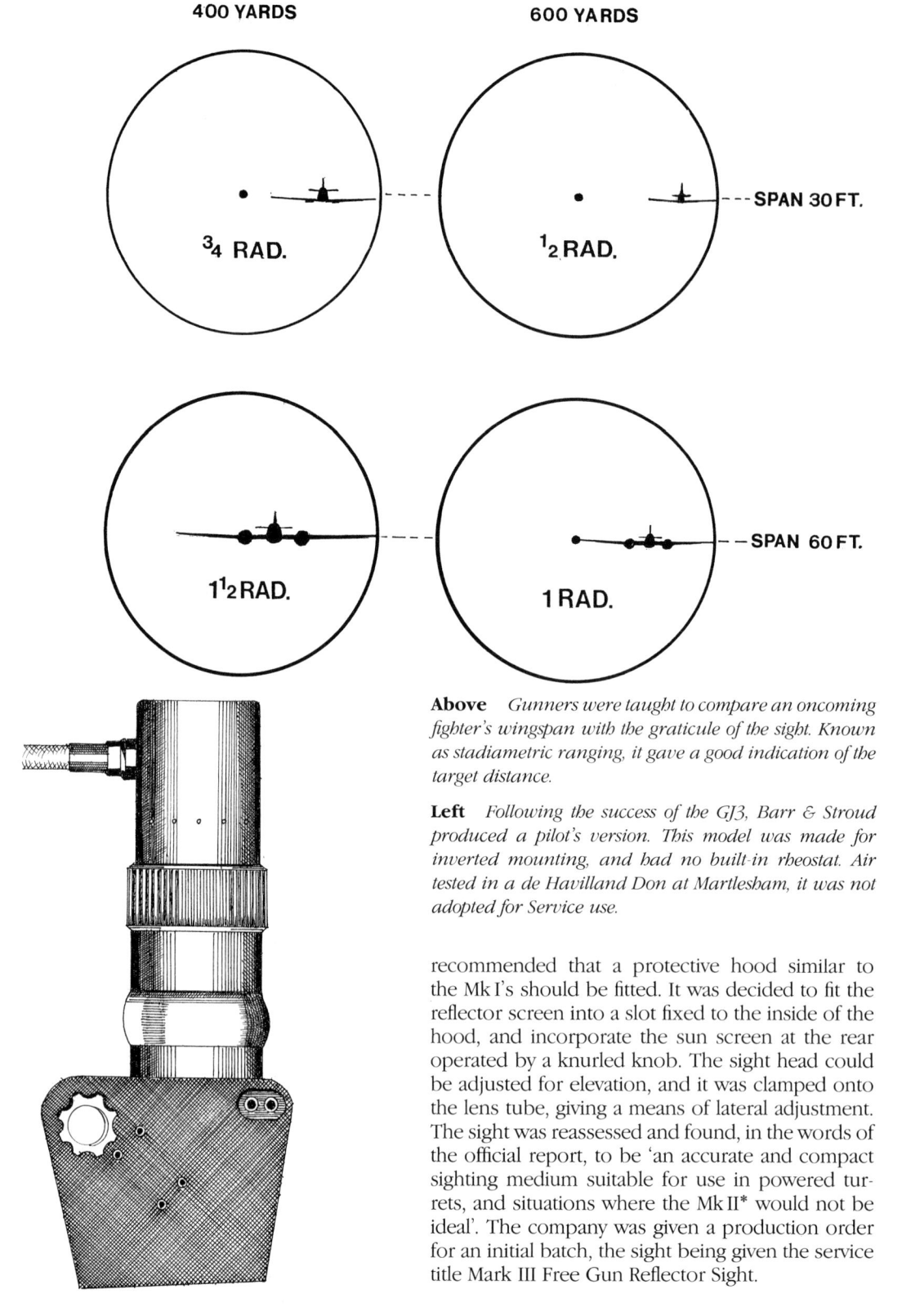

¾ RAD.   ½ RAD. — SPAN 30 FT.

1½ RAD.   1 RAD. — SPAN 60 FT.

Above *Gunners were taught to compare an oncoming fighter's wingspan with the graticule of the sight. Known as stadiametric ranging, it gave a good indication of the target distance.*

Left *Following the success of the GJ3, Barr & Stroud produced a pilot's version. This model was made for inverted mounting, and had no built-in rheostat. Air tested in a de Havilland Don at Martlesham, it was not adopted for Service use.*

recommended that a protective hood similar to the Mk I's should be fitted. It was decided to fit the reflector screen into a slot fixed to the inside of the hood, and incorporate the sun screen at the rear operated by a knurled knob. The sight head could be adjusted for elevation, and it was clamped onto the lens tube, giving a means of lateral adjustment. The sight was reassessed and found, in the words of the official report, to be 'an accurate and compact sighting medium suitable for use in powered turrets, and situations where the Mk II* would not be ideal'. The company was given a production order for an initial batch, the sight being given the service title Mark III Free Gun Reflector Sight.

THE GUNSIGHTS

Left This experimental GJ3 was fitted with a stadiametric ranging system. It was not accepted as the dials could not be operated with gloved hands. (Barr & Stroud)

Above On the Mk IIIL, the reflector angle could be depressed for aiming rocket projectiles from 0° to 5°. The size of the hood and screen was enlarged on this version of the L.

Below A Mk IIIA* reflector sight in a Wellington FN5 tail turret. The vee-shaped sight mounting was linked to the gun elevation system by rods and connecting arms.

REFLECTOR SIGHT. Mk.IIIA. FOR FREE GUN.

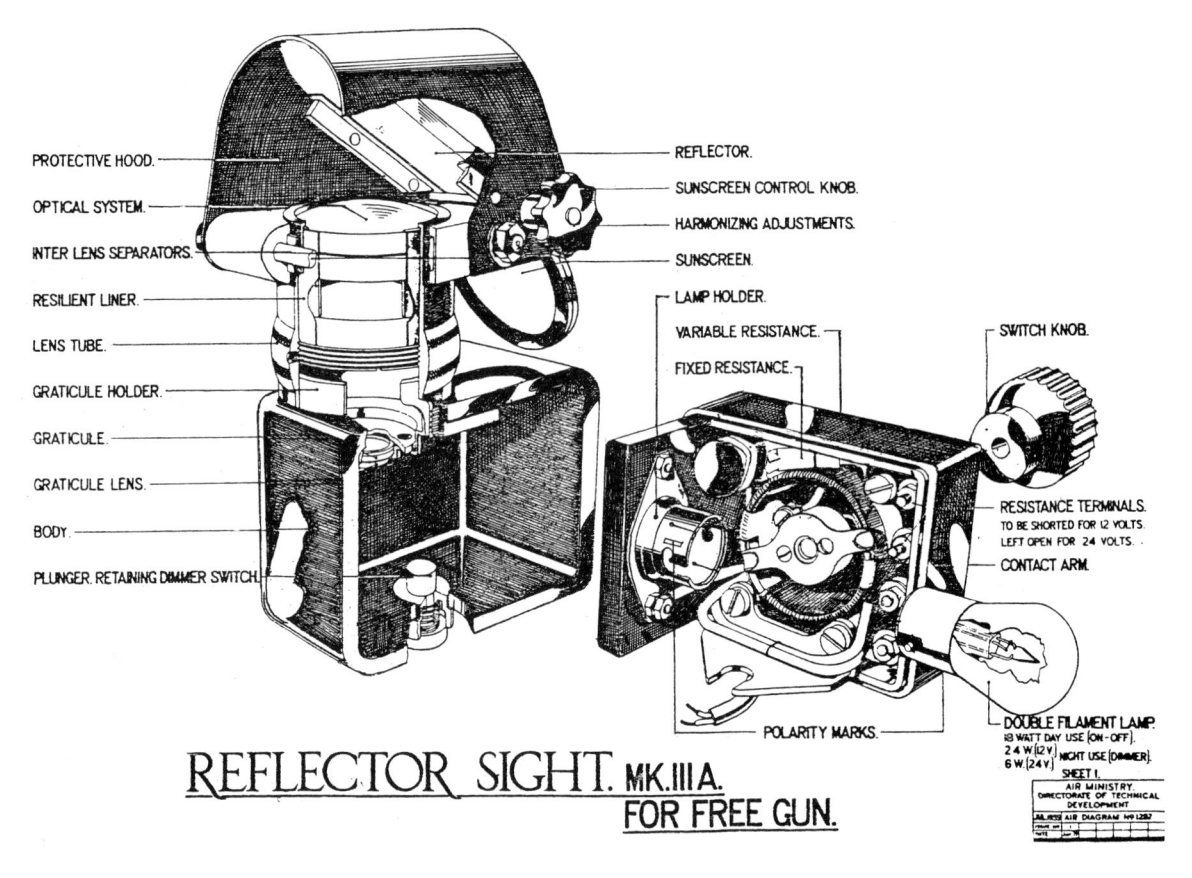

Labels on diagram:
- PROTECTIVE HOOD.
- OPTICAL SYSTEM.
- INTER LENS SEPARATORS.
- RESILIENT LINER.
- LENS TUBE.
- GRATICULE HOLDER.
- GRATICULE.
- GRATICULE LENS.
- BODY.
- PLUNGER. RETAINING DIMMER SWITCH.
- REFLECTOR.
- SUNSCREEN CONTROL KNOB.
- HARMONIZING ADJUSTMENTS.
- SUNSCREEN.
- LAMP HOLDER.
- VARIABLE RESISTANCE.
- FIXED RESISTANCE.
- SWITCH KNOB.
- RESISTANCE TERMINALS. TO BE SHORTED FOR 12 VOLTS. LEFT OPEN FOR 24 VOLTS.
- CONTACT ARM.
- POLARITY MARKS.
- DOUBLE FILAMENT LAMP. 18 WATT DAY USE (ON-OFF). 2.4 W.(12V.) NIGHT USE (DIMMER). 6 W.(24V.)

AIR MINISTRY. DIRECTORATE OF TECHNICAL DEVELOPMENT. AIR DIAGRAM N°1287 SHEET 1.

Above *Section through the Mk IIIA*.*

Left *A Mk IIIN with neither hood nor dimming screen, used on daylight operations.*

When production got under way, the company's lens-making department was fully extended, the Mk III and Mk II pilot's sight (described later) requiring many high-quality lenses. As demand increased, sub-contractors were appointed for some of the work. The Glasgow team then decided to produce a version of the Mk III with stadiametric ranging, but it was found to be difficult to operate with heavy flying gloves and was dropped.

The Mk III series proved to be one of the most successful reflector sights ever produced. Virtually every turret in RAF Bomber Command used it, and it was adopted by the US Navy as the Mark 9, produced in America by the Woolensak Optical Co. of Rochester, NY, and in Australia by the Kriesler (Australasia) Pty Ltd. (This model featured an anti-vibration lampholder.) As well as turret use, it was fitted to free-mounted Vickers K guns, and, being less obtrusive than the Mk II*, was used as a pilot's sight on many multi-engined aircraft. The final rede-

The Mk IIIA* was widely used by the USAAF as the US Mk 9. The gunner's 'No. 2' was brought back from Africa in a B17!

sign was the incorporation of a tilting head mechanism for rocket firing.

The production versions were:

Mk III	First version, no inbuilt adjustment, 24/12 v
Mk IIIA	Die-cast, modified adjustment 24 v, short hood
Mk IIIA*	Main production version, some bakelite housing, 24 v
Mk III G	Tilting head for rocket firing use
Mk III L	Tilting head with hood
Mk III N	Hood deleted
Australian Mk III	Spring-damped bulbholder
US Mk 9	With or without adjustable head

In most turrets the sight was mounted at the gunner's eye level and connected in elevation to the gun cradle by rods and cranks (in rotation the sight moved with the turret). Nash & Thompson FN4 tail turret gunners often used the sight as a convenient handhold when getting into their seats, which could put the sight out of alignment, so a prominent 'Hands Off' sign was attached.

The Mk IIIA* was used on many free gun mountings, and on heavy-calibre installations such as the Molins gun on the Mosquito XVIII. The most recent use was probably the twin Hispano mountings on South African Shackleton IIIs.

Prismatic sights

The reflector sight was not universally accepted as ideal: during trials at Northolt and Farnborough, pilots complained that the graticule was hard to see against bright cloud, and could not be held in view in a tight turn. These complaints were usually made by older pilots used to the black outline of the ring and bead, and as they were of a higher rank, a conference was held at Farnborough on 12 October 1938 to discuss alternative systems. The Armament Department of the Royal Aircraft Establishment suggested a sight similar to the Aldis, but using prisms to reduce the size for use behind the windscreen. The RAE were given an order for two such sights, the G1 type 'A' pilot's sight, and the type 'B' for free guns and turrets. These were to have a black-line graticule and a facility for night illumination.

Messrs Ross Ltd produced four type 'A' sights, to be given comparative tests with reflector sights in a Gladiator. The type B was produced for use in turrets, but the small (20 mm/0.78 in) eyepiece proved a drawback, and it was soon replaced by the Mk III reflector sight. All work on prismatic sights was finally abandoned in 1940. The G1 was used as the optical head of the Mk I gyro, and some Bristol B.I turrets used it, but from 1940 onwards the

A type G prismatic sight.

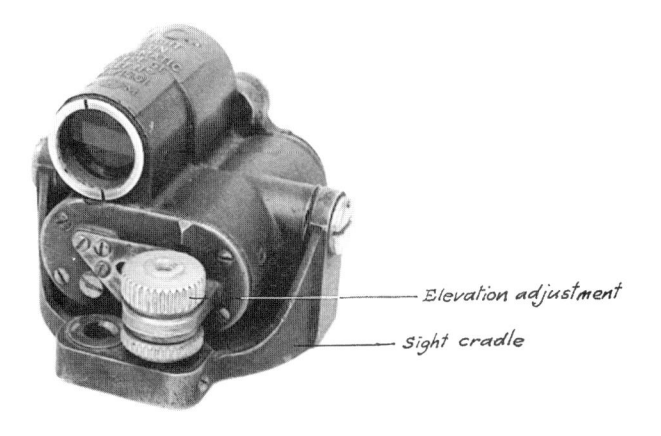

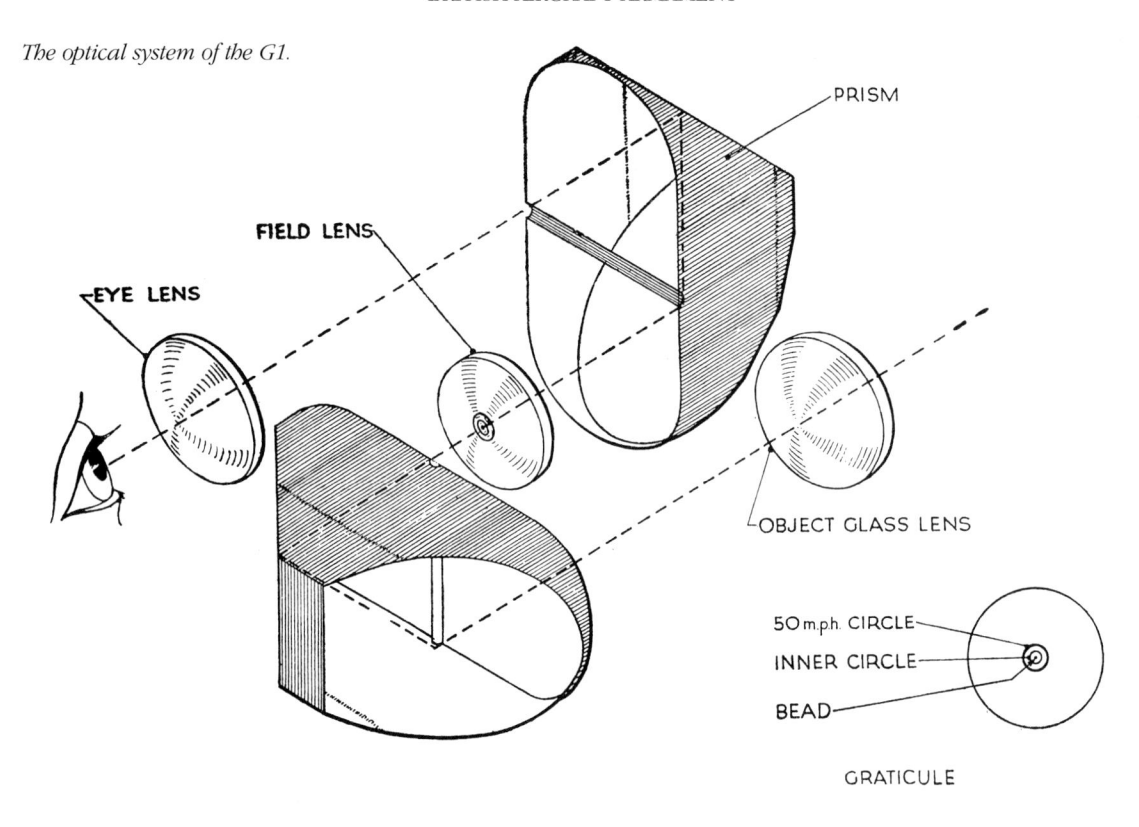

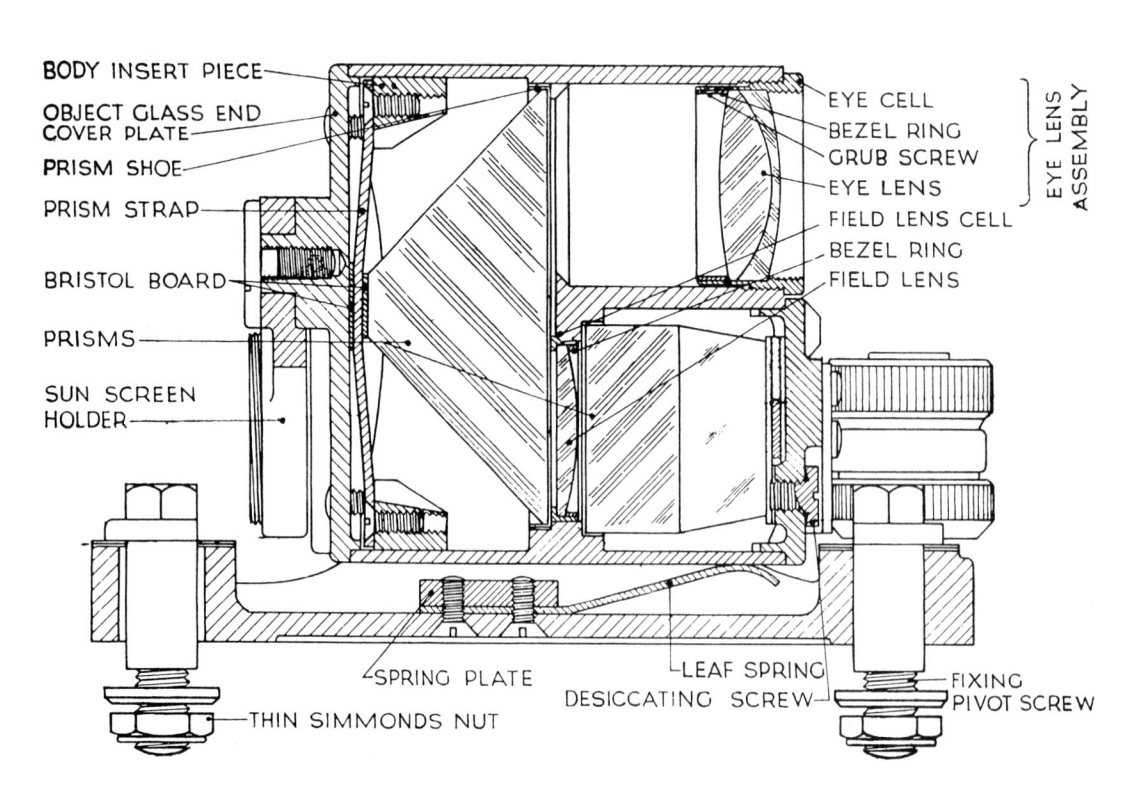

The optical system of the G1.

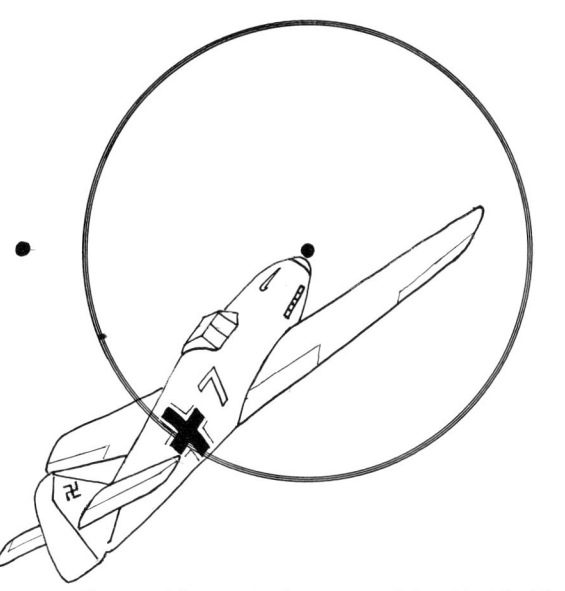

Above *The graticule pattern of the G1 sight. The spots were used for ranging.*

Right *A G1 sight mounted on a Bristol B.I turret. The gunner placed his chin on the pad to steady his aim.*

reflector sight became standard apart from the ring and bead sights used on some free-mounted guns, as mentioned elsewhere.

The G45 camera gun

In July 1939 the G45 camera gun was issued to armament schools and fighter units. Designed and manufactured by the Williamson company of

The Williamson type 29 replica Vickers K gun for free gun training.

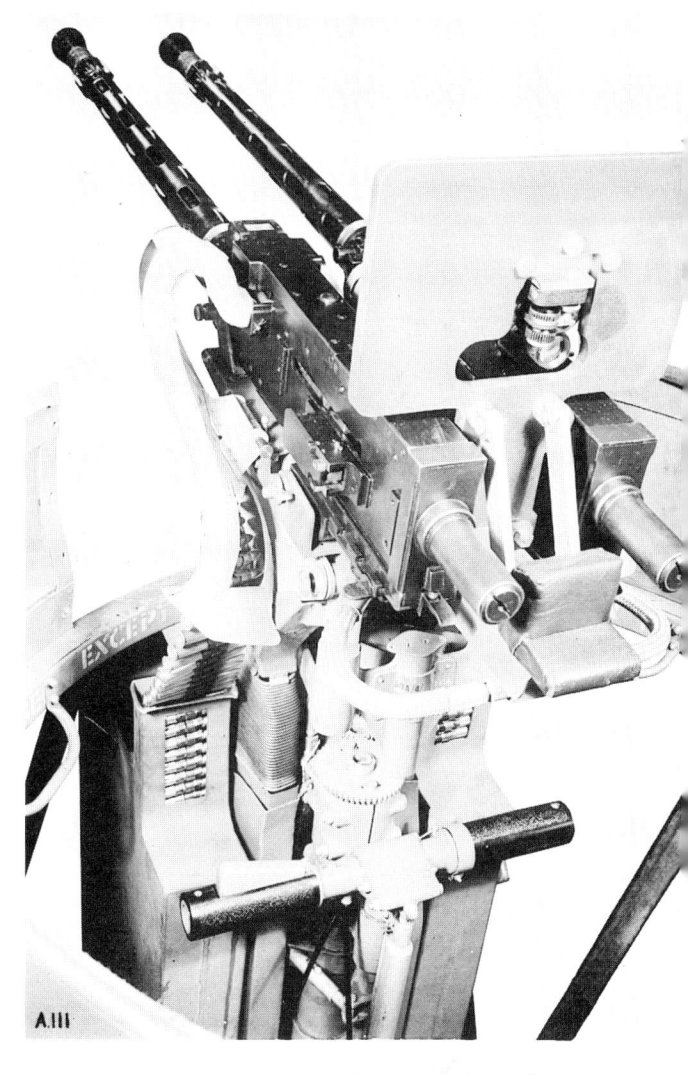

London and Reading, it was developed from the earlier G42B which had been in service for some years. The G45 used 16 mm Orthochromatic film supplied in 25 ft (7.62 m) lengths. Frame speeds could be regulated to 16, 18 or 20 per second, these speeds corresponding to the rates of fire of the Lewis, Vickers K and Browning. It was supplied in two versions, the fixed gun type with a short lens unit, and a long lens for other use. Williamson also supplied a replica Vickers K gun, the type 29, with a camera mount and reflector sight bracket.

The G45 proved to be an essential aid to aerial gunnery, enabling a trainee to be shown the results of his 'shooting' after an exercise and be advised on any improvements needed. A footage recorder was provided in the cockpit or turret, wired from contacts in the camera. In the centre of the recorder was a 'sunny or cloudy' switch wired to the aperture of the camera. Early problems with condensation and moisture were eased when heating elements were fitted to the lens and camera body, although they were never fully overcome. The G45 was fitted as standard on Fighter Command aircraft, but, mounted in the leading edge close to the guns, the lens was often obscured by smoke from tracer and vibration affected film clarity. It was controlled by an electrical switch operated by the gun-firing pneu-

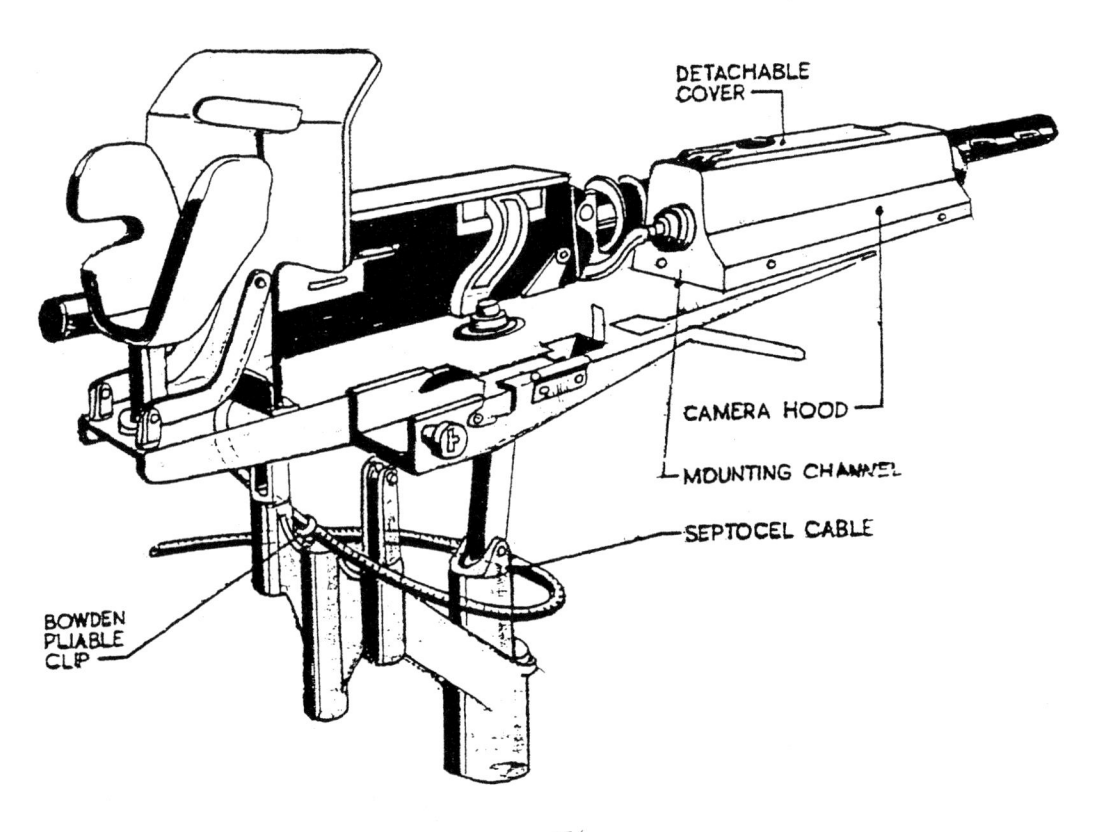

Above *Film footage recorder.*

Below *Camera mounting in a Bristol Blenheim B.I turret.*

Right *Camera position in a Spitfire wing.*

Middle right *The G45 camera gun showing film cassette access.*

matic system of a fighter, or from the turret electrical firing unit. The film was loaded into a cassette, which could be inserted either from the top or side of the housing. The G45 was one of the most practical means of weapon aiming training, and the gun camera hut in gunnery training schools was in constant use. A purpose-made cine-projector made by Specto Ltd of Windsor was used to show the films. It could show frame stills or slow motion shots of the trainees' performance.

Barr & Stroud GM series fixed gun sights

Although the GD5 pilot's sight had proved to be quite effective, the windscreen projection system had caused problems with double imaging, and it

The Barr & Stroud GM1 fixed gun sight. This was the prototype with a perspex range/base indicator. (Barr & Stroud)

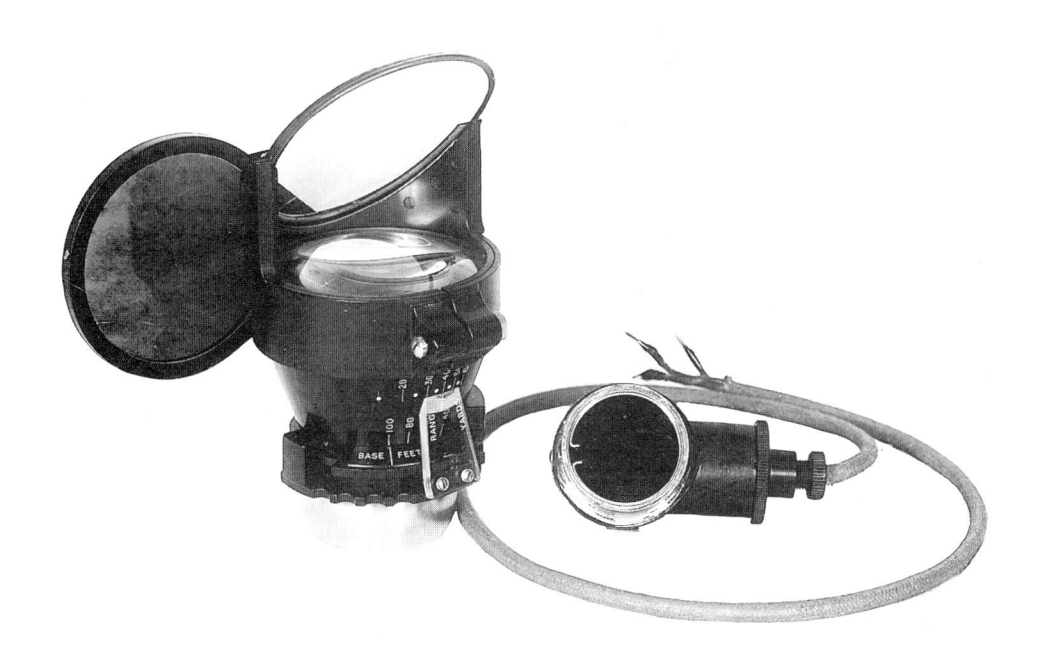

had proved difficult to manufacture. With this in mind, the company was asked to design a sight with an integral reflector which would be more suitable for series production. In late 1935 the prototype of the new sight was submitted for evaluation. The optical department produced a lens with a numerical aperture of $F/O = .68$, through which a large circular graticule was projected onto a circular glass reflector screen 3 in (75 mm) in diameter. The graticule was bisected by a cross, the horizontal bar of which was broken in the centre, the gap being set by (1) a knurled ring which turned a perspex pointer to various range settings, and (2) an adjustable ring which turned an indicator to wingspan in feet. The internal mechanism then set the gap according to the required range. A central dot was added for a further aiming point. The pilot first set the span dial to the known wingspan of his prospective target, then the range dial to the maximum

Left *The GM2, adopted for use in 1938 as the standard RAF fixed gun sight and designated the Fixed Gun Reflector Sight Mk II. (Barr & Stroud)*

Below *The Srb-a-Stys optical sight offered to the Air Ministry in 1938. It was brought to London by a representative of the Czech company after Germany had marched into the Sudetenland. The offer was declined with thanks. It would appear that it would have been even more hazardous than the Mk II in a rough landing. This is the only known picture of the sight.*

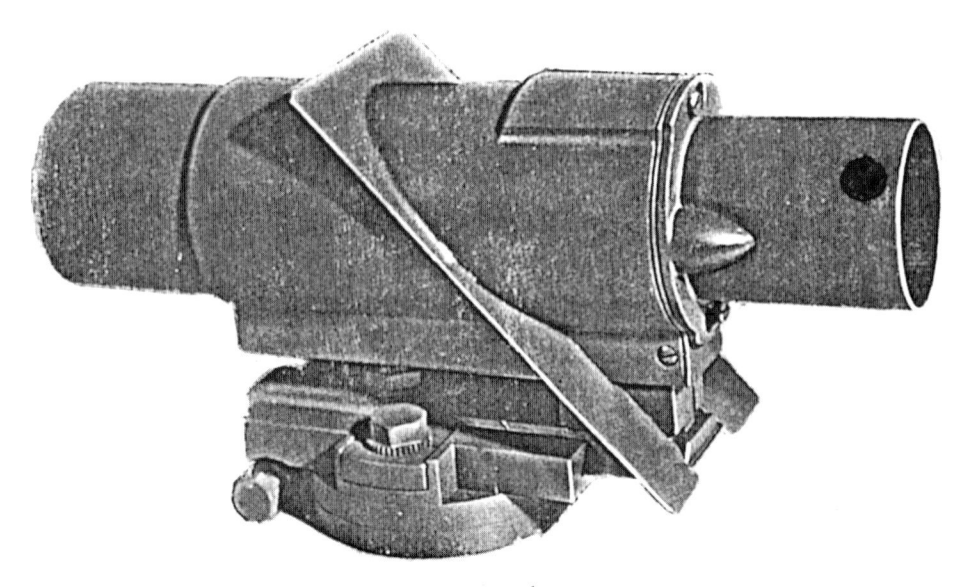

for accurate fire. When the target coincided with the gap it was within range. The radius of the graticule ring gave the deflection allowance for hitting a target crossing at 100 mph (161 km/h). The sight, designated the GM1, was illuminated by a half-silvered 12 v lamp in a quick-release holder at the base of the sight body.

The prototype was tested at Farnborough, and then at Martlesham in a Bristol Bulldog for air firing tests. The trials reports led to various modifications. A substantial rubber pad was fitted to protect the pilot from injury in the event of a rough landing, the range/base setting was modified to two similar knurled rings with their own scale and indicator, and the lamp changing was also made easier. The blue-tinted swing-in sun screen was found to be effective in high brightness situations although the orange graticule was not perfect in some conditions. The modified sight was designated the Barr & Stroud Type GM2, and was accepted as the standard fixed gunsight of the RAF, being known as the Reflector Sight Mark II. It was patented in 1937, and the first sights of an initial order of 1,600 reached some Gloster Gladiator squadrons in 1938.

The Srb-a-Stys sight

When Germany occupied the Sudetenland, a Swedish optician working for the Czech Srb-a-Stys Company was sent to London on a secret mission to offer a new aircraft gunsight design to the British. The instrument resembled a large-diameter shortened Aldis sight which needed to be inserted through the pilot's windscreen. This would have posed problems with thick bullet-proof windscreens. Although the sight was tested and found to be quite accurate, the offer was declined and the agent, Mr George Vogel, returned to Prague with a letter of thanks to the Czech Company. Meanwhile, Gladiator pilots tested their new sights.

War clouds gather

At the time of the Munich crisis in 1938, increased orders were placed for the Mk II for the new Spitfire and Hurricane fighters just coming into production. Even with 24-hour working, Barr & Stroud were already at full stretch, and as all British companies with the required production facilities were committed to other defence contracts, the Air Staff asked the company to find an overseas licensee. This resulted in an agreement being signed with C. P. Goerz of Vienna. Drawings and a Mk II sight were sent to Vienna, and Herr Neuman of Goerz concluded a production contract with an Air Ministry representative. In early 1939, with 55 sights

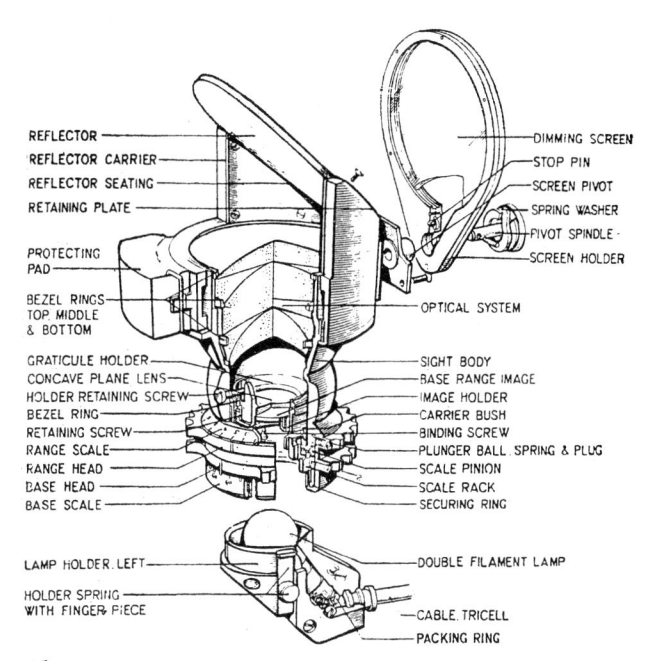

Above *Section through the Mk II reflector sight.*

Below *The range/base system of the Mk II. When the required range and base (wingspan) had been set, the gap in the horizontal line gave the maximum range at which to open fire.*

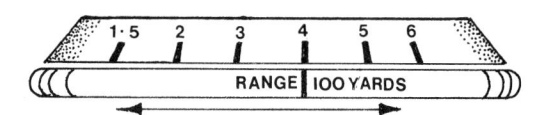

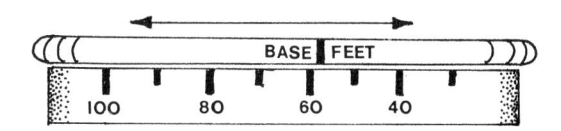

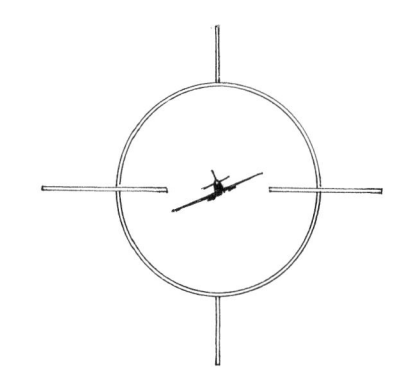

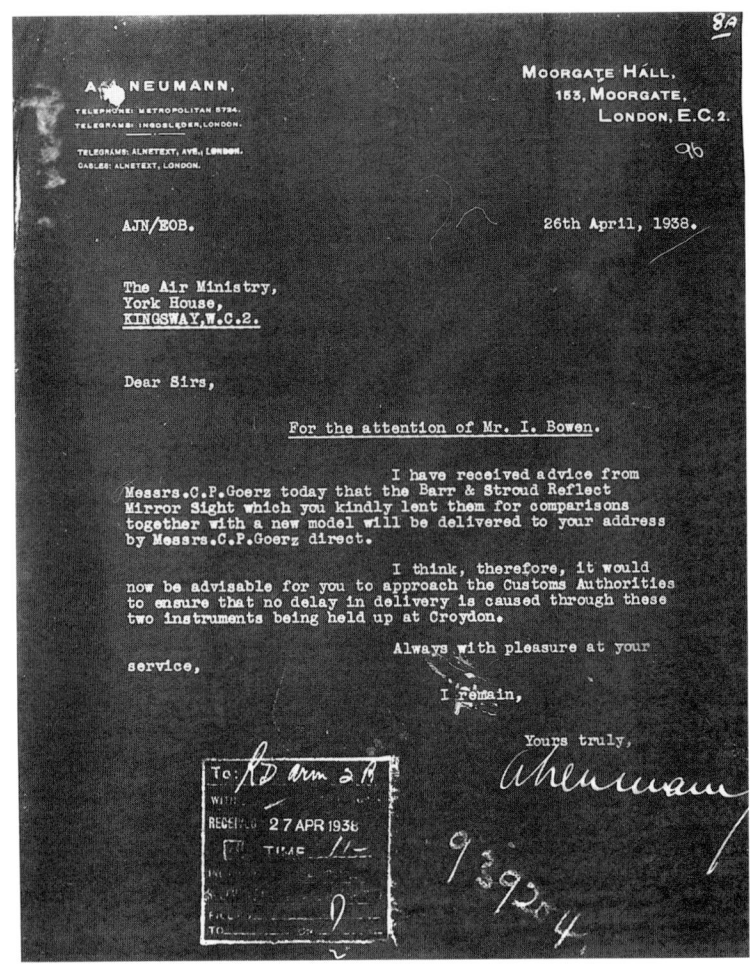

> A. NEUMANN
> MOORGATE HALL,
> 153, MOORGATE,
> LONDON, E.C.2.
>
> AJN/EOB.
>
> 26th April, 1938.
>
> The Air Ministry,
> York House,
> KINGSWAY, W.C.2.
>
> Dear Sirs,
>
> For the attention of Mr. I. Bowen.
>
> I have received advice from Messrs. C.P. Goerz today that the Barr & Stroud Reflect Mirror Sight which you kindly lent them for comparisons together with a new model will be delivered to your address by Messrs. C.P. Goerz direct.
>
> I think, therefore, it would now be advisable for you to approach the Customs Authorities to ensure that no delay in delivery is caused through these two instruments being held up at Croydon.
>
> Always with pleasure at your service,
>
> I remain,
>
> Yours truly,

Above *Typical correspondence between Goerz and the Air Ministry.*

Left *The Mk II* replaced the Mk II in 1941, the circular reflector screen being replaced by a rectangular one. The sun screen was also deleted.* (Barr & Stroud)

delivered, the *Anschluss* pact was signed by Germany and Austria. However, Air Ministry fears of cancellation were groundless and Goerz were only too pleased to honour the contract, 700 sights known as the GM2 Mk III being delivered before the outbreak of war in September. These gunsights made by the 'enemy' were invaluable, as production at Glasgow could not satisfy the needs of Fighter Command. In early 1940 the situation was eased when the Salford Electrical Co. began production under licence. The sight was used in most British fighters of 1938–43, although as mentioned earlier, some Beaufighters and Mosquitos used the Mk III* turret sight for fixed guns.

The first change in the design was made in 1941, when the circular reflector glass was replaced by one 4.5 in (114 mm) square, the circular design hav-

THE GUNSIGHTS

A Mk II* reflector sight mounted in a Spitfire.

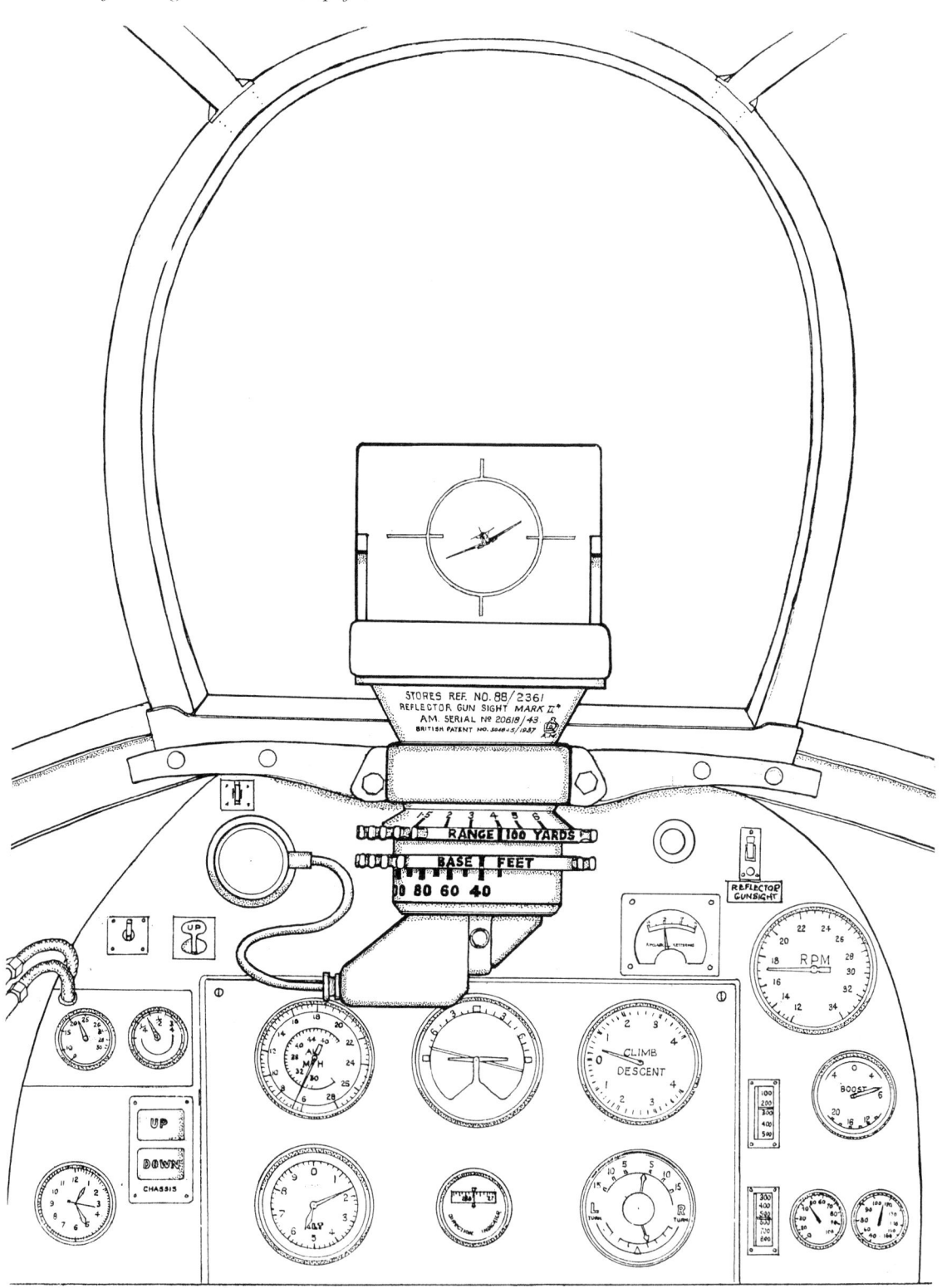

When rockets came into widespread use, a special version of the GM2 was designed in co-operation with Farnborough, to allow for the increased gravity drop of these projectiles. The reflector screen was made to tilt forward by the pilot from 0° to 5° depression, according to airspeed and the known drop of the missiles being used. This modification, which involved the replacement of the sight head, became known as the type 1 Mk II conversion, and the sight then became the Mk IIL.

The Beamont modification

S/Ldr R. P. Beamont commanded a squadron of Hawker Typhoons engaged on low-level attacks on German installations in occupied France. Low clouds and poor visibility were often encountered, and it was found that, even if the gunsight lamp was turned right down, the large ring and range bar tended to obscure tracking and target acquisition. Beamont decided to cut down the graticule pattern on his MK II* sight, so squadron armourers blanked off the ring and bars, leaving only the centre dot visible. The smaller aiming mark proved a big success: Beamont could discern much more without the large orange glow. On subsequent operations he realized that it would be even better if the glass reflector screen was removed and the graticule projected onto the windscreen, the rake of which was 45°. The following day he used the sight in an attack on an installation in the Lille area, and found he had a better field of view than ever before, so he had all 609 Squadron's sights altered to the same configuration. When the news of this modification reached the authorities, terse instructions were issued to remove the unauthorized alterations forthwith. Eventually it was agreed that in low light conditions the system had its advantages, and the spot was easily seen in full daylight conditions if the brightness was fully turned up.

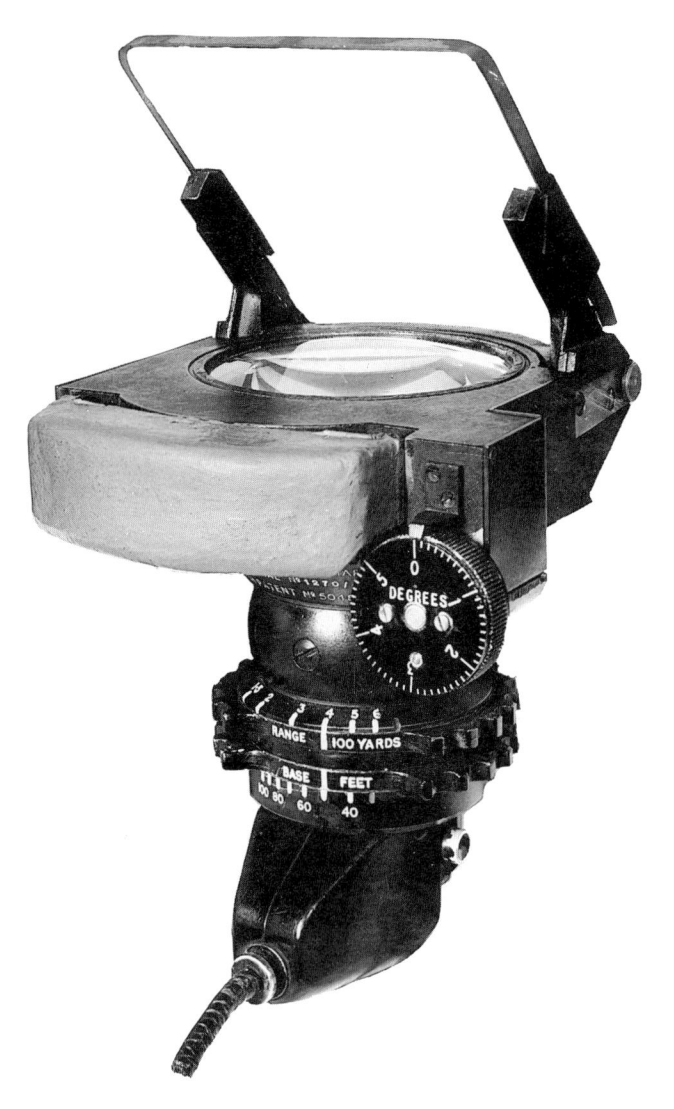

The Mk IIL sight head could be depressed for firing rocket projectiles. (Barr & Stroud)

ing been found to have slight optical aberrations. This modification was designated the Mark IIs, later Mk II*, fixed gun reflector sight. Existing sights were retrofitted with new sight heads, and the circular sight heads are now highly prized by collectors. The dimming screen was discontinued, as it was said to be rarely used.

In 1941 the US Navy adopted the GM2, designated the Navy Mk 8. It was decided not to use the stadiametric range base system, the graticule consisting of a ring and centre dot. Licence agreements were signed with Bausch & Lomb and Bell & Howell for production in the USA. The British Mk II* was also used in P-51 Mustang aircraft of the USAAF.

Type I Mk III projector sight

Beamont's modification was taken up by Farnborough and a new version of the Mk II* sight was designed with the reflector screen removed and the graticule modified to a plain 100 mph (161 km/h) ring and dot. The two control rings were replaced by a single ring which moved the graticule forwards or backwards, raising or lowering it on the windscreen. A dual-purpose graticule was devised consisting of an adjustable ring for rocket firing and a dot for the guns. An angle of depression scale was fitted, but it was found that a white celluloid strip, on which a pilot marked settings for various dive angles with a marker, was more practical. A detach-

THE GUNSIGHTS

TYPE I MARK III PROJECTOR SIGHT (PRODUCTION VERSION)
A. ANTI-DAZZLE SHROUD
B. ½°° DIVISIONS
C. CELLULOID STRIP FOR SPECIAL SETTINGS
D. SIGHT ADJUSTMENT CONTROL RING

The Mk III projector sight, which was used on Typhoon and Tempest aircraft. The graticule was projected onto the windscreen.

able shroud was fitted over the sight head to prevent images of the sun being reflected into the pilot's eyes.

Trials revealed several shortcomings in this model, which was known as the Type I Mk III projector sight: the graticule was dim and allowed very little eye movement. However, a simpler modification was agreed in which the graticule reverted to the single dot only, without a reflector glass. A production order was issued for this version, which was first officially used when the Hawker Tempest V entered service in February 1944. Similarly altered Mk IIs were also used by night-fighter and Typhoon squadrons. Beamont's modification was further improved when Grade A armoured windscreens became available. These were optically correct, and prevented the double image which sometimes occurred if the lamp was turned up too high (shades of the GM5). In a special night-fighter version tested by the Fighter Interception Unit in November 1942, a green diffuser cell was inserted between the lamp and the graticule to enable tracer to be seen to a maximum range at night. Another idea tested by the Air Fighting Development Unit consisted of an optically correct reflector glass screen the same size as the windscreen fitted 1.5 in (38 mm) behind it, which gave the pilot much more eye movement.

More than 84,000 GM2 sights were manufactured, mostly by Barr & Stroud, the production of lenses alone amounting to over 1 million units. The production versions were as follows:

Mark II   Oval reflector with sun screen
Mark IIs   Rectangular reflector with no sun screen
Mark II*   Similar to the IIs, slightly different dimensions
Mark IIL   Adjustable sight head (0–5°) for rocket firing and 40 mm guns
Mk III   Produced by Goerz of Vienna.
Type I   Mark I (Projector) open/shut graticule one ring only
Type I   Mark III (Projector) dot graticule
Type I   Mark 8 (Projector) standard range/base graticule

The last reflector gunsight produced by Barr & Stroud was based on the GM2, but incorporated automatic adjustment for elevation and azimuth (horizontal movement). It was designed for the large anti-shipping missile of 1945 code-named 'Uncle Tom'. Radar signals were fed into a computer which energized flexible shafts driving actuators which tilted or slewed the reflector glass to the left or right, giving a correct point of aim. Approximately 150 of these intricate instruments were produced, but the Uncle Tom system was cancelled when the war ended.

Bomber Command tracer ranging

Though tracer was known to be a misleading aid, on 8 April 1940 the Farnborough Gunnery Research Department issued a report to Bomber Command gunnery leaders entitled 'The use of tracer ammuni-

BRITISH AIRCRAFT ARMAMENT

METHOD OF DEFLECTION ESTIMATION USING G. MK.IV TRACER. RS. IIIA.

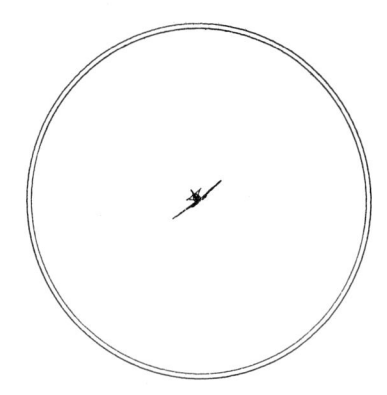

When target is seen, keep it on the bead, identify, and remember the fraction of the ring at 600 yds.

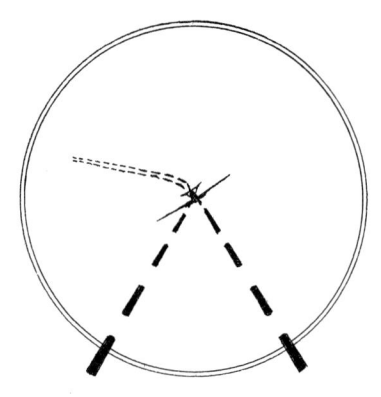

Fire a short burst at 600 yds. keeping target in sight, note position of end of trace in ring.

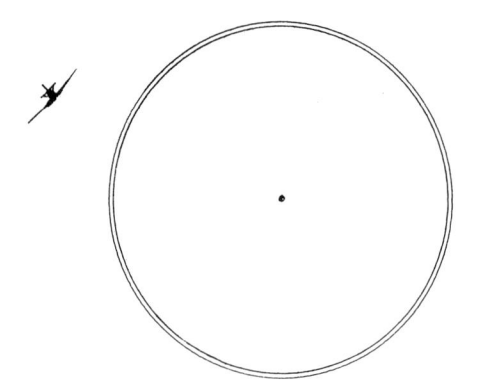

Set bead 'ahead' of target so that target is in same position relative to ring as end of trace was.

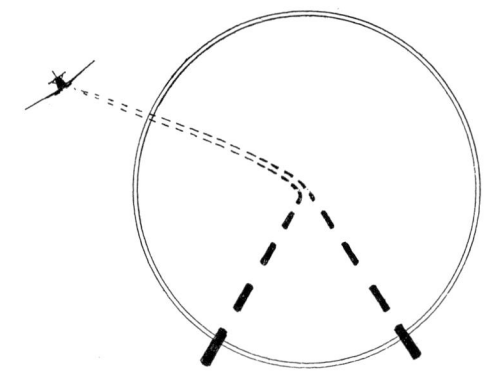

Fire burst, keeping target on the end of trace, until range has fallen to 400 yds.

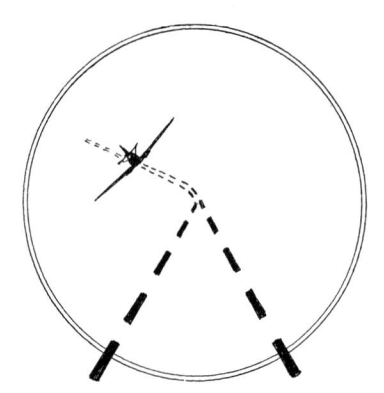

When range is 400 yds., set target half-way between bead and end of trace.

When range falls below 150 yds. ignore trace and fire point blank.

THE GUNSIGHTS

Opposite *The tracer method of deflection estimation.*

tion as an aid to air sighting'. The reason for this was an unforeseen problem encountered by rear gunners. It was found that, when a fighter approached on a 'curve of pursuit' (a turning attack, keeping his guns aligned on the bomber), it was very difficult to make a correct allowance for deflection. This was because a tail gunner had nothing to tell him whether an attacking fighter was moving across his line of fire or not. The gunner had the impression of being suspended in space.

After many novel ideas had been tried, it was found that the solution was to use tracer. The tracer then in use was the Type G Mk IV, which burned to 600 yd (549 m). The theory was that this would help solve problems with both range and deflection. The advice given was:

> As the fighter approaches at long range fire a burst, keeping the centre dot aligned on the target. You will see the tracer rise to the centre of the ring, remain in a cluster, then move sideways in the direction from where the fighter has come. This gives a clear indication of the allowance required. Adjust your line of fire in the opposite direction of the movement.
>
> Commence firing when the target is just over half-way along the trace, then maintain your aim and keep firing until the range closes to 150 yd, then fire point blank ...

The amount of deflection was set out in a directive entitled 'Zone system of sighting allowance'. It gave a series of allowances, such as: 'When a fighter is seen in the sector between 10° and 30° round dead astern, allow two rads (radii of the graticule circle). This may seem complicated, but failure to grasp such advice could mean the loss of an aircraft and crew. Any fighter pilot engaging an aircraft with a skilful rear gunner did so at his peril. A Sunderland on convoy patrol off Norway was attacked by a group of JU 88s which dived on the flying-boat in pairs. After an hour two of the Junkers had been shot down, and the Sunderland returned to its base at Invergordon.

Various types of tracer were designed by ICI Kynoch. One type burned red to 400 yd (366 m), then a brilliant red, before ending in a puff of smoke. This was to warn the pilot that his ammunition was nearly spent. The G Mk III did not trace until 200 yd (183 m), to avoid giving away the position of the aircraft. The Luftwaffe did not use ball ammunition in its 7.92 (rifle-calibre) guns: ammunition comprised tracer, AP and incendiary rounds.

Over-the-nose sighting

The Spitfire, like many fighters, gave the pilot a poor view ahead over the long nose cowling. When an enemy was engaged from behind, the point of aim had to be slightly above the target to allow for bullet drop, and when the target was climbing away the pilot needed a lead angle above his quarry. In these

Spitfire over-the-nose sighting system, designed to overcome the blind spot created by the cowling.

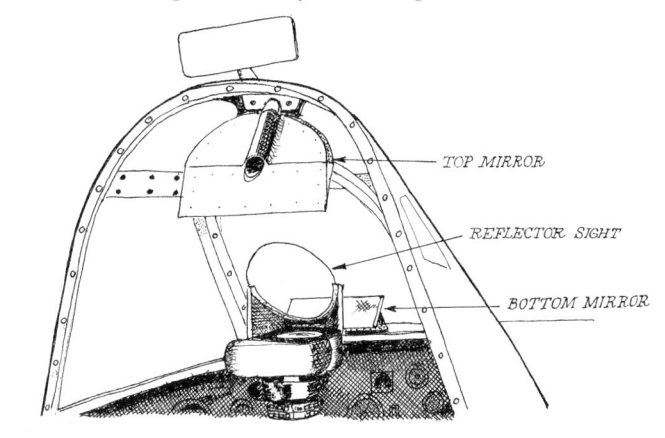

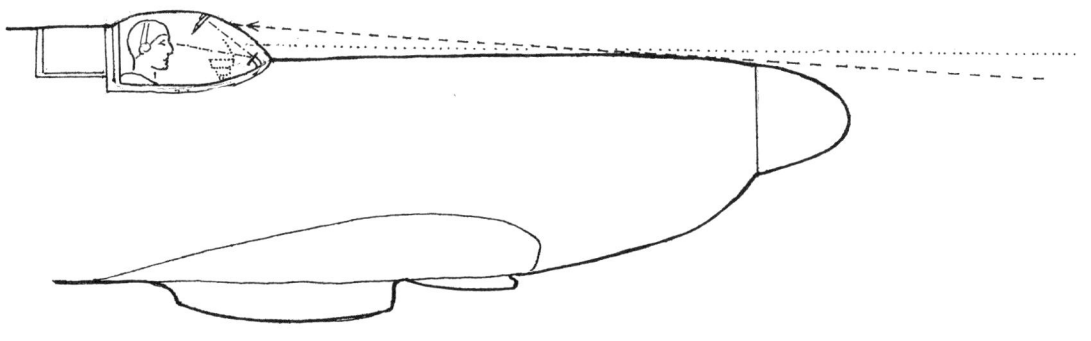

situations the pilot had to ease the nose down and observe the target, then pull the nose up slightly and press the gun button.

In early 1942 a periscopic device was tested at the Air Fighting Development Unit at Duxford with a view to overcoming this problem. A Spitfire II (converted Mk I *K9830*) was fitted with two mirrors, one mounted just behind the gunsight, and the other, much larger, fixed inside the top of the windscreen looking forward into the blind area. It gave the pilot vision into the blind zone, but it was difficult to keep the target in view once the nose was raised, and was not a practical proposition for the extra few degrees of vision gained. The officer in charge of the tests duly reported these findings and the device was not accepted for Service use.

Wartime ring and bead sights

Even in the Second World War the ring and bead sight was used operationally by RAF gunners, because it was not practical to use reflector sights in some positions. Gunners manning side hatch positions in flying-boats, observers in naval aircraft, and extra hand-operated guns in bomber aircraft all used such sights. They were usually the 2 in type, which were far less prone to damage than the prewar 4.5 in. Observers in the Beaufighter TF X had three ring and bead sights on the single Browning in the cramped rear cockpit. They used whichever gave a bead on the target.

Fairey Battle light bombers took a leading part in trying to stop the German advance into France in

The triple ring and bead of the Beaufighter TF X observer's mounting.

Fairey Battle observer using a 2 in ring and bead sight on a Vickers K gun.

Twin Vickers K guns in a Catalina, each with its ring and bead.

May 1940. Their armament, against high-performance Luftwaffe fighters, was identical to that of the 1917 two-seaters. The observers had a single Vickers K gun, and rarely had a chance to align their sights onto their opponents, who decimated them.

The gyro sight

The gyro gunsight at last solved the most difficult problem faced by pilots and air gunners, that of finding the correct angle of deflection or 'lead' needed to hit the target when it was moving across the line of fire. Even experienced fighter pilots had great difficulty in overcoming the problem. Indeed, the world's top-scoring fighter ace, Eric Hartmann, revealed that his secret was 'Get in close – down to 25 metres – then you don't need to worry about deflection shooting!'

When the gyro sight was perfected, no guesswork was needed. The sight presented an aiming mark which automatically allowed for range, deflection and gravity drop of the projectiles. As early as 1917 the basic theory of using a gyroscope as an aid to deflection shooting was propounded by W/Cdr L. J. Wackett RFC. With him on the same station was Capt. (later Prof. Sir Melville) Jones, who was to be a leading member of the Farnborough gyro team 25 years later. The idea of using a gyroscopically controlled sighting system was later suggested by Dr L. B. C. Cunningham of the RAF Education Branch. In 1936 he pointed out that mathematics could be used to solve any problem of dynamics; as an example he put forward the theory that even the complicated problem of air-to-air deflection shooting could be solved mathematically 'by using a gyroscope to offset the sight line from the gun line through an angle determined by the rate of turn of the sight line'. Although his pupils could not understand the implications of his theory (most of them had no idea what the angle of deflection was anyway), his observations did not go unnoticed.

In 1938, with the international situation worsening, the Air Staff arranged an exercise at Northolt in which all aspects of fighter combat were to be assessed. The Committee for the Scientific Survey of Air Defence, headed by Henry T. Tizard, appointed observers from Farnborough to advise on points of scientific importance which might otherwise be overlooked. Camera guns were fitted to many of the aircraft, and a squadron of Gladiators had just been fitted with some of the first batch of Mk II reflector sights. When exposed films from the camera guns were examined, it was found that although the pilots often got into position to dispatch their victims, very few had any idea of the amount of deflection needed.

A report on the findings was duly presented to Tizard, stating:

> Although the new Spitfire and Hurricane fighters performed well, their effectiveness would be vastly improved if some means could be found to predict the amount of lead required to hit the target accurately. Many of the gun camera films proved that if the combats had been in earnest the enemy would have escaped unscathed.

The Air Staff were deeply concerned, and the Director of Farnborough was instructed to investigate the possibility of a predictor gunsight, suitable for use in fighters and the gun turrets of bombers. The problem was to be given priority over all other work, no expense was to be spared and the utmost secrecy was to be observed.

Dr Cunningham's theory depended on the fact that a gyroscope will resist any rotation of its axis. If a gyro is clamped onto a rod on which is fixed a ring and bead sight, any attempt to follow a crossing target will be resisted by the gyro. The degree of resistance will depend on the target's crossing speed. A gyro sight would have to present the marksman with a sight line held back by the gyro, whilst the line of flight and guns would be in front

Layout of the Mk I gyro sight installation.

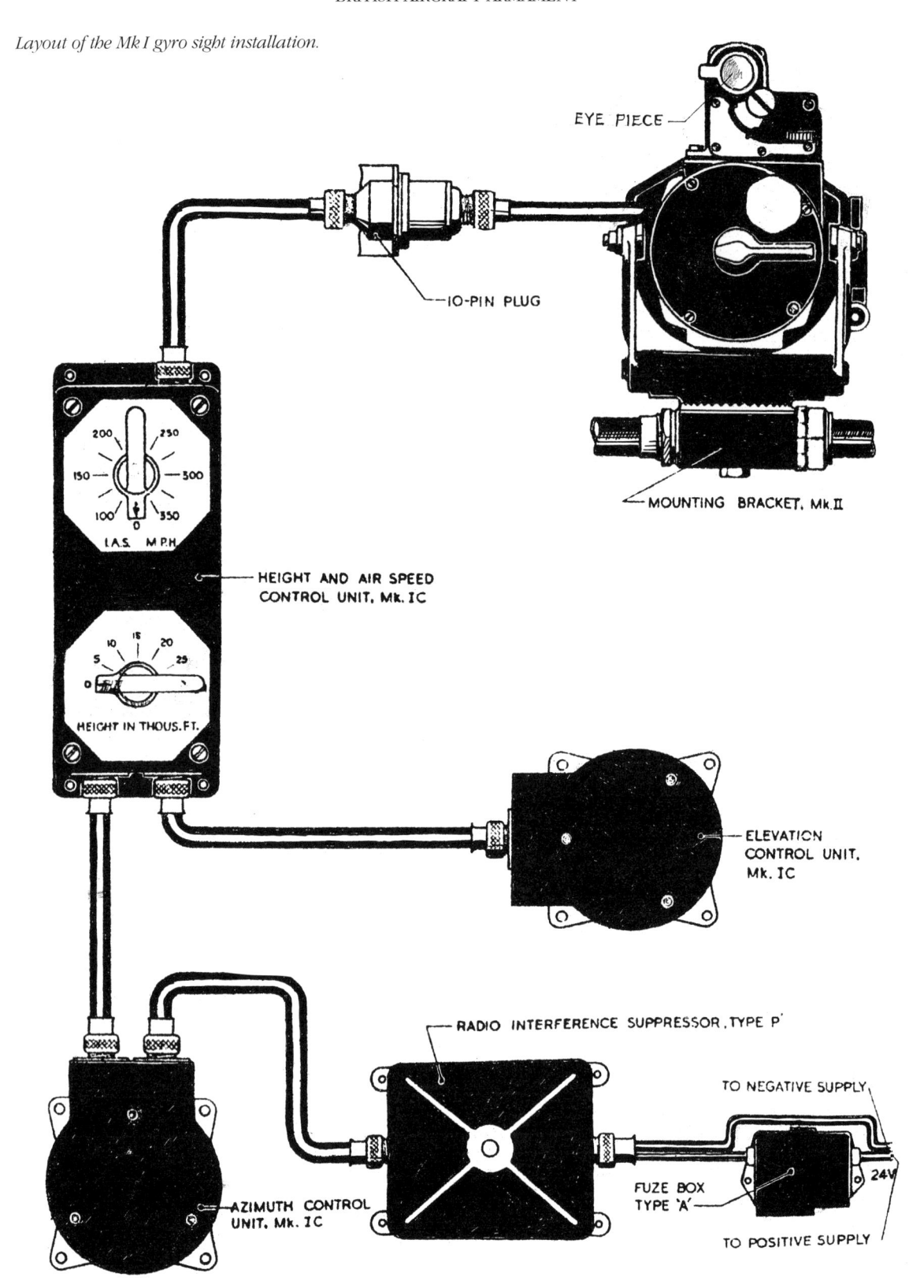

THE GUNSIGHTS

of this sight line. In other words, if the pilot kept to the sight line indicated by the gyro, his guns would automatically point correctly, in direct relation to the rate of turn.

By October 1939 two experimental sights of slightly different design had been fitted to a Hurricane and a Wellington at Boscombe Down. A sight was fitted in the normal position in front of the pilot in the Hurricane and in the Frazer-Nash FN 25 under turret of the Wellington. The results of the trials were promising, although the sights were rather primitive. By the summer of 1940 the RAE Director was able to report to the Air Staff that a potentially operational sight, the Mk I GGS (gyro gunsight) was ready for testing. It was given the code name Type 6 mechanism.

The Type 6 contained an electric motor driving a rotor mounted on a stem, the end of which actuated a linkage. Also fitted to the stem was a saucer-shaped aluminium dome surrounded by four electro magnets. The motor ran at a speed of 4,000 rpm, so the gyro resisted any angular movement of the housing. Such a movement tilted the axis of the stem and the linkage moved a celluloid strip in the optics of a G1 prismatic gunsight. A black ring was

Above *GGS Mk I produced by Elliot Bros (London).* (Elliot)

Below *Sectional drawing of the 'works' of the GGS Mk I. The optical system was a G1 prismatic sight, which proved too small.*

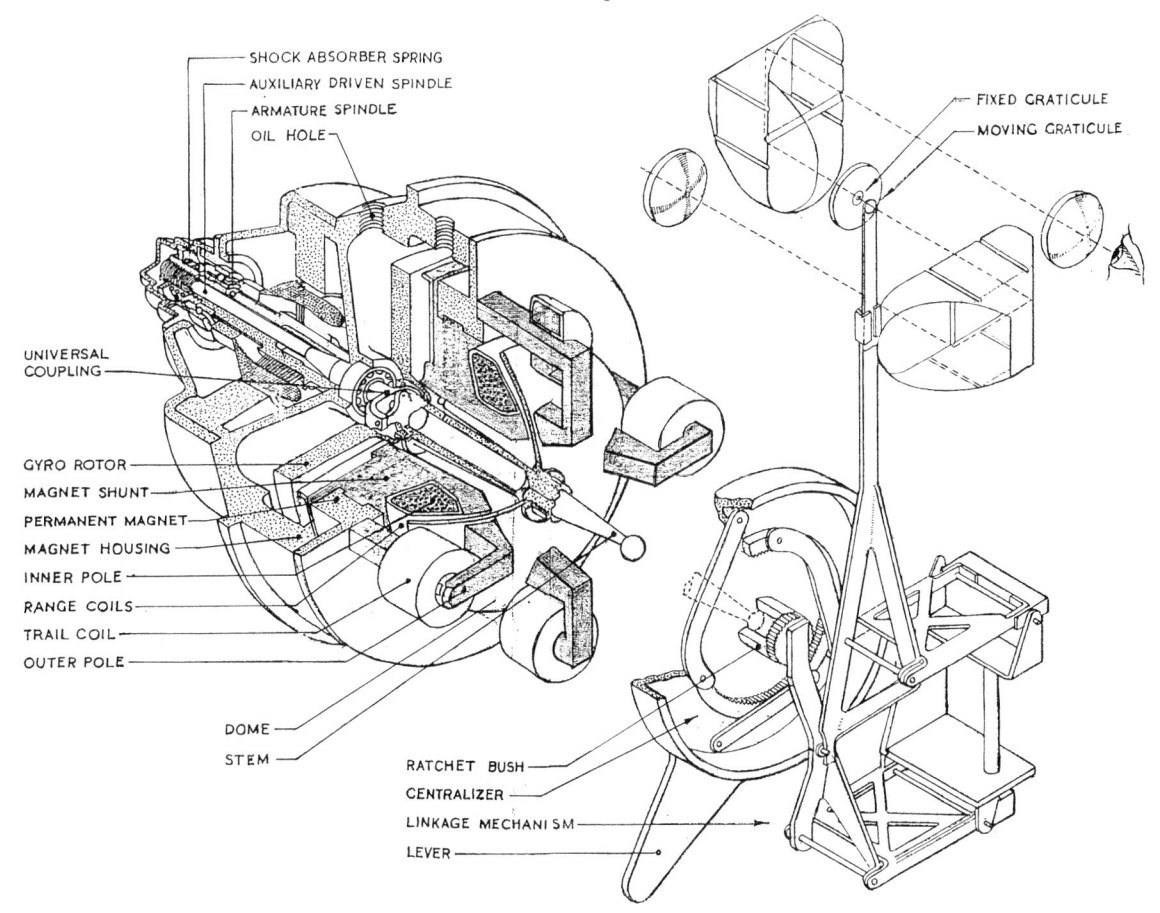

engraved on the end of the strip, so that the gunner saw two graticule rings, one fixed, the other on the celluloid (this will be explained later).

In principle, the Mk I GGS was very similar to a rate-of-turn indicator which also used a gyro. In this, as the aircraft turned, the gyro held back the needle of the instrument, this deflection indicating the force of the turn with a needle on a dial. In the GGS the gyro resistance was used to deflect the ring on the celluloid. The electro magnets allowed for host aircraft speed and height: the thinner the atmosphere, the less the bullets drop.

As soon as the Air Staff heard that the design had been completed they made arrangements to rush it into production, the prime contractors being Ferranti and Elliot Bros (London) Ltd. But the Director of the Gunnery Research Unit advised:

> It is the considered opinion of the Gunnery Research Unit that the true value of the sight cannot be proved without operational trials, and before full production is authorised we respectfully advise that a number of Squadrons should be equipped to enable the pilots, after all the advice that can be given to them, to find for themselves to what extent it aids them in actual combat.

Meanwhile, the Air Staff's fears about inaccurate shooting had been borne out. Intelligence summaries of combat reports quoted pilots who could not understand why their fire seemed to have no effect, even when a long burst was fired from an ideal position.

In early 1941 Farnborough produced the first pre-production batch of Mk I gyro sights and a Spitfire and Defiant were flown into the airfield to be fitted. The sight was rather bulky, and difficult to fit into the turret of the Defiant. The experienced trials pilots reported an almost magical performance. As they turned into the attack on the stooge aircraft, two circles were seen in the eyepiece, the one lagging behind being the aiming point, the leading circle being the direction in which the guns were pointing. It was found a little difficult to locate the target in the small eyepiece, and the circles wandered during a high 'g' turn, but the correct deflection angle was presented.

Ferranti (Edinburgh) began low-rate production in April 1941, and Spitfire and Hurricane fighters from operational squadrons tested the sights in interceptions of German raids during July and August. After six weeks the AOC Fighter Command, Sir Sholto Douglas, took his eagerly awaited report (dated September 1941) to Whitehall. It stated:

> The initiative and interest shown in these trials has been far from satisfactory, owing to the fol-

This GGS was one of the first installed in a Spitfire. It was a considerable hazard to the pilot in a rough landing.

THE GUNSIGHTS

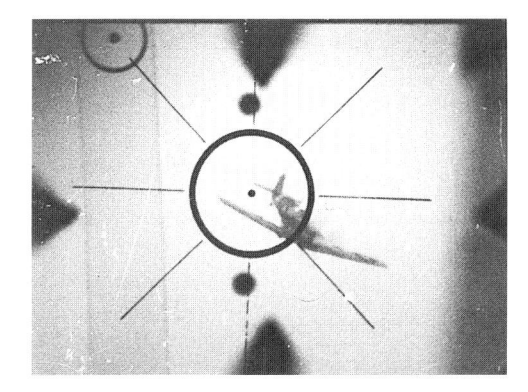

Gunner is using the wrong aiming mark, he is also having trouble manipulating his turret to follow the target correctly. The sketch below shows the ideal line of sight and target tracking

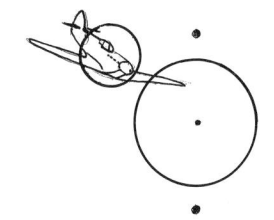

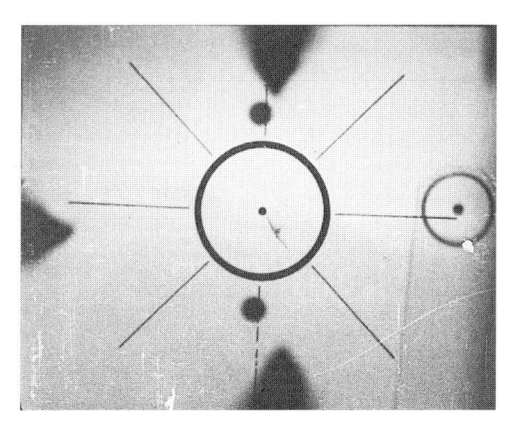

Gunner is again using the fixed graticule as his point of aim, he should be using the moving graticule as shown below. Note the faint outline of the celluloid strip carrying the moving graticule.

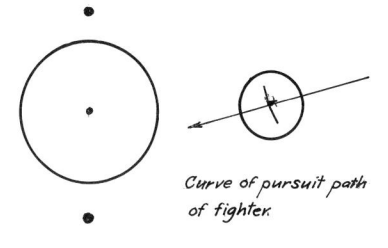

Curve of pursuit path of fighter.

Assessment of a turret gunner's aim; serious faults shown by a sight recording camera fitted to a Mk I GGS.

lowing reasons: the difficulty of carrying out Service trials of gunsights under operational conditions, and the reluctance of pilots to use experimental sights in a life-and-death situation. Also, the restriction that the sight should not be used where the possibility exists that it might fall into enemy hands has been a drawback. However, sufficient information has been obtained to enable the following conclusions to be arrived at:

1. The hard and angular nature of the sight in the position it occupies constitutes a danger of facial injury to the pilot in the case of a forced landing.

2. The amount of eye freedom is very limited. It is necessary to place one eye on the small eyepiece, so it is not possible to observe other aircraft which would otherwise be seen within the normal field of vision.

3. The sight is too sensitive. It requires some device to damp the violent changes in position of the moving graticule caused by bumps, and alterations in the rate of turn. Rates of turn above rate 3 cause the graticule to disappear entirely, and on reappearance it takes one or two seconds to settle down. It is therefore considered that in its present from the gyro sight Mk I is not a practical proposition for the operational requirements of Fighter Command. It is recommended that consideration should be given to the development of a reflector sight embodying the principle of the gyro sight, in which the disadvantages referred to above would be eliminated. The value of training pilots in the correct amount of deflection to be applied in air fighting has been considered, and it is proposed to allocate a number of these sights to Operational Training Units of this Command.

Bomber Command had also been testing sights fitted to the rear turrets of Wellingtons. Reports from gunnery officers were more enthusiastic than those of Fighter Command. Once they had mastered the technique, turret gunners showed a 50 per cent improvement in marksmanship over those using Mk III reflector sights. However, as there was no illumination of the graticule, the sight could not be used at night. The restricted nature of the small eyepiece was also mentioned, and the unstable nature of the graticule was a drawback.

Following these reports, the Air Staff had to postpone full-scale production. This was doubly disappointing, as the Spitfire V was being outperformed by the new Focke-Wulf 190, and Bomber Command losses due to German fighters were mounting. Limited production continued to give trainee pilots and gunners practice in deflection shooting; some Coastal Command squadrons decided to use the sight operationally.

It is perhaps worth outlining again the principles by acting the part of a trainee turret gunner. You first turn on the switch on the end of the body. In a few seconds the gyro will run up to 4,000 rpm. To prevent unnecessary wear, the sight should be switched on only when hostile aircraft are expected. You then set the aircraft's height and speed on a control box on the left side of your turret. Looking through the eyepiece you will see two black circles. The larger graticule is fixed and indicates the direction in which the guns are pointing. The smaller ring is the point of aim computed to allow for deflection and bullet trail. When the turret is stationary and the guns pointing in any direction other than astern, the moving graticule will be seen as being displaced from the fixed ring by the four electro magnets. When you turn your turret to follow a target, the gyro will make the moving graticule lag behind in relation to the rate of turn or rotation. This lag, added to the bullet trail allowance, gives the point of aim required to hit the target – in other words, you can't miss – provided you can manipulate your turret controls accurately. The range of your sight is fixed at 300 yd (274 m), which has been found to be the optimum – you merely place the graticule round the target. The sight is protected from misting by a filter of silica gel, which dries the air as it enters an inlet at the bottom of the sight. Two adjusting screws harmonize the guns with the sight. As the elevation screw is turned part of the anti-vibration cradle is tilted, causing the line of sight to be elevated or depressed. Similarly, as the traverse screw is turned, the line of sight is rotated in azimuth.

The Farnborough team worked urgently to overcome the problems noted in the Fighter Command report. The first and most serious fault was the means of display. It had been decided to base the display of the Mk I on the optics of the GI to save time, but the choice of this system had precluded night use, and the small eyepiece was far too restrictive. The obvious answer was to insert a moving graticule into a reflector sight, as suggested by Sholto Douglas.

The Mk II GGS

Farnborough devised a solution which was one of the simplest yet most effective inventions of the war: a mirror was fixed to the end of the gyro and made to reflect an illuminated graticule onto the reflector plate. This graticule moved to the correct position allowing for deflection, and also incorporated a ranging facility. The graticule consisted of a ring of six small diamonds, the diameter of which could be set to correspond with the target span. The type of enemy aircraft was set on a dial, enabling the sight to calculate the range. The reflector screen was a large glass plate 4.75 in (112 mm) × 2.5 in (61 mm). Looking into the screen, the operator saw two illuminated graticules. The one on the left was a fixed ring graticule which could be used if the gyro system failed, its main use being to harmonize the guns with the sight. In the right half of the screen was the

The Mk IIC gyro gunsight.

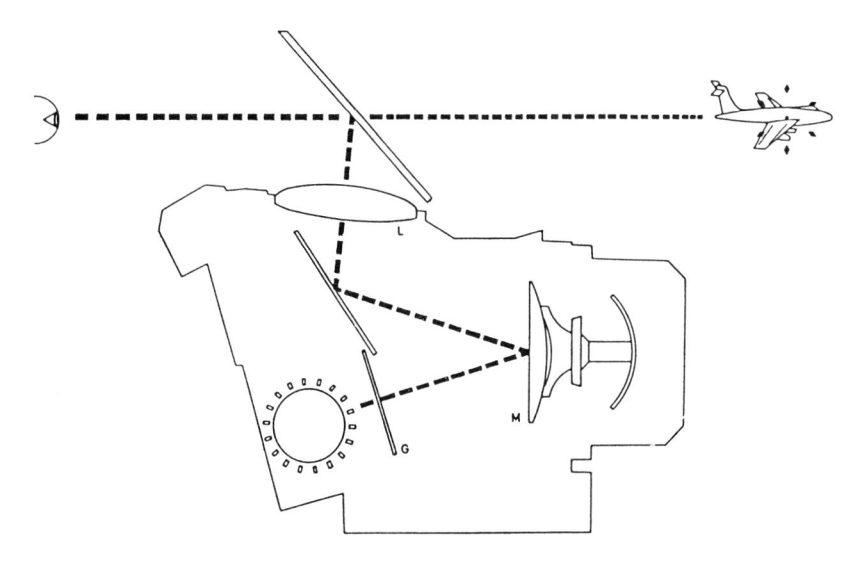

The optical system of the GGS. The lamp shines through the graticule (G) onto the gyro mirror, which reflects the image of the graticule onto the fixed mirror (M), which reflects it through the lens (L) onto the pilot's reflector screen.

gyro-controlled ring of six diamonds. The diameter of the ring was adjusted by foot pedals in the turret version of the sight, and by a twist grip on the pilot's throttle lever on the fighter type. Both graticules could be dimmed for night use, or used singly by

Right *The rotor motor, dome and mirror.* (Ferranti)

Below *Stadiametric ranging system, which set the span of the target into the mechanism.* (Ferranti)

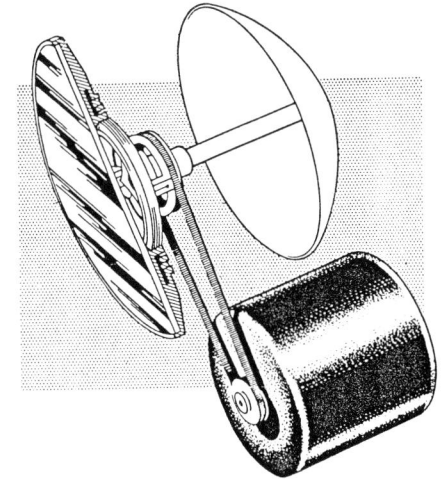

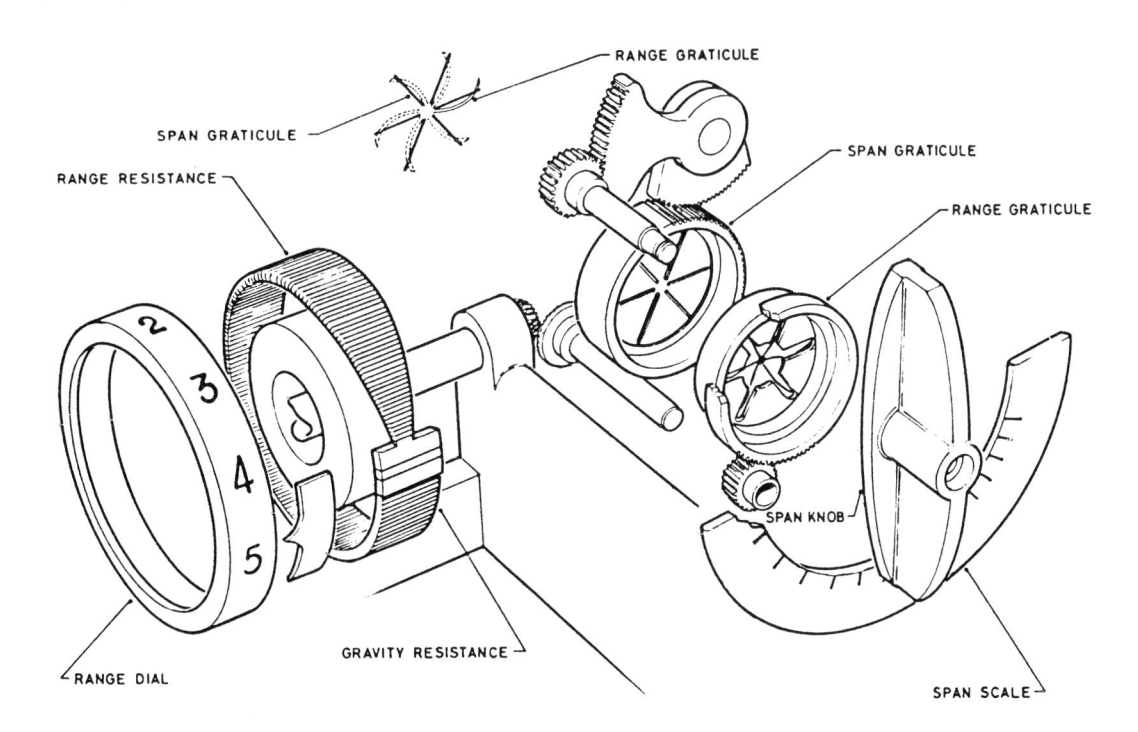

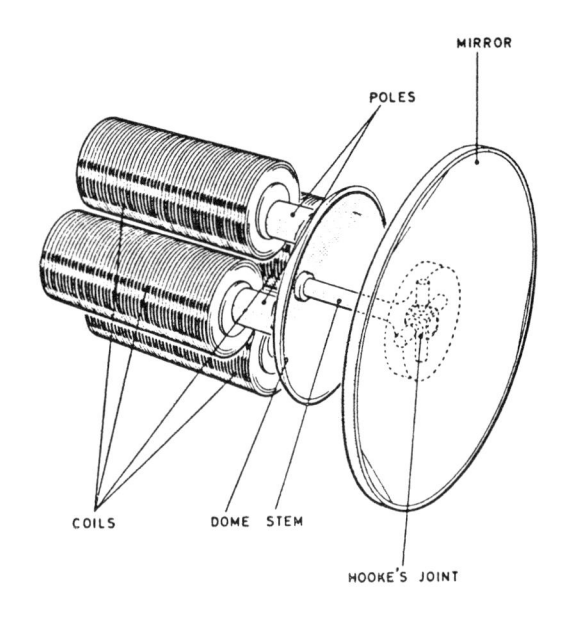

switching either on or off. The height and speed setting unit of the Mk I was found to be ideal, and could not be improved. The first Mk II sights made at Farnborough were tested by the Armament Research Unit in July 1943, and it was clear that the problems had been solved.

The new sight had had the undivided attention of some of the most able brains in the country, some of those taking part under Sir Melville Jones being A. A. Hale, B. Sykes and G/Capt. Ford. Ferranti played a central role in the design and development of the complex gyro and electrical components. The sight was to be known as the GGS Mk IIC (turret)

Left *The four coils in the rotor unit injected the allowance for range and the ballistic trajectory of the ammunition.*

Below *The Mk IIC gunner's installation showing the various components.*

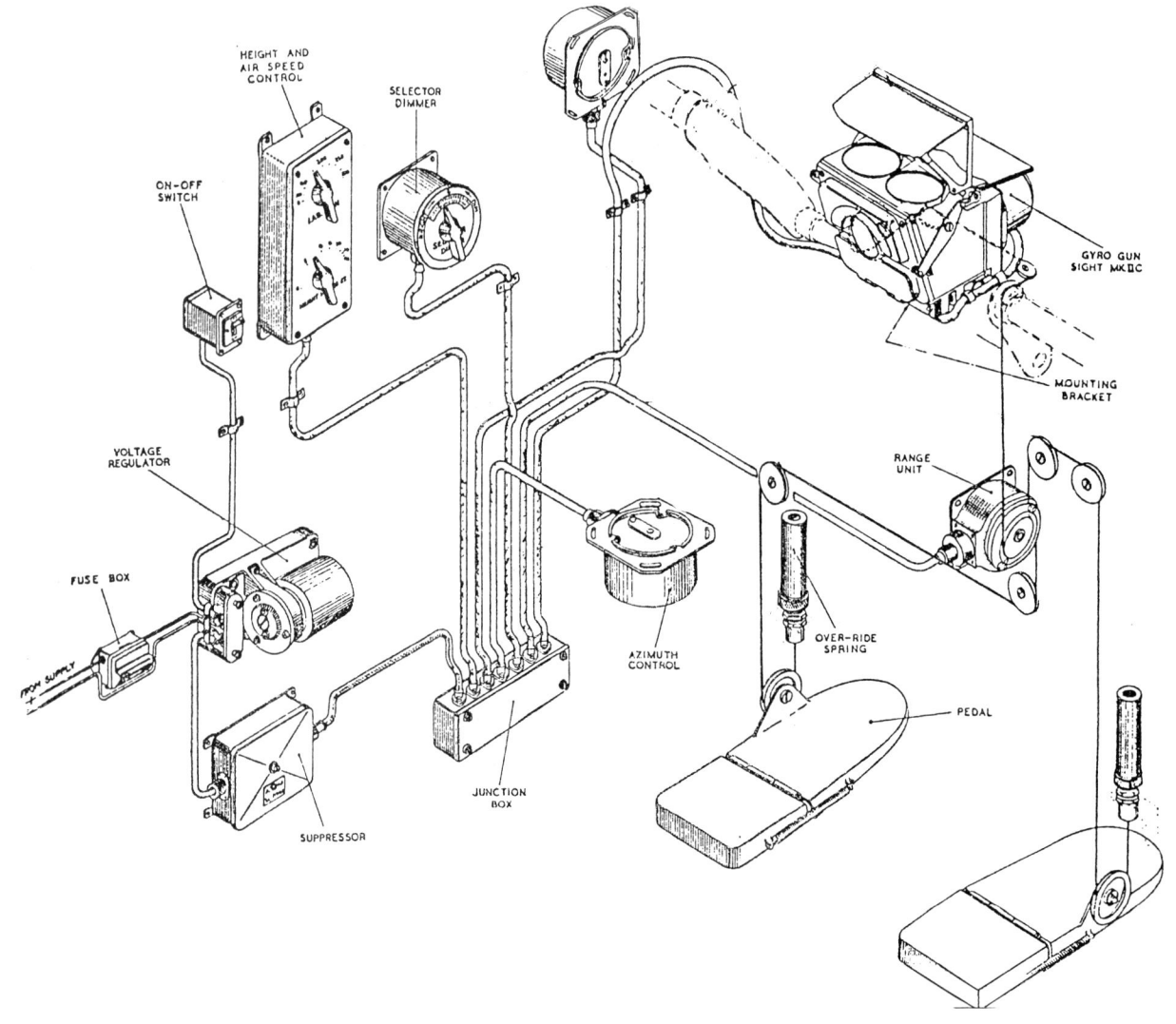

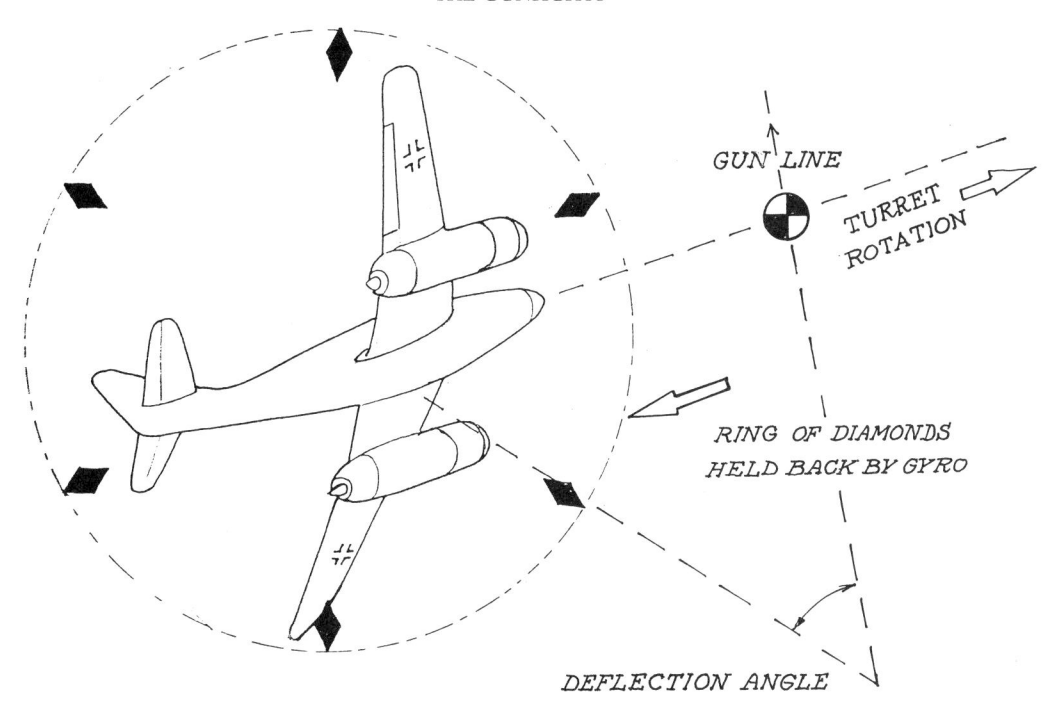

Automatic deflection angle presented by the GGS.

and Mk IID (pilot). Deliveries from a purpose-built factory near Edinburgh began in late 1943.

The sights seemed to possess almost magical qualities. As an ex-Battle of Britain pilot stated: 'I look back on previous combats where the enemy escaped more or less intact, and realize that I could most certainly and easily have destroyed it if I had been using a good gunsight.' A demonstration was also staged for two pilots of the USAAF. One reported:

> I believe this sight would improve gunnery at least 100 per cent. Shooting is at the moment for most pilots purely guesswork. A pilot cannot guess with this sight, due to this I am sure that at least the lower bracket of pilots (75%) will improve their shooting to the level of the best gunnery shots now, and the best ones can do even better. It is easy to handle, and there is no situation it cannot handle as well as the GM2, and in most cases (90%) it will do a lot better.

The second pilot reported:

> Speaking from the point of view of the day fighter, I would say that the Mk IID gyro gunsight is definitely the answer to our problem with deflection shooting. We are proving daily that the average pilot cannot do deflection shooting, even with small angles, accurately with a fixed sight. I think that the sight should be put into production immediately and fighter squadrons equipped with them as soon as possible.

Bomber Command were also very keen to receive

Sight recording camera fitted over the hood of a Mk IID (Ferranti)

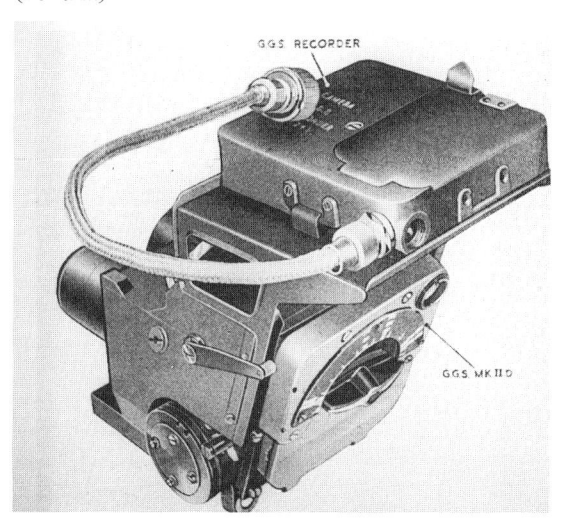

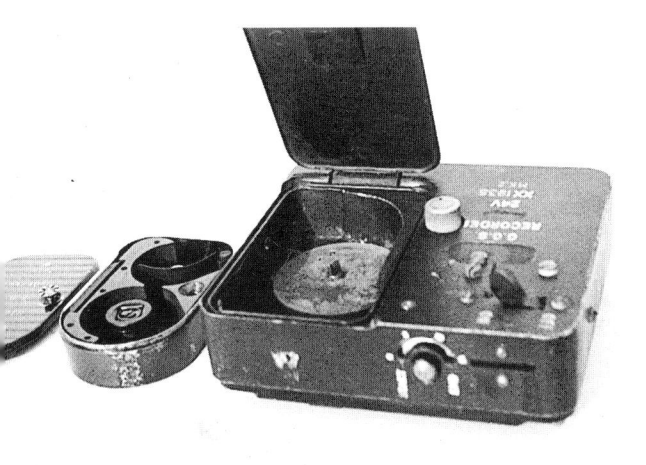

The electrically driven camera operated when the gun button was pressed. The film holder (left) was loaded with 16 mm film.

A Mk IID in a Hawker Hurricane, set to engage an FW 190 at 350 yd (320 m).

the Mk IIC version. It was first issued to gunnery schools and operational training units where lectures and air-to-air practice were quickly arranged. The gunnery schools had been very pleased with the Mk I sight, which had proved invaluable for instructional purposes. A compact 16 mm sight-recording camera had been produced, and gunnery training was much improved owing to the fact that no ammunition needed to be fired, and a record of accuracy could be shown to the trainee. As with the Aldis sight, word soon spread round the operational squadrons and the new sight was eagerly awaited.

At first there was concern over the possibility of the sight falling into enemy hands, and there were restrictions on gyro-equipped aircraft flying over enemy-held territory, but as they became more numerous this rule was relaxed, and the Luftwaffe began to suffer from the attentions of an enemy who could suddenly fire with uncanny accuracy. Not that all fighter pilots accepted the gyro sight with enthusiasm at first, for it required a fair degree of dexterity: select graticule brilliance, set graticule presentation, set span level, then once the target is presented align the ring of diamonds to the enemy span. No such preparation was needed on the Mk II, but as the pilots gained experience the early scepticism vanished, and results bore witness to the gyro's effectiveness. The US Navy and Army Air Force formally accepted the sight, and production commenced in America where it was designated the Mk 18 (Navy) and K-14 (USAAF). In Canada, Semco Instruments produced a naval sight more robust than aircraft versions and with two dimming screens to counter glare off the sea. Otherwise the 'works' were identical to the Ferranti model.

Perhaps the reader can imagine himself seated in the Frazer-Nash FN.20 tail turret of a Lancaster, flying in daylight over Germany in early 1945, when an enemy fighter is seen to dive to the attack.

> The heart misses a beat, but then you realise that only you can save the aircraft and crew, you are in effect in full control. First you inform the captain and crew of imminent attack, and tell the pilot which way to break away to give the enemy maximum deflection. You will have already set the height and speed on the dials of the control box mounted horizontally at hand level to your right. You identify the fighter and set its wingspan on the span handle. You now operate the left pedal which opens the ring of diamonds to the maximum range setting. As your attacker closes in, you keep him inside the ring until he fills the ring;

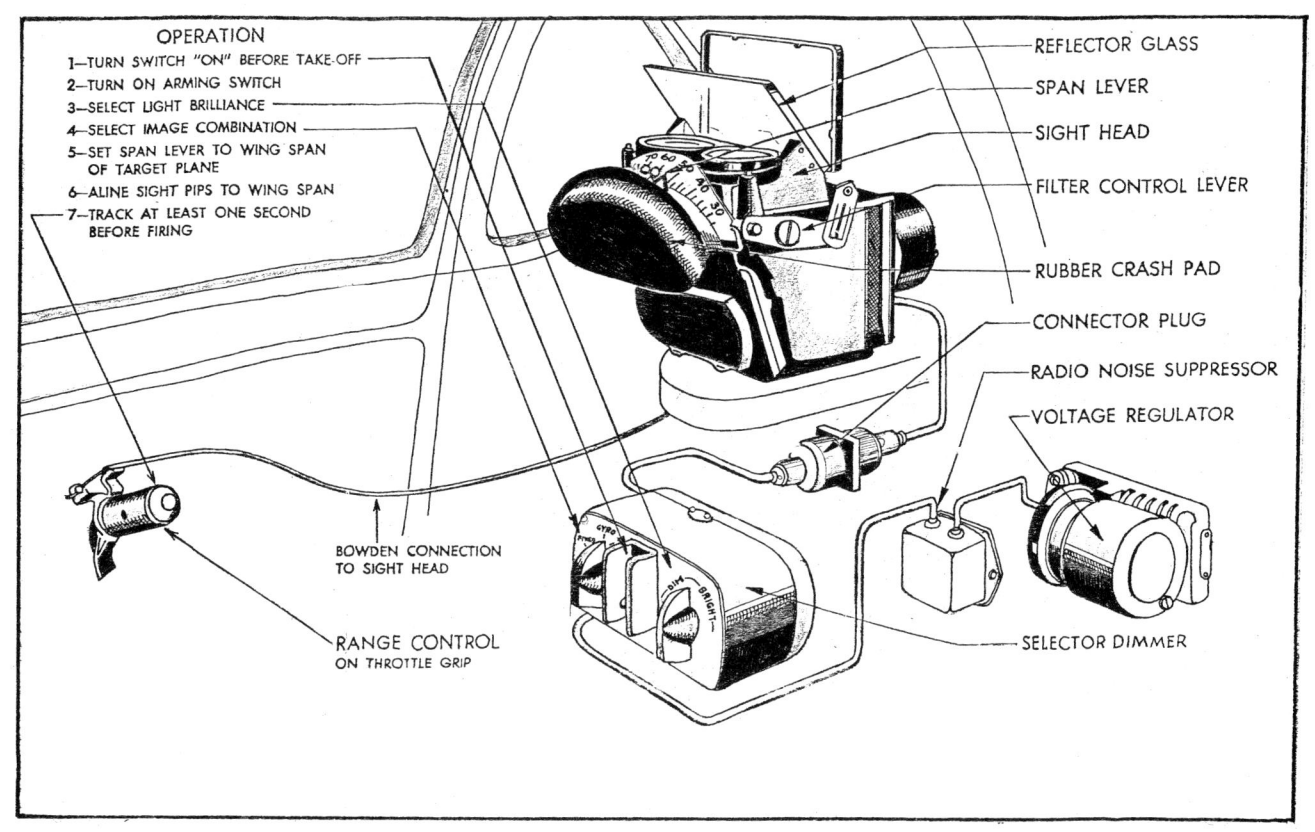

OPERATION
1 — TURN SWITCH "ON" BEFORE TAKE-OFF
2 — TURN ON ARMING SWITCH
3 — SELECT LIGHT BRILLIANCE
4 — SELECT IMAGE COMBINATION
5 — SET SPAN LEVER TO WING SPAN OF TARGET PLANE
6 — ALINE SIGHT PIPS TO WING SPAN
7 — TRACK AT LEAST ONE SECOND BEFORE FIRING

Labels: REFLECTOR GLASS, SPAN LEVER, SIGHT HEAD, FILTER CONTROL LEVER, RUBBER CRASH PAD, CONNECTOR PLUG, RADIO NOISE SUPPRESSOR, VOLTAGE REGULATOR, SELECTOR DIMMER, BOWDEN CONNECTION TO SIGHT HEAD, RANGE CONTROL ON THROTTLE GRIP

Above *The American version of the GGS, the K14, showing its operating sequence.*

Right *The graticule selector/dimmer unit. The fixed graticule was used mainly for gun harmonizing, but was also a standby in case of gyro failure.*

he is now framed and within range – open fire with a four-second burst. As long as you can keep him centre in the ring your bullets will be striking home. If he keeps coming in, he will appear to get larger: depress the right pedal to control the ring and keep his wingtips touching the diamonds. Track the target accurately and smoothly at the same time as closing the range with the pedals. When the target reaches 200 yd the graticule will not get smaller, but the sight will still be accurate. Keep the right foot pressed and aim at the part of the fighter where hits will do the most damage.

A fighter pilot using the Mk IID would not use pedals to control the diamonds but a twist grip on his throttle. The operator could select various combinations of illumination. With Gyro Night, only the gyro graticule was visible, and the range was set to 180 yd (165 m) irrespective of pedals or twist grip. This was the usual maximum range at night.

Several types of German aircraft were marked on

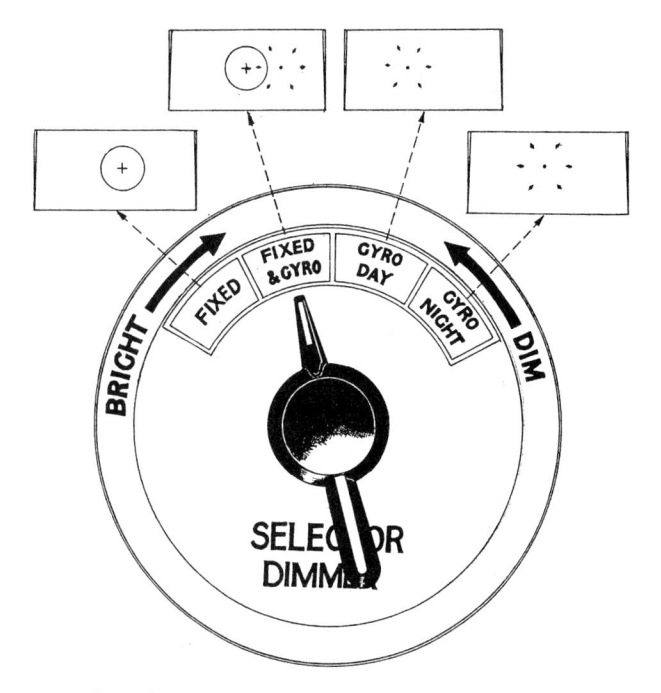

The 'Fixed' position was used for gun harmonizing or as a standby position. 'Fixed and gyro' used for sight alignment. 'Gyro Day,' used for day combat. Gyro Night, – range fixed at 150 yards, and graticule brightness adjustable by moving outside ring.

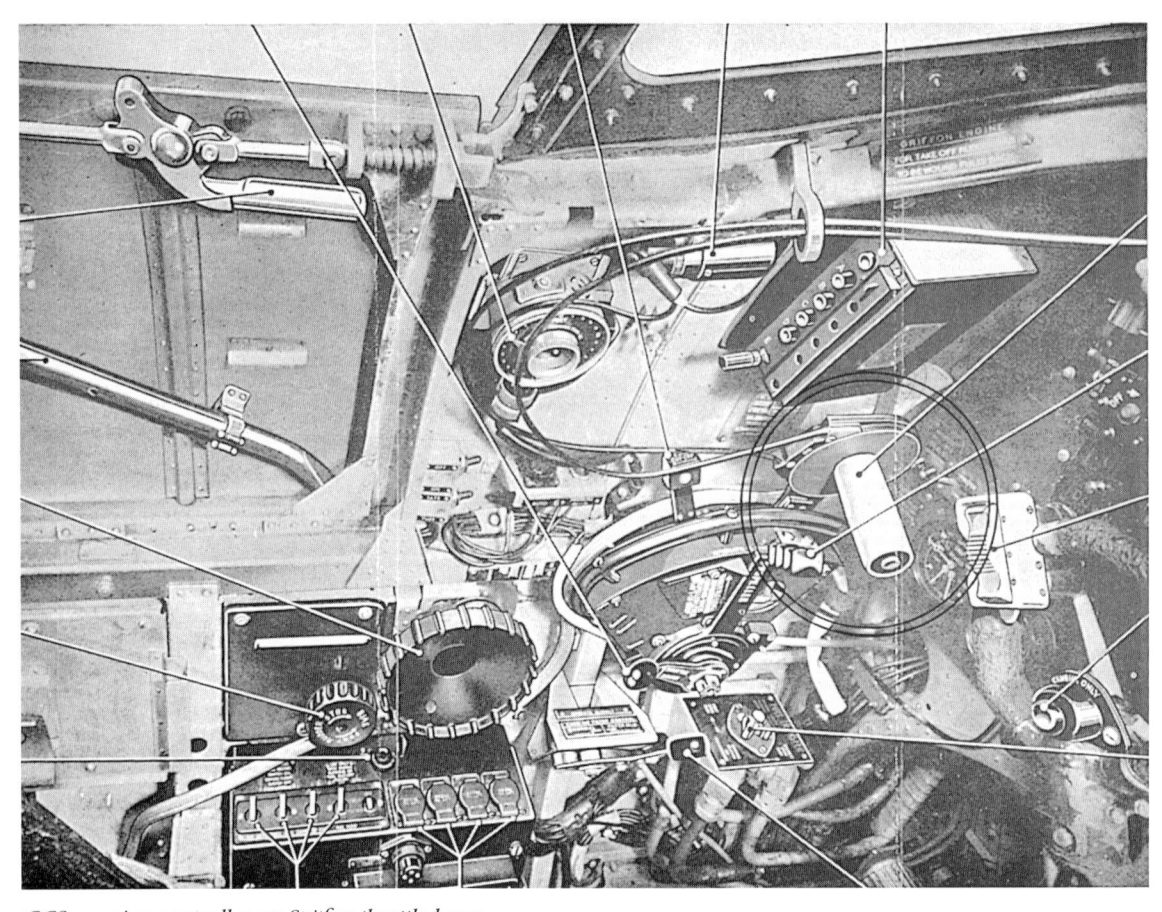

GGS ranging controller on Spitfire throttle lever.

the dial, and aircraft identification became even more important to pilots and gunners. Gunners were taught to recognize the frontal silhouettes of German and Allied aircraft instantly. Anyone who failed the aircraft recognition test badly often failed the course.

By 1943 Frazer-Nash gunnery trainers were built on most Bomber Command airfields, housed in an igloo-shaped building known as Instructional Building Type A. In the centre was a turret supplied with hydraulic and electric power. The pupil, in flying clothes, manned the turret, and a gunnery instructor sat at a projector with which he could simulate a curve-of-pursuit attack by a fighter with a spotlight on a realistic sky painted on the hemispherical ceiling. The gunner manipulated his controls with the sound of engines in his earphones. As he aligned his sight on the moving target, a tiny cross indicated his point of aim. It was easy for the instructor to assess the gunner's ability, and some gunners who had flown on operations were found to be well below standard and were sent on refresher courses. On some of the old bomber bases the 'igloo' is the only building still recognizable, having defied demolition.

Another type of simulator housed in a larger building was used mainly at gunnery schools. The instructor sat on a platform below the trainee and projected the image of a moving fighter onto a parabolic screen some 25 ft (7.62 m) in diameter. The turret projected the two graticules onto the screen, giving an excellent indication of accuracy.

Details of Mk II GGS gyro sight production
GGS specification received from A. M. to Ferranti Ltd, Edinburgh, February 1943
Site for new factory purchased December 1942
Building commenced February 1943
Factory opened June 1943
First production sight 30 November 1943
Quantity production commenced February 1944
Output by March 1945: 1,000

GGS training simulator at the RAF Central Gunnery School. The instructor projected an image of an attacking fighter, the trainee manipulated a GGS which projected the diamond graticule and a gun boresight ring.

Other companies were also involved in production: Barr & Stroud supplied lenses and produced a small quantity of complete sights, Salford Electrical Company produced gyro sights to Ferranti drawings, and other concerns carried out sub-contract work.

Ventral turret sights

After suffering a high loss rate in the early months of the war, the Blenheim light bomber was fitted with a rearward-firing under turret, the FN54A, as described in Volume 1. The twin Browning guns were aimed by a periscopic sight, the Type A. This consisted of an optical tube with a mirror at each end, a graticule and angle-of-drift scale. The two operating handles each had a trigger, and the graticule was a black circle. The field covered was 20°

The Type A periscopic sight used on the Blenheim rearward-firing under turrets.

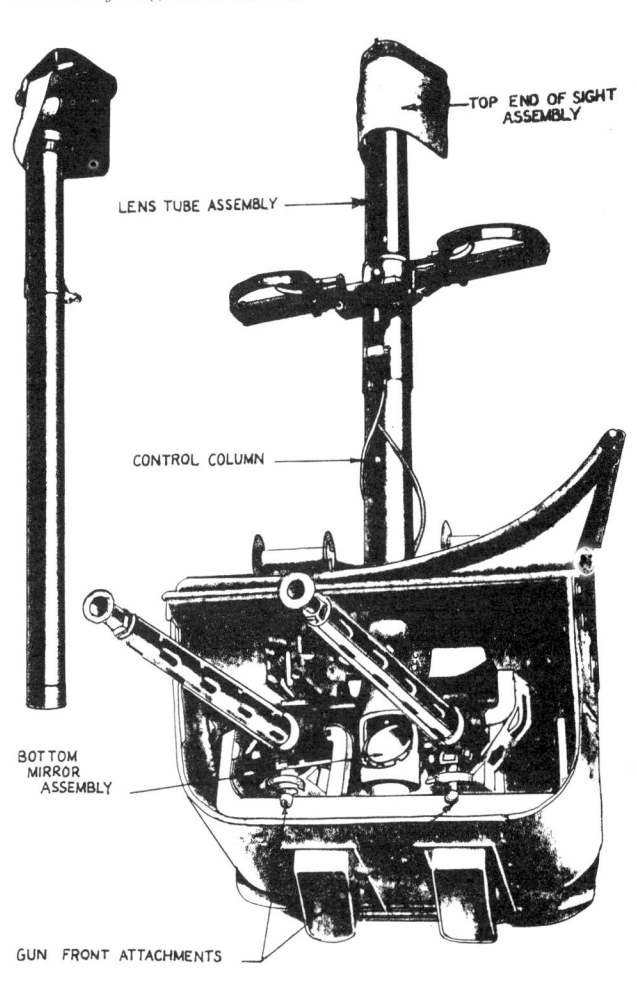

From a labour force of 100 in July 1943, Ferranti employed 950 at peak production in October 1944.

	Number produced	
	1944	1945
February	8	400
March	110	1,000
April	200	1,100
May	250	
June	370	
July	380	
August	420	
September	540	
October	700	
November	720	
December	600	

BRITISH AIRCRAFT ARMAMENT

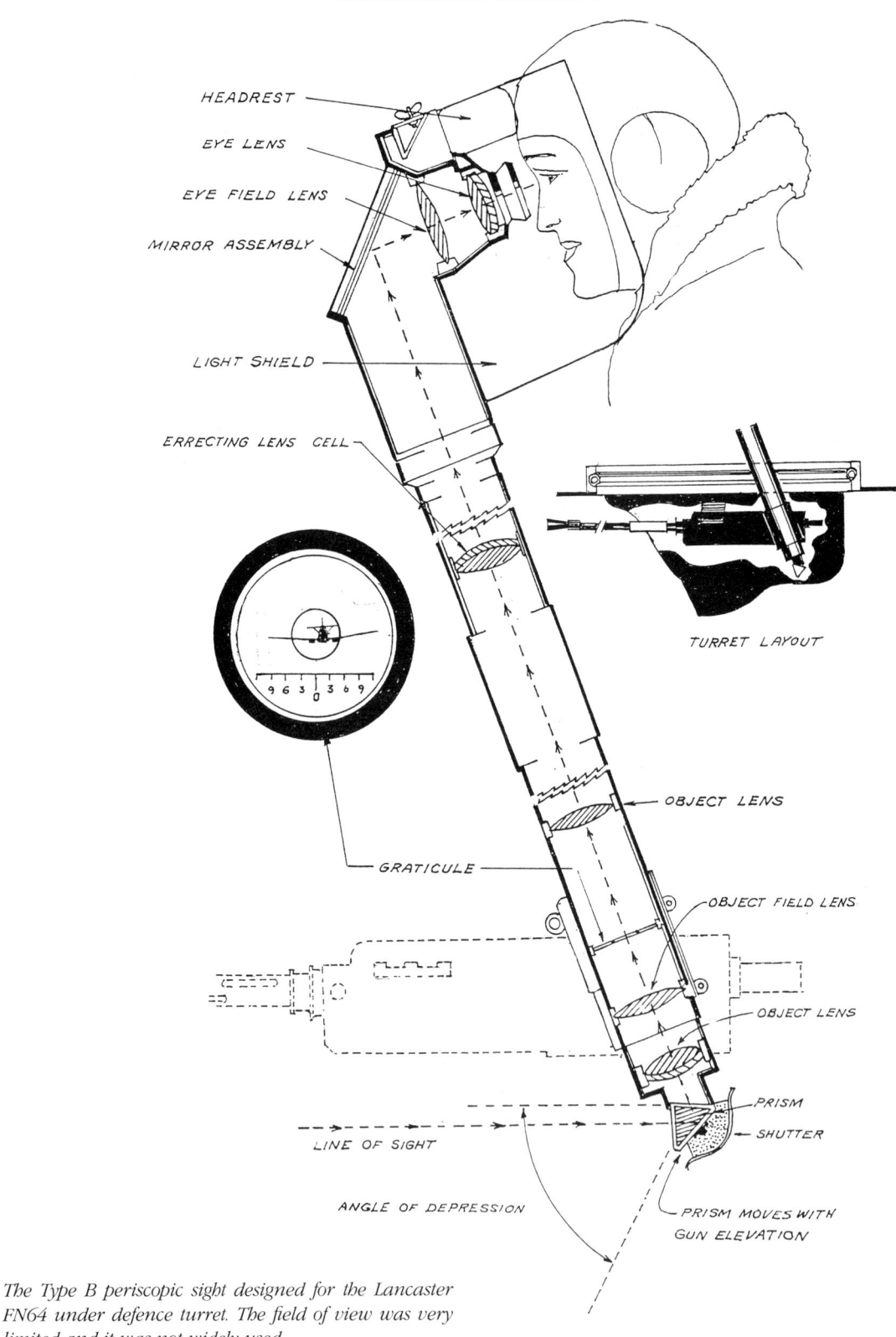

- HEADREST
- EYE LENS
- EYE FIELD LENS
- MIRROR ASSEMBLY
- LIGHT SHIELD
- ERRECTING LENS CELL
- GRATICULE
- OBJECT LENS
- OBJECT FIELD LENS
- OBJECT LENS
- PRISM
- SHUTTER
- PRISM MOVES WITH GUN ELEVATION
- LINE OF SIGHT
- ANGLE OF DEPRESSION

TURRET LAYOUT

The Type B periscopic sight designed for the Lancaster FN64 under defence turret. The field of view was very limited and it was not widely used.

each side, and 17° depression. A more elaborate sight was fitted to the Frazer-Nash FN64 fitted to some Lancasters. This, the Type B, featured a full optical system culminating in a prism linked to the guns in elevation. The turret was powered by a small version of the Nash & Thompson hydraulic system. As in all prismatic systems, the gunner's vision was limited to the narrow FOV through the lenses.

American gunsights

Computing sights

Coastal Command and some special-duty squadrons used Boeing Fortress IIs (B-17F/G) and Consolidated Liberator IIIs, whose turrets used Sperry sights, which gave the gunner a computed impact point allowing for deflection. These were mechanical rather than gyro-controlled, but effective. The K-3 was used in the Sperry dorsal turret of the Fortresses. The Sperry ball turret used the K-4, in which the sight head was at the bottom of the housing. The K-3 was too large for the Martin dorsal turret of the Liberator, so the K-9 was developed in which the sighting head was remote from the computing unit.

Above right *The American K-4 computing sight used on Sperry ball turrets. This was the 'upside down' version of the K-3 used on mid upper turrets.* (USAF)

Right *The view through the K-3. The ring and bead was the standby sight.* (USAF)

Below *The Sperry K-9 turret sight in which the computer was remote from the sight head.* (Pete Sanders)

THE SPERRY K-9

The target span was set on a dial. Looking through the sight the gunner saw a horizontal line, which he aligned with the target, and two upright range bars with which, using a pedal, he bracketed the target. Providing he tracked the target smoothly and aligned the graticules, the target would be hit. These sights used the rate-by-time system, measuring angular velocities with respect to the host aircraft. They were effective in level flight, but rough air or evasive action by the bomber caused considerable inaccuracy.

Left *US K-13 sight used on waist guns on some Fortresses and Liberators.* (Pete Sanders)

Below *American ring and bead sights used on US aircraft operated by the RAF.*

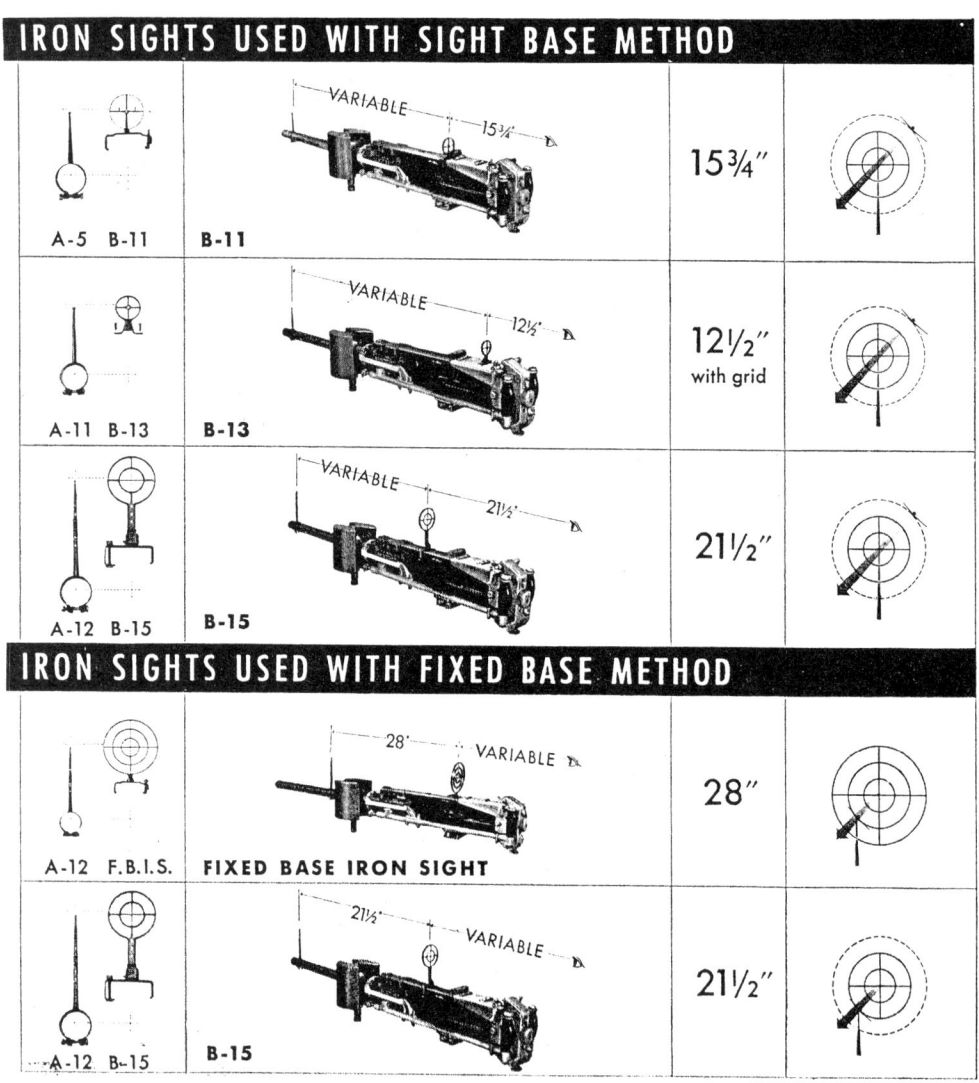

IRON SIGHTS USED WITH SIGHT BASE METHOD

Sight	Gun	Span	Sight Picture
A-5 B-11	B-11	15¾"	
A-11 B-13	B-13	12½" with grid	
A-12 B-15	B-15	21½"	

IRON SIGHTS USED WITH FIXED BASE METHOD

Sight	Gun	Span	Sight Picture
A-12 F.B.I.S.	FIXED BASE IRON SIGHT	28"	
A-12 B-15	B-15	21½"	

THE GUNSIGHTS

The waist guns of some Fortress IIs used the K-13 compensating sight, which was bulky but gave the gunner the correct lead angle for rate of rotation.

American ring and bead sights

The flexible guns fitted to some US aircraft operated by the RAF were aligned by an assortment of ring and bead, or 'iron' sights, as the Americans called them. They were designed to use 35 mm (1.37 in) radius rings in two ways: the sight base system, where the eye to ring distance was fixed, or the fixed base method, in which the distance between the ring and bead was fixed, as shown in the illustration.

American reflector sights

When the USA entered the Second World War, the sights used in their turrets were all of US design and manufacture, but several British fixed gunsights

Right *The US Navy Mk 8 reflector sight, a licence-built version of the British Mk II without the ranging mechanism.* (USAF)

Below *The N-6 reflector sight used on US turrets.* (USAF)

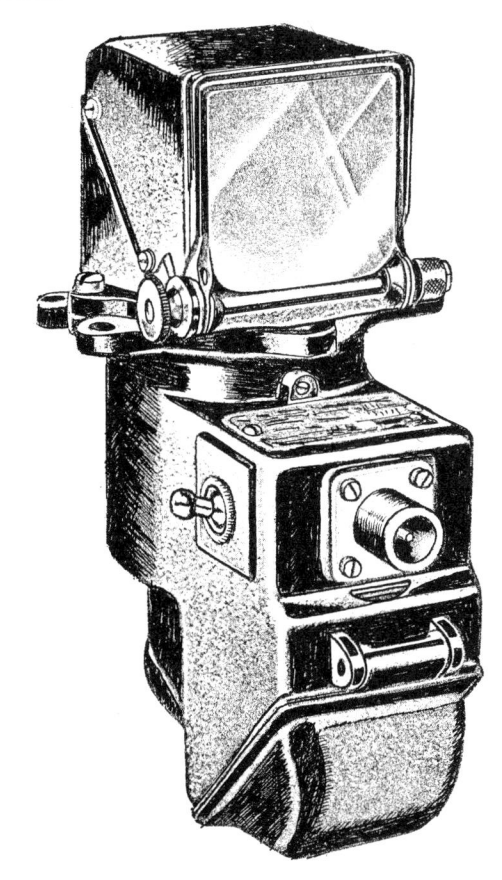

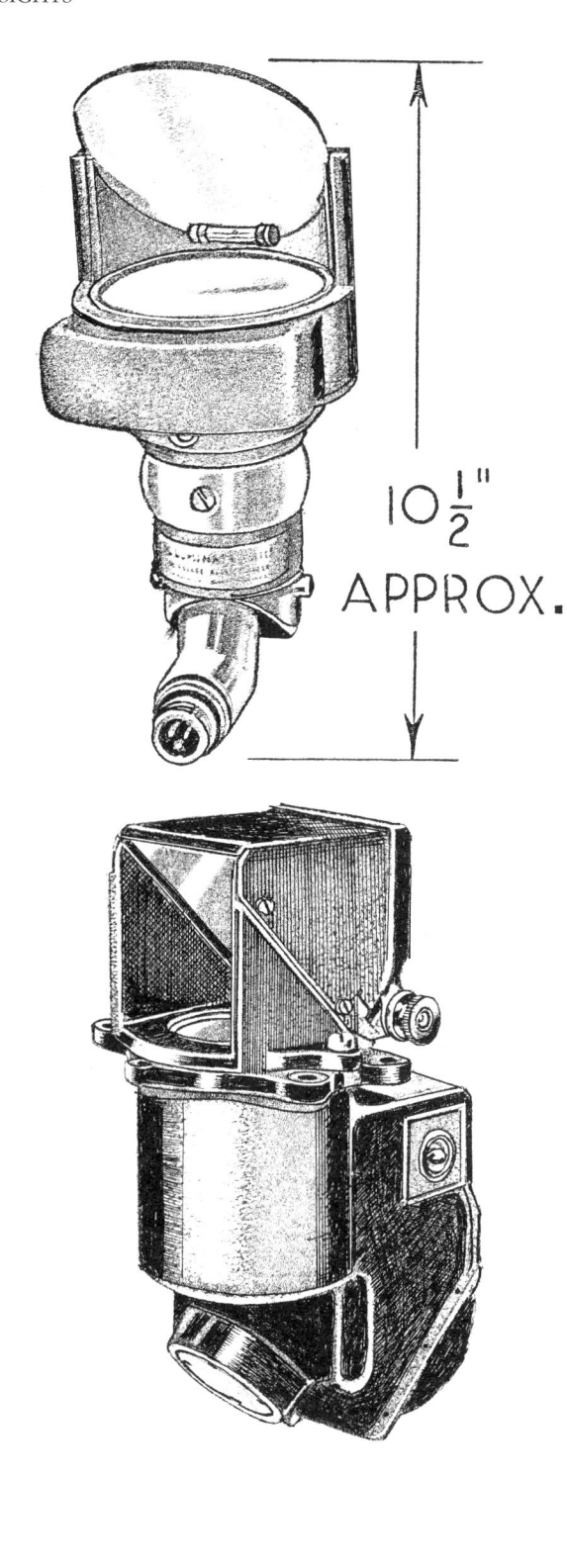

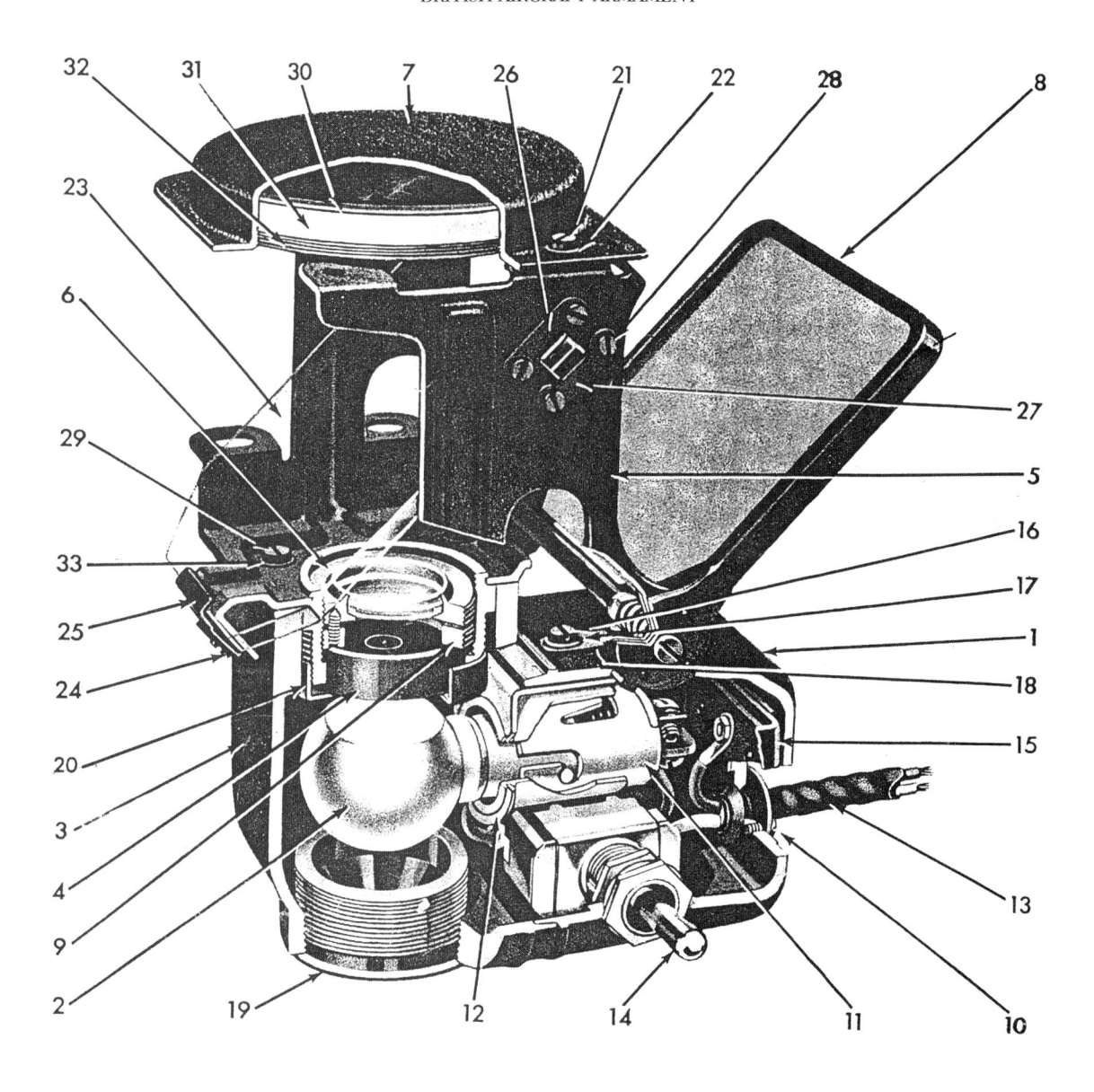

1. Lamp House Assembly—Complete
2. Lamp—Mazda
3. Lamp House—Stationary
4. Reticle Assembly—Complete
5. Frame and Reticle Retainer Assembly
6. Lens Assembly—Sealing
7. Mangin Mirror and Holder Assembly
8. Filter and Frame Assembly
9. Seat—Reticle mount
10. Screw—No. 40 x 5/16 fillister-head
11. Socket—Lamp
12. Holder—Socket
13. Cable Assembly
14. Switch
15. Spring—Latch
16. Screw—No. 4-48 fillister-head
17. Washer—No. 4 internal lock
18. Washer—Plate
19. Cap
20. Nut—Reticle mount retaining
21. Screw—No. 4-48 fillister-head
22. Washer—Medium spring lock
23. Plate—Viewing
24. Screw—No. 4-48 fillister-head
25. Clip and Pad Assembly—Lower
26. Clip and Pad Assembly—Retaining
27. Clip—Retaining
28. Screw—No. 3-56 fillister-head
29. Screw—No. 5-40 fillister-head
30. Spacer
31. Mirror—Mangin
32. Ring—Mirror retaining
33. Washer—No. 5 internal lock

were later adopted by the USAAF. The following designs were used by the RAF in Lease-Lend aircraft:

Opposite *The American N-8A used the retiflector system, in which the graticule was projected up onto a Mangin mirror at the top of the sight body, from where it was reflected down onto the semi-reflective reflector screen.* (Ross Whistler)

Number	Type	AM Ref.	Comments
N-2	Reflector	N/A	Used on A-14, A-18, BC-1, P-47B, P-26, P-35.
N-3A	Reflector	108B/12	Used on P-38, Martin & Emerson turrets, P-47C & D.
N-3B	Reflector	108B/68	Used on P-51B, P-51C.
N-6, N-6A	Reflector	108B/19	Most widely used turret sight, with single-ring graticule 35 mm (1.37 in) radius.
N-8, N-8A	Retiflector	108/24	B-17 and B-24 used retiflector principle, where the graticule was projected onto an upper mirror and down to the reflector screen.
N-9	Reflector	N/A	Some P-51Ds.
L-8			Spirit level on sight head.
L-3	Reflector	N/A	Made to fit late P-38 cockpits.
Navy Mk 8	Reflector	108B/20	British Mk II without stadiametric ranging, used on P-47, Avenger and some P-51s.
Navy Mk 9	Reflector	108B/21	Based on British Mk IIIA.

The Washington remote control sighting system

The Boeing B29 Superfortress entered RAF service in 1950, when 88 of these aircraft, known in the Service as the Washington, formed the UK's main striking force until the arrival of the V bomber.

A gunner using the pedestal sight of a B29 Washington. Signals from these remote sighting stations were relayed to the computing system which controlled the guns and made the correct allowance for deflection, range and 'own speed'. (USAF)

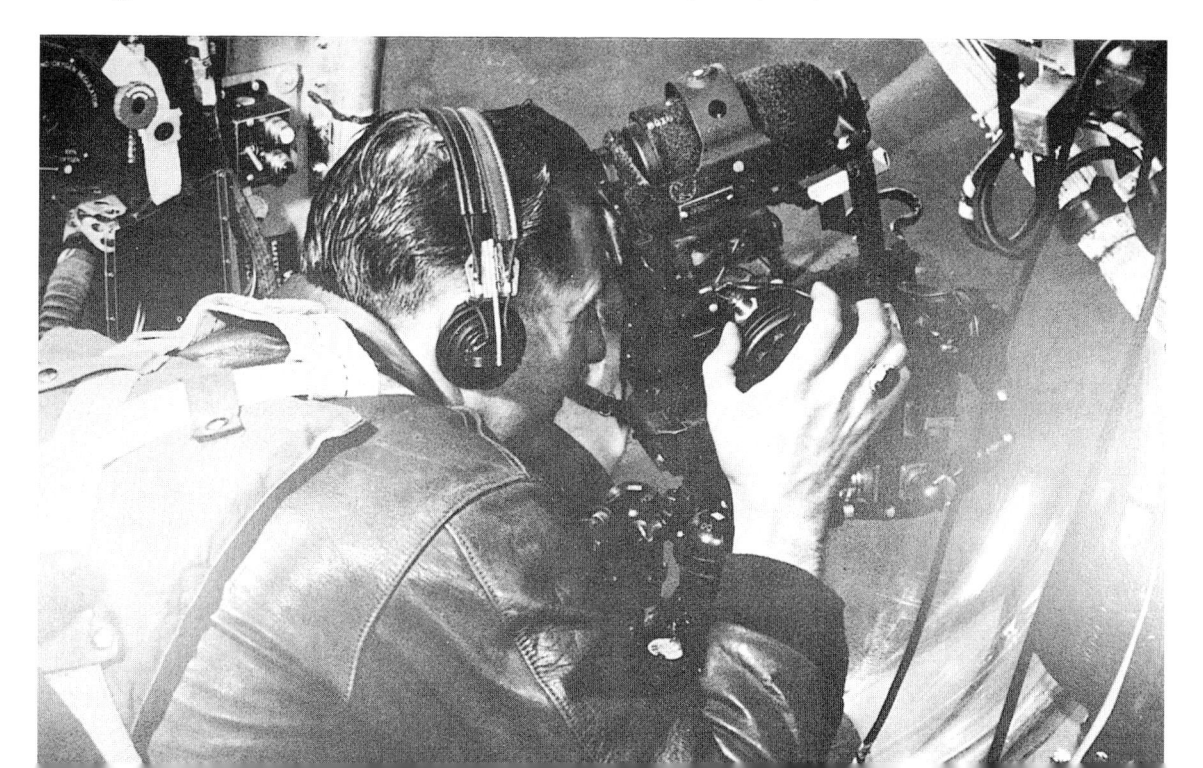

BRITISH AIRCRAFT ARMAMENT

Sighting procedure of the RCT system.

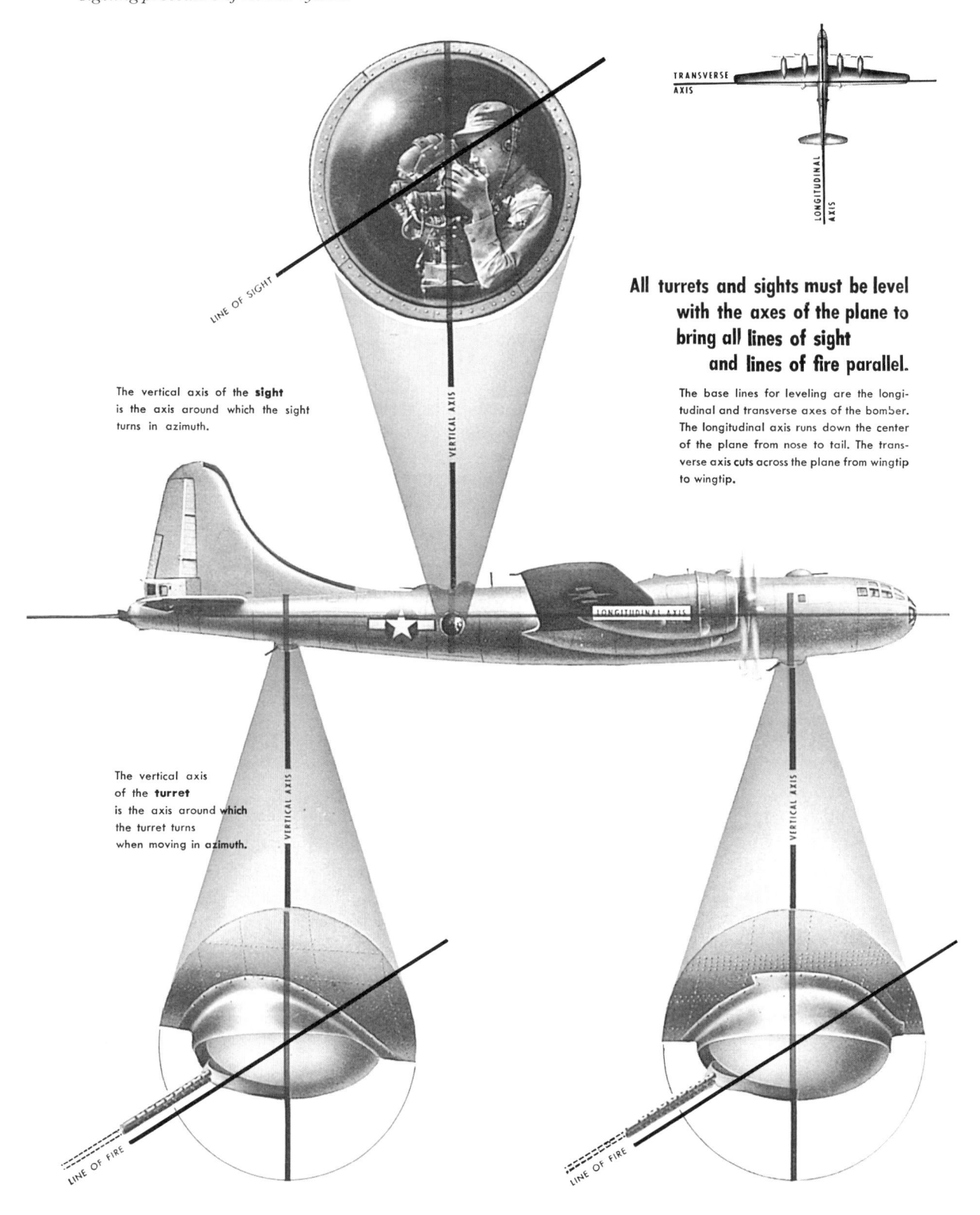

All turrets and sights must be level with the axes of the plane to bring all lines of sight and lines of fire parallel.

The vertical axis of the **sight** is the axis around which the sight turns in azimuth.

The base lines for leveling are the longitudinal and transverse axes of the bomber. The longitudinal axis runs down the center of the plane from nose to tail. The transverse axis cuts across the plane from wingtip to wingtip.

The vertical axis of the **turret** is the axis around which the turret turns when moving in azimuth.

184

THE GUNSIGHTS

Layout of the Washington RCT turret control system. (USAF)

These advanced bombers were fitted with remote turrets controlled by five sighting stations. Each station contained a controller which incorporated a reflector-type sight linked with a computer and four Selsyn signal generators. As the gunner tracked a target, the Selsyns transmitted information to receivers in the turret which controlled servo motors on the gun control system. When the gunner acquired the target, he set the aircraft type on a dial and adjusted a ring of spots on his reflector screen to correspond to the target's size, which gave the computer the range. As he tracked the target, the computer analysed the rate of movement which enabled it to predict the lead angle. The Selsyn signals were processed through the computer, which adjusted the signals so that the guns pointed to a position in space where the target would be hit.

This system, known as the Remote Control Turret System (RCI), was produced by the General Electric Co. It needed to be very accurately harmonized and set up, and gunners required a five months' training course to become operational. The system is further explained in Volume 1.

Axis gunsights

The Luftwaffe used the excellent Revi series of gunsights, the most used fixed-gun version being the Revi C/12D, which was the German equivalent to the Mk II reflector sight. There were many sub-types of this sight, one of which (the Revi C/12F) featured an angled second reflector at the rear of the main sight screen, which is similar to the AGLT collimator screen and was probably used for radar input.

The German Revi C/12D fixed gun reflector sight mounted in a Bf 109G.

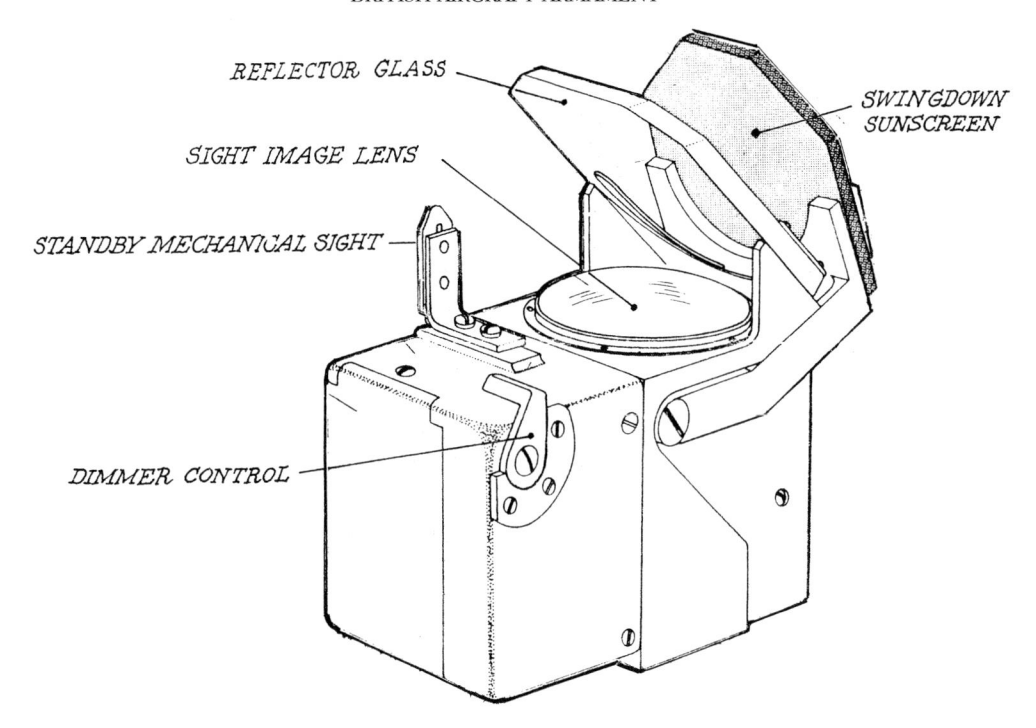

Labels on diagram:
- REFLECTOR GLASS
- SWINGDOWN SUNSCREEN
- SIGHT IMAGE LENS
- STANDBY MECHANICAL SIGHT
- DIMMER CONTROL

Above *The Revi 16B free gun and turret sight, the equivalent to the British Mk IIIA*.* (Bundesarchiv)

Right *The Japanese type 98 pilot's sight, which bore a strong resemblance to the Revi series. Note the large standby ring and bead.*

The German equivalent to the GIIIA* free gun reflector sight was the Revi 16B. Most German reflector sights were fitted with standby ring and bead sights in case of electrical failure.

The German type VE22 compensating gunsight.

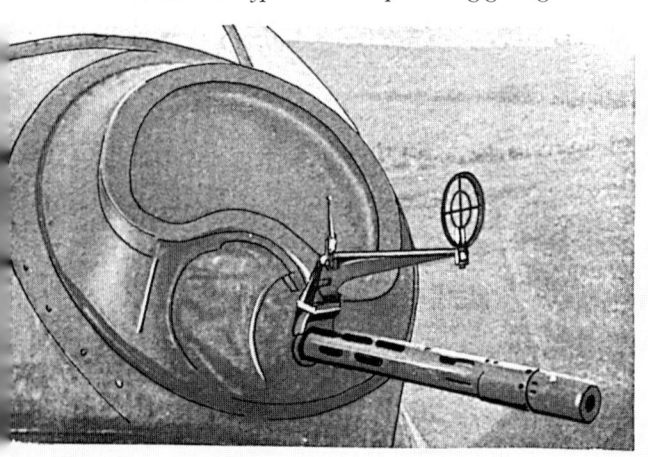

THE GUNSIGHTS

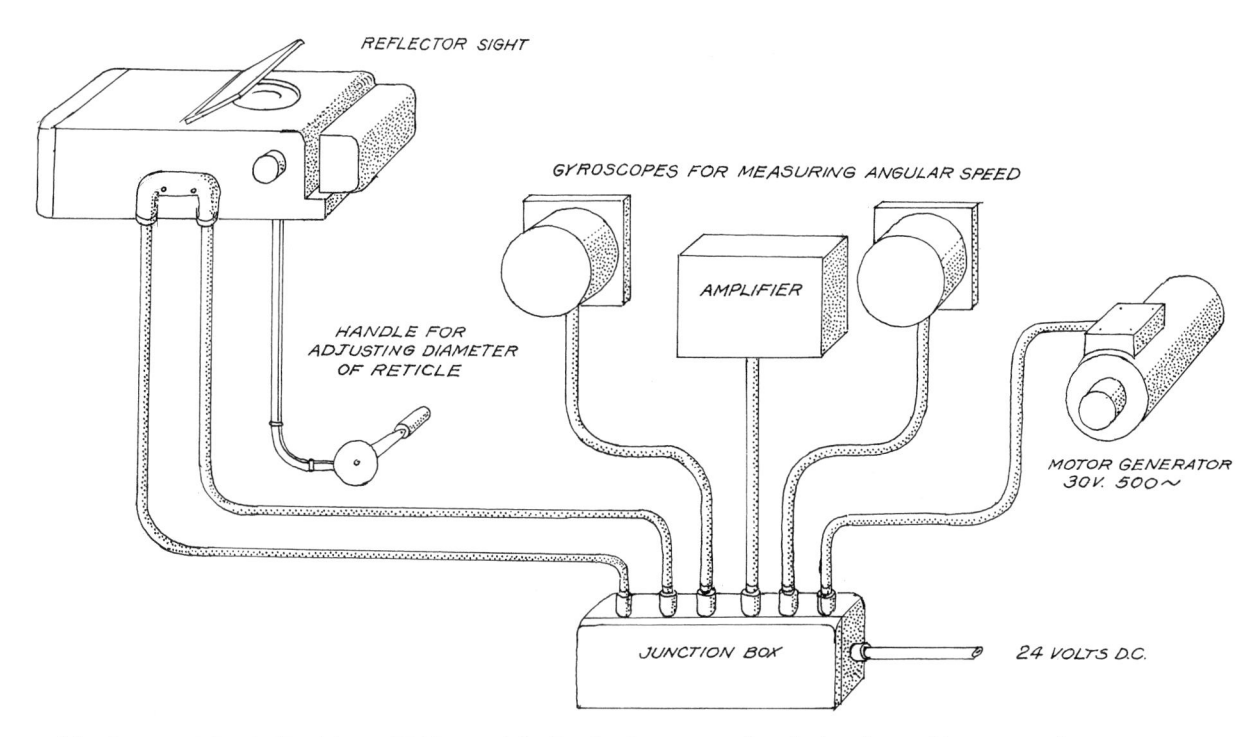

The German Askania 'Eagle' type EZ42 gyro sight. Production was too late for it to be used in any numbers.

Another free gun sight used by the Luftwaffe was the VE series, which consisted of a ring and bead unit fixed to a housing geared to the angle of aim relative to dead astern. They were used on rear defence, hand-operated guns, and were effective against a fighter coming in on a pursuit curve attack.

Both Russian and Japanese reflector sights bore a remarkable resemblance to Revi sights, the Japanese type 98 and Russian ASP3 series being in widespread use during the Second World War.

The German Askania EZ42 Eagle gyro sight
The Askania 42 gyro sight was developed by Askania in 1944, production beginning in November. The system used two rate gyros, one for lateral, the other for vertical angular speed. The gyros controlled two servo motors which moved a mirror to give the required lead angle. The EZ42 project was considered to be the most important Luftwaffe programme of the closing months of the war, operational trials having shown that if the adjustable ranging graticule was lined up with the target's wingspan, and the cross lines were on the target, the target would be hit. However, by the end of hostilities only 400 sights had been delivered, relatively few of these being used operationally.

The Thompson sight

In 1941–2 the Royal New Zealand Air Force base at Wigram had a fitter armourer, Cpl. John R. Thompson, who was described by his sergeant as 'possessing a mind which can only be described as an encyclopedia of mechanical movements, whose ability as a designer and craftsman probably has no peer. Some of his ideas appear fantastic, but he seems to be able to put his ideas into practice.' Thompson, who had already invented various gunnery devices, devised a sight which gave the gunner a point of aim which predicted the angle of deflection. In theory, the firer matched the target's path and speed with a moving graticule pattern seen through a screen similar to that of the Mk II reflector sight. The gunner controlled the moving series of graticule sections engraved on an endless belt of heat-resisting 16 mm film, the direction of the graticules being aligned with the apparent path of the target. After a short period of tracking, when the target was made to move towards the centre of the sight, the guns were fired when the target reached a line marked across the graticule movement. The author has Thompson's specification in which every detail of operation and manufacture is meticulously described. He produced the drawings and proto-

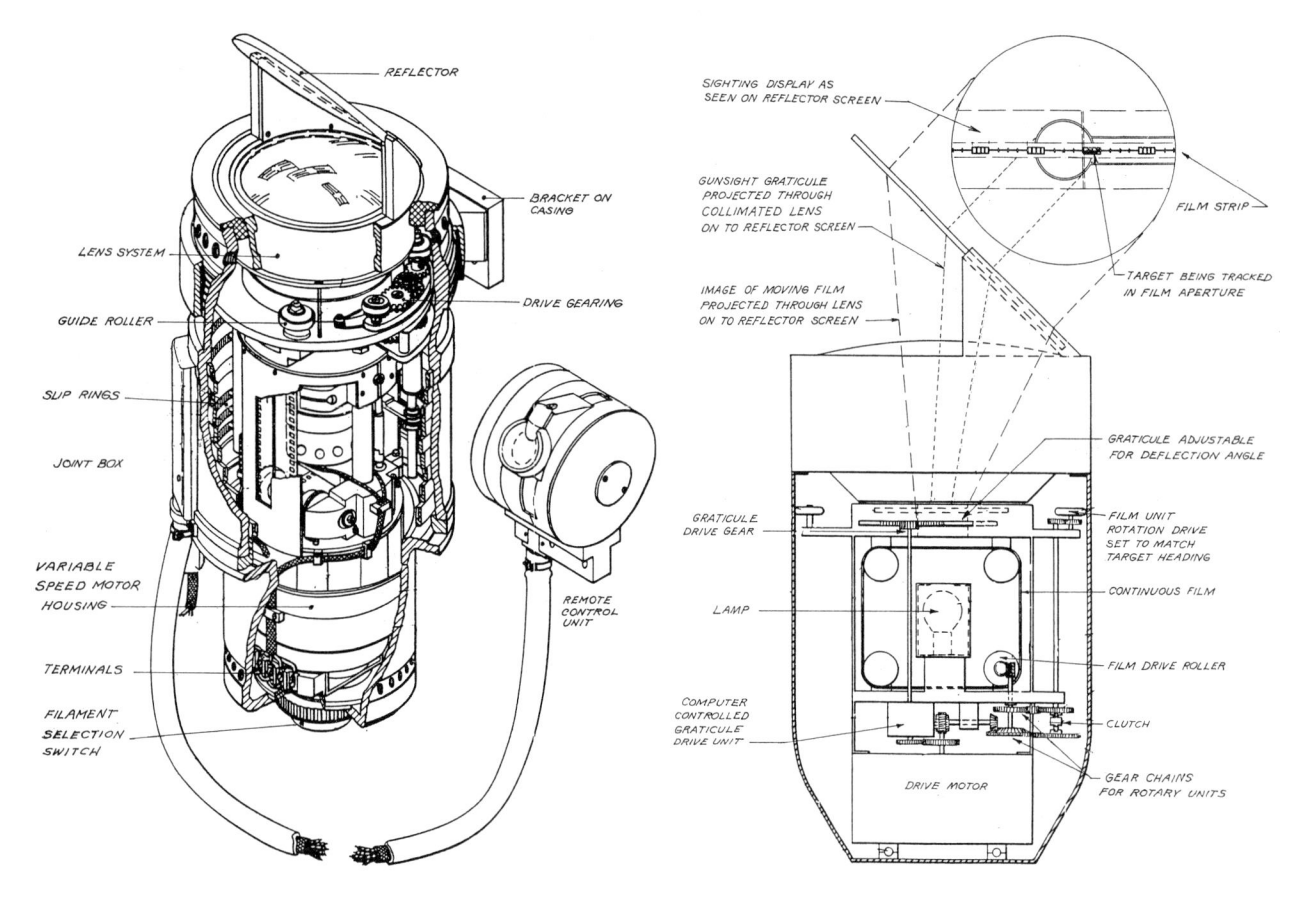

The compensating gunsight invented by Cpl. John Thompson of the RNZAF. This used a film strip moving at a set speed, which was made to coincide with the speed of the target. (RNZAF Wigram)

The moving film/graticule system of the Thompson sight. The target was aligned with the graticule, which could be adjusted to the crossing speed. The moving film then gave the correct deflection. (RNZAF Wigram)

type in nine weeks for £50 (the Mk IIC gyro sight was developed in 12 months and cost £750,000).

Any technical invention of this kind is often met with scepticism by the authorities, and Thompson's was no exception. It was only after a desperate letter to the Chief of the Air Staff dated 26 May 1943 that he was summoned to Britain with his invention, which was evaluated at Farnborough. Unknown to Thompson, the gyro sight was nearing completion, but his sight was thoroughly tested by the Armaments Section. The relevant parts of their conclusions were as follows (in memo Arm. No. 73):

The sight presents to the gunner a collimated graticule which consists of a broken line which can move across the field of view in the direction of its length at a rate which can be controlled by the gunner. The orientation of the travelling line can also be controlled by the gunner, in such a way that it always travels through a central point in the field of view to which the guns are organised. No arrangements for including trail and gravity drop allowance are included.

The sight therefore demands: the setting of guns and graticule pattern to cause the target to fly towards the centre, matching of the apparent graticule speed and the apparent target speed, tracking the target with the aiming point thus determined and judgement of the firing range with the stadiametric aid of breaks in the graticule pattern.

The defect of the principle is that the angular velocity of the target is determined some time before firing, and is determined relative to the gunner's aircraft which must, therefore, fly straight and level if large errors are to be avoided.

The operations required of the gunner could present him with difficulty under combat conditions, and the addition of trail and gravity drop allowance would further complicate what is already an involved mechanism.

This sight is a good attempt at the solution of the problems of controlling aircraft guns, and the engineering skill displayed in the design of the prototype is high. The arrangement is, however, outdated by other developments in Great Britain and in the USA: in particular the British GGS Mk IIC sight is more compact, is likely to be simpler to use, is referred to space axes rather than own-aircraft axes, and includes trail and bullet drop. In view of the fact that this development is in production, further work on the Thompson sight is not recommended.

It is recommended, however, that this inventor be encouraged to consider further problems, since the skill and thought shown by his prototype sight is of a high class.

Thompson returned to New Zealand where he continued to work as an engineering officer in the RNZAF. Until 1942 no such sight had been developed by any nation. Thompson's achievement must rank amongst the most remarkable of the war. The official assessment was probably correct, but if the gyroscopic sight had not materialized there is every possibility that the Thompson sight would have been adopted by the RAF. Before the sight was sent to Britain it was demonstrated to a group of visiting operational pilots of the RNZAF, including W/Cdr Wells, who described it as 'the answer to a fighter pilot's prayer'.

Airborne Interception Radar Gunsighting

Well before the war it had been realized that if daylight bombing attacks were successfully countered, a future enemy would switch to raids under the cover of darkness. When 'radiolocation' was seen to be feasible, a design team was set up at Bawdsey in Suffolk to investigate the possibility of airborne radar detection for night fighters. This project, under the direction of Dr E. G. Bowen, produced the first successful airborne radar sets. The first equipment was limited by having a frequency of 1.5 m (1.64 yd), which gave a very wide beam and limited range. However, these first sets were successful in detecting intruders. After the invention of the cavity magnetron by Messrs Randal and Boot, and the Klystron oscillator, it was possible to produce a radar of 10 cm (3.94 in) frequency, giving a narrow beam of 30°, which was less prone to ground returns.

In March 1941 the first airborne centimetric radar was tested in a Blenheim by Alan Hodgkin of the TRE and Edwards of GEC, the prototype system being known as the AIS. In autumn 1941 sets had been delivered for Service trials by Beaufighters for the FIU (Fighter Interception Unit) and 100 sets were ordered from GEC with Nash & Thompson scanners. The first production model was the Mk 7, the first squadron being equipped in March 1942. The first kill was recorded in April 1942, followed by 100 more in the following months of the year.

The Mark VIII

These radars were designed for two-man operation: the radar operator viewed the screen and gave the pilot instructions for the interception. A modification to the earlier Mk V was the 'windscreen project', which consisted of a unit which projected the image of the radar display up onto the pilot's windscreen. It was seen as a step forward and the system was tested at Farnborough, but it was not adopted for service. However, an improved version was designed for the main production version, the Mk VIII. This proved successful but was not adopted for use at that time. The official description of the projector system read:

> The windscreen projection endeavours to combine the four main functional aids into a single co-ordinated picture showing the pilot what is happening, and enable him to carry out the interception by its sole aid.
>
> The whole picture is therefore projected ahead of the aircraft, and observed by the pilot looking through his windscreen. The pilot sees the position of the enemy aircraft and its range, his own position relative to it, and his attitude in relation to the horizon and a pre-set heading, the whole being encompassed by the gunsight circle. In this way, many difficulties experienced while carrying out a night interception are hoped to be successfully overcome.
>
> Firstly, the pilot's attention will not be divided between several instruments in his cockpit, concentration will thereby be enhanced at the expense of less strain and time lags reduced. Secondly, the pilot will be in the most favourable condition to achieve a visual since his dark adaptation will be the best possible for the existing

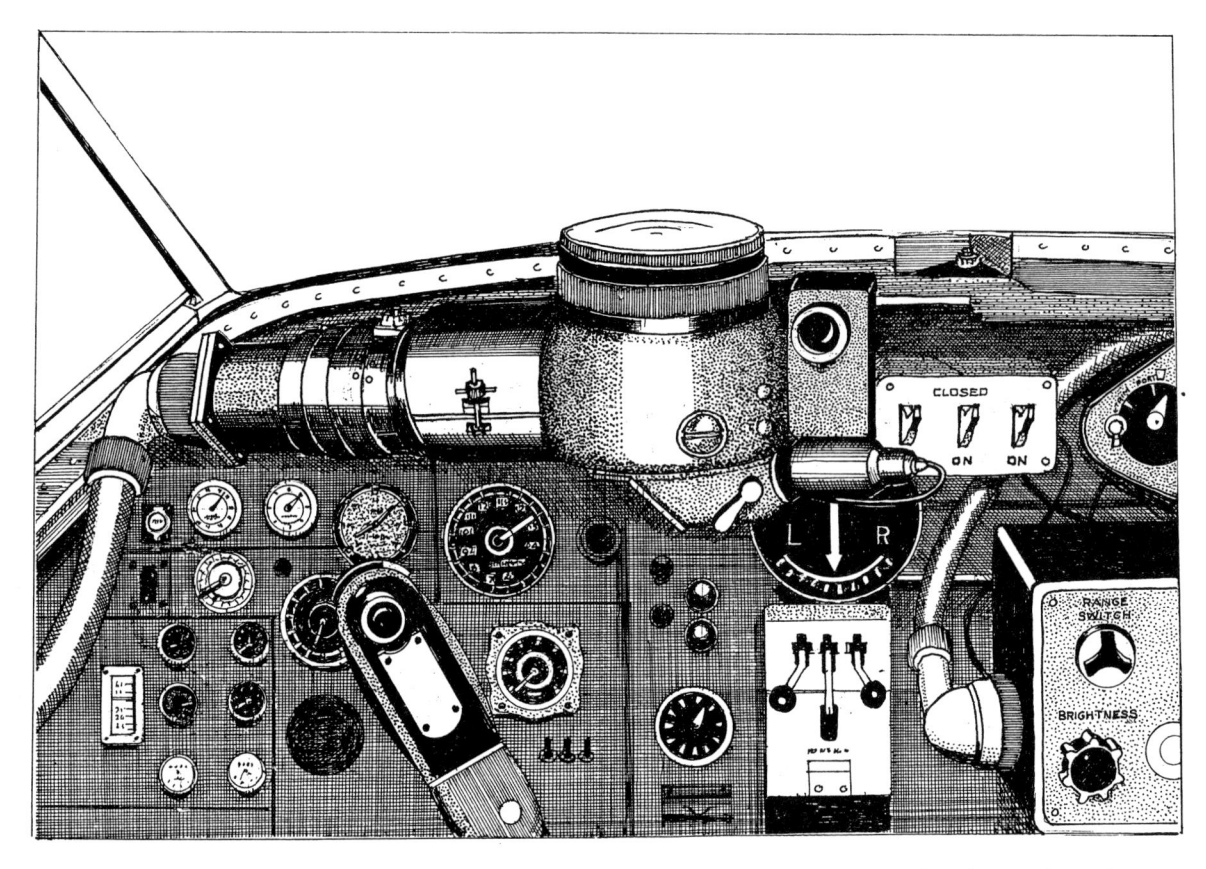

Above *Drawing showing the position of the AI Mk VIII in a Mosquito night fighter.*

Below *The projector unit of the AI Mk VIII, showing the CRT, gunsight, lampholder and main body.*

night conditions, his eyes will be focused to infinity. Additionally the visual will be aided by the fact that the pilot will be looking out in the direction of the enemy aircraft. Lastly no change in the pilot's eye position will be required throughout the whole interception, since the gunsight ring is included in the picture.

A particular development of the system has been carried out for use with AI Mark VIII to provide the pilot with blind flying information.

Method of operation of the AI Mk VIII projector system

When the set was switched on the pilot adjusted the tube brightness and background control; the radar scanning system then began to operate. When a target was detected an electronic spot or 'blip' appeared on the screen, showing the position in relation to the pilot's field of view. The aircraft was then brought into a position where the blip was central. When the range had decreased to 7,500 ft (2,286 m), small horizontal lines or 'wings' began to grow on either side of the blip. The pilot then had to reduce speed to avoid overshooting the target. From this time, depending on conditions, the target

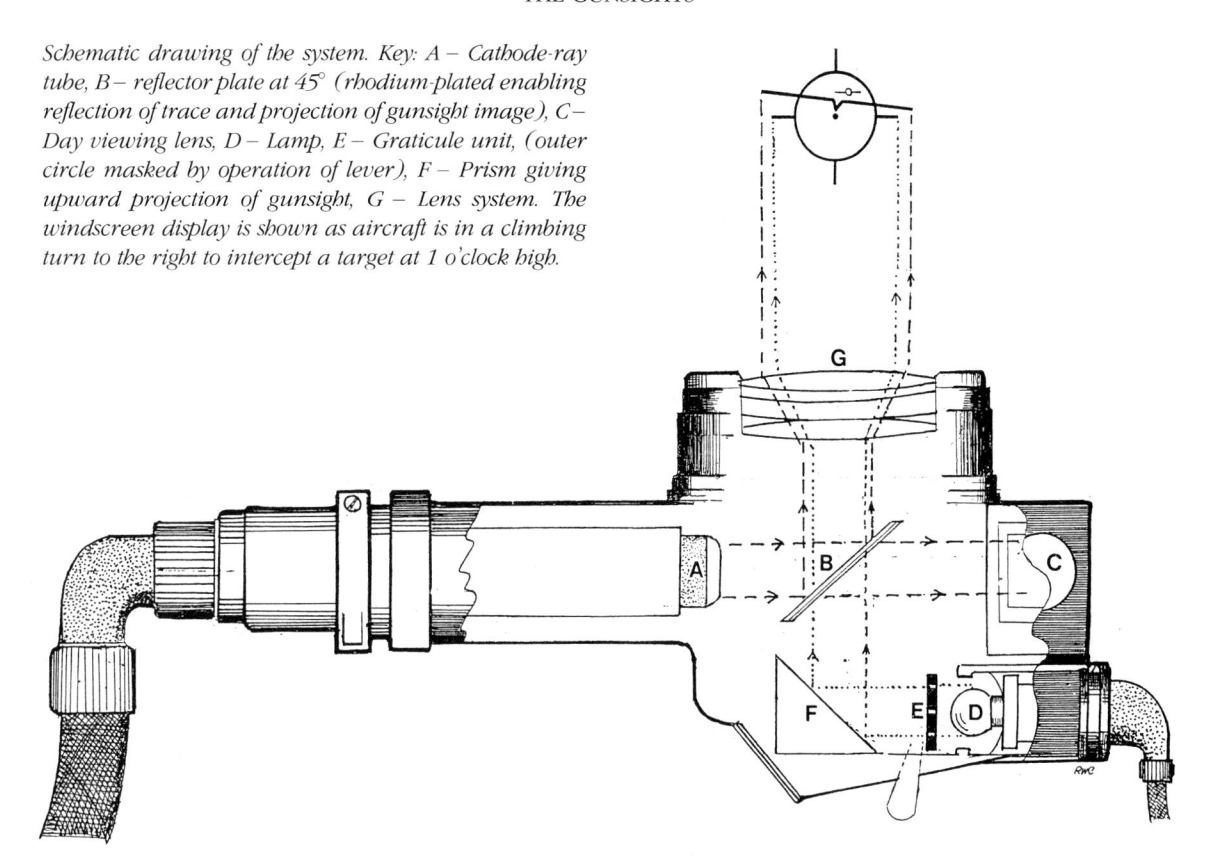

Schematic drawing of the system. Key: A – Cathode-ray tube, B – reflector plate at 45° (rhodium-plated enabling reflection of trace and projection of gunsight image), C – Day viewing lens, D – Lamp, E – Graticule unit, (outer circle masked by operation of lever), F – Prism giving upward projection of gunsight, G – Lens system. The windscreen display is shown as aircraft is in a climbing turn to the right to intercept a target at 1 o'clock high.

was likely to be seen. The gunsight would have been switched on during the search period, and adjusted for brightness. When the target was seen the radar image was turned off, leaving only the gunsight graticule visible. The pilot then opened fire.

Other facilities available from the system was IFF (Identification Friend or Foe) and a heading indicator consisting of a vee in the horizon position line which could be set to a beacon or specific heading. Identifiable ground returns noticed during the operation of airborne radar sets led to the development of blind navigation by Dr Bernard Lovell which later became the H2S navigation and blind-bombing system.

As can be seen by the above description, this system could be regarded as the ancestor to the modern head-up display. The information projected onto the windscreen consisted of an artificial horizon, target position and range, and a gunsight graticule. The gunsight consisted of a lamp in a quick-release housing illuminating a 100 mph (160.9 km/h) ring and dot graticule. Interposed between the lamp and graticule unit was a hemispherical shutter, pivoted and rotated by a lever, which, when rotated, shrouded the ring, giving a spot graticule only (as in the Beamont modification). The graticule image was projected through a prism and a stadiametric lens system onto the windscreen, and a dimming rheostat was fitted close to the unit on the instrument panel. Pilots who used the Mk V and VIII equipment suggested that as the directional gyro heading indicator (pre-set heading) was rarely used, a more useful aid would have been a presentation of airspeed – an essential ingredient of night interception.

Airborne Gunlaying in Turrets (AGLT)

With the success of air interception radar for night fighters, it was decided to develop a system which would provide rear gunners with a means to locate fighters making a radar-assisted attack. In early 1943 the Air Staff held a conference to advise on a specification for such an installation, resulting in the following suggestions:

The transmitter/receiver to be mounted at the

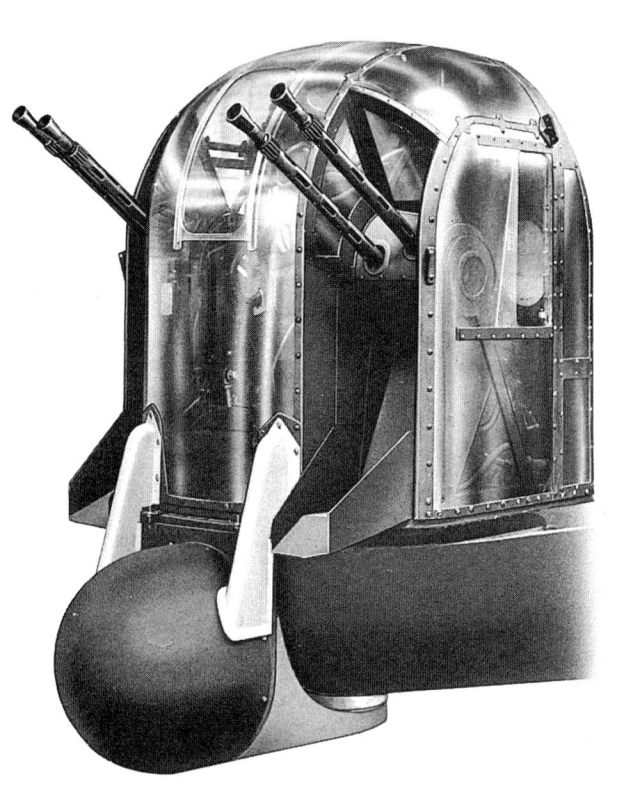

AGLT 'Village Inn' seen here fitted to a Lancaster FN121 tail turret. The scanner under the dome moved in elevation with the guns. (Frazer-Nash)

navigator's position to save weight in the tail of the aircraft. The design should include an automatic ranging facility, which would be relayed into the Mk. IIC gyro sight. A small scanning aerial should be mounted in such a way as to follow exactly the movement of the guns. The display screen would consist of a small cathode ray indicator mounted adjacent to the sight screen of the GGS.

A research team under Dr P. I. Dee was formed, the design leader being Dr Alan Hodgkin and the code name 'Village Inn'. The official Air Ministry title was Airborne Gunlaying in Turrets (AGLT).

AGLT scanning

After various experimental designs, the scanning system was agreed upon. It consisted of a small dipole antenna mounted near the focal point of a 16 in (406 mm) parabolic reflector. The antenna was mounted eccentrically and rotated by an electric motor at 2,000 rpm through a cone angle of 30°. The unit was to be covered by a perspex radome and fixed to the lower rear section of the turret, rotating with it and being connected to the gun elevation system by parallel linkage. The antenna was fed by a 9.1 cm (3.59 in) transmitter remote from the turret, radiating half-micro-second pulses at a prf (pulse repetition frequency) of 660 per second. Limits of operation were 85° to each side and 45° above and below horizontal. When the gunner searched for fighters, the scanner followed, and any approaching aircraft gave a 'blip' on the screen.

AGLT display

The first design for the gunner's radar display was a small cathode-ray spot indicator mounted to the left of the gunsight, in which the displacement of a green spot from the centre gave the target bearing and elevation. Tests revealed that the gunner, who was trying to get a visual on his attacker *and* manipulate the turret controls, had to evaluate two screens: his sight and the radar indicator. It was found that a better solution was a single screen which incorporated both the radar spot and the sight graticule. After several ideas had been tried, one of the team devised a means of projecting the radar spot onto a semi-transparent mirror fixed behind the reflector screen of the GGS. This consisted of an upright enclosed cathode-ray tube mounted on the right side of the sight. The tube display was projected upwards through a prism onto the mirror. This was to become known as the AGLT collimator. The gunner was provided with a control box to adjust the brightness and focus of the spot, and a selector for controlling the signal.

AGLT ranging

In the Mk IIC GGS, target range was fed in by foot pedals, with which the gunner adjusted the diameter of the circle of diamonds to correspond with the target span. The AGLT was designed to make this process automatic. Most of the radar 'black boxes' were in the navigator's compartment, and the range was read by the navigator. Briefly, the ranging system worked as follows: Two integrating valves which emitted timed pulses were controlled by a strobe generator. The timing was fractionally different for each valve, and the difference in echo time was fed into a balancing circuit, which controlled a ranging motor, which controlled a shaft on which were three variable resistances. At the end of the shaft was a range dial reading 0–1,400 yd (0–366 m). As range information was fed from the balancing circuit, the motor turned the shaft and the

THE GUNSIGHTS

Above An AGLT collimator, now a very collectable item.

Right AGLT collimator installation. The target 'blip' was projected through the prism onto a small angled reflector plate which was fixed behind the GCS screen. (Frazer-Nash)

A. VIEW FROM TOP
B. RADAR BLIP PROJECTED THRO' PRISM
C. CATHODE RAY TUBE
D. PRISM
E. MK.IIc GYRO SIGHT
F. A.G.L.T. COLLIMATOR

range reading was relayed by the navigator over the inter-com. The three resistances triggered: 1. An aural warning of an impending attack (similar to the Monica tail warning system); 2. Control information for the strobe generator; and 3. Continuous ranging information for the predictor section of the gunsight.

A pursuing fighter was ideally detected at 3,000–4,000 ft (914–1,219 m). As it approached, the range would be read off by the navigator. The gunner would aim his guns to keep the 'blip' lined up with his sight graticule, and when he was told that the range was under 400 yd (366 m) he would open fire, whether the target was visible or not. The weak point was the need for a second member of the crew to tell the gunner when his target was in range. To overcome this weakness, the blip was given 'wings' (as in the A1 Mk 8), which gave the appearance of an approaching fighter. When the range was under 500 yd (457 m) the 'span' of the image grew rapidly. This new facility was controlled by a fourth resistance on the range shaft.

As with all new technology, limitations were discovered. The first was 'spot wander' or 'jitter', caused by echo fading when the signal reacted to the propellors of the fighter. Another was the time

Right A Frazer-Nash FN82 turret equipped with AGLT. The bracket on the left was for a Type Z indicator designed to detect infra-red I.F.F. projectors to be mounted in the nose of Bomber Command aircraft. (Frazer-Nash)

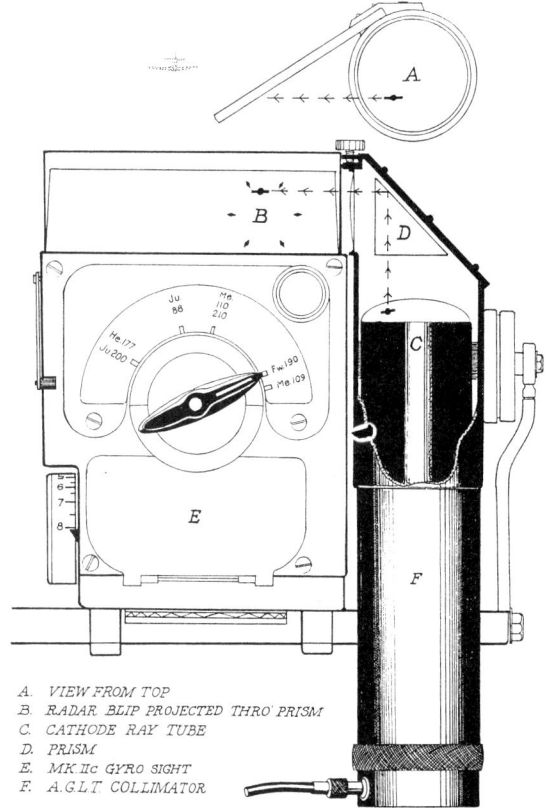

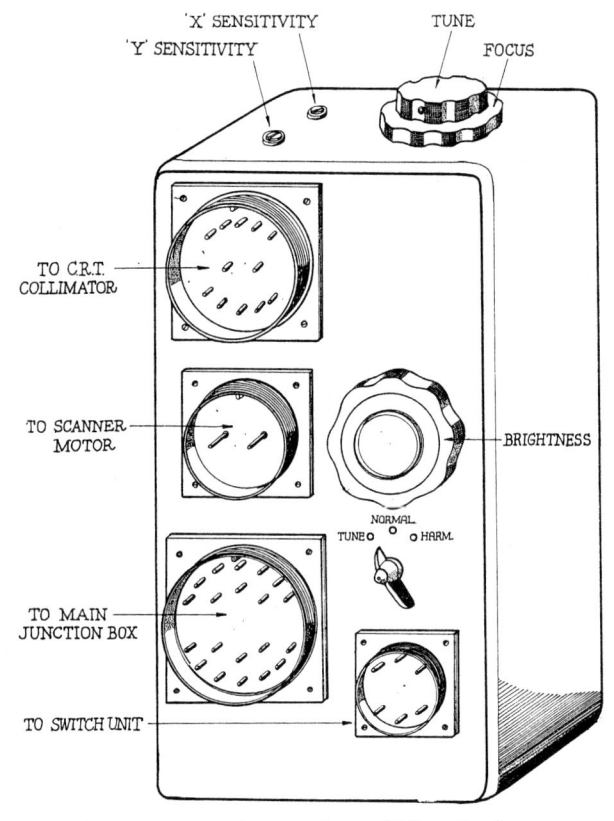

Above *Gunner's control unit, 'Village Inn'.*

Below *The Mk III AGLT scanning system. The dish scanned the rear hemisphere until a target blip was received, then it locked onto the signal.*

lag of 0.5 sec before the blip took up its final aiming position. As the antenna rotated with the turret, if the gunner traversed in a jerky manner the blip would wander aimlessly for a fraction of a second. If two aircraft were present within the 30° cone, the spot would wander between one target and the other.

To overcome the serious problem of friendly aircraft appearing on the screen, a method of identification was devised. Twin infra-red lamps, appearing as small rings, were fitted into the nose cone of bomber aircraft, and the gunners of AGLT-equipped turrets were provided with infra-red detectors mounted on the left side of the sight arm. Known as Z-type IR detectors, they picked up the infra-red emissions from the lamps, which were programmed on an auto timer to the signal of the day.

When the AGLT equipment was introduced, a self-destruct unit was fitted to prevent it falling into enemy hands. Before a gunner left the aircraft on or over enemy territory, he lifted a red flap on the device and pressed a red button, which activated a 10 sec timer.

Several Avro Lincolns were fitted with AGLT. These were improved Mk I units mounted on the Boulton Paul Type D turrets, armed with 0.5 in guns.

AGLT Mk III

A Mk II version of the system was designed incor-

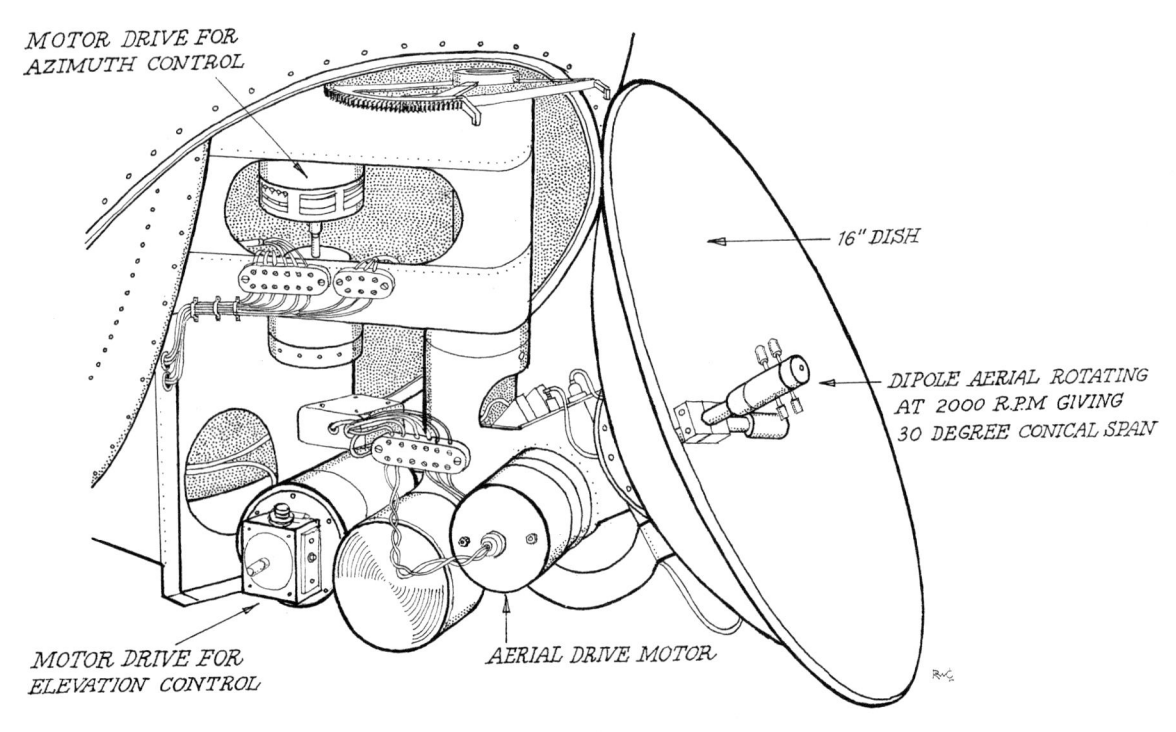

Emerson Electric blind-tracking radar fitted to Lancaster KB805 for trials. The installation, the EM-APG-8, was not accepted for use by the RAF.

porating several modifications, but this was not proceeded with. The Mk III, however, was put into limited production. In this version the scanner was not linked to the turret movement: when a target was located the scanner locked onto the echo, and the information projected onto the collimator. This enabled the gunner to dispense with the continual searching movement of the turret.

Testing was carried out by the Telecommunications Research Establishment at Defford. The aircraft used were Lancasters *ND712* for the Mk I and *JB705* and *LL737* for the Mk III. After lengthy development the Mk III was finally put into production. The first operational squadron to be fitted with AGLT was 101 at Ludford Magna in the autumn of 1944. Soon after this, 49,159 and 635 Squadrons were converted. These units found difficulty in operating the equipment, and the scanner drive mechanism gave trouble. Dr Hodgkin's team continued to improve the system but, pending the elimination of the problems, large-scale conversion of turrets was postponed. In September 1945 proving trials were carried out by 113 Squadron, but by this time there was no great urgency as hostilities had ceased.

Emerson APG-8

Running concurrently at Defford was the evaluation of an American turret radar system. In 1944 a Canadian-built Lancaster, *KB805*, had been sent to the St Louis factory of Emerson Electric, where a Model 3 tail turret had been installed equipped with the APG-8 blind-tracking radar. Defford found that the US system was not quite so effective as the British design, and in fact only 15 APG-8s were produced before the project was shelved.

Remotely sighted gun turrets

In May 1942 the Air Ministry placed an order for a comprehensive remotely controlled turret system for Bomber Command aircraft. The main contractors were Boulton Paul, who would produce the turrets, or barbettes as they were to be known, and British Thompson Houston, who had experience of remote control systems for ground use. The gunner would occupy a sighting station in the tail, where he would have an ideal field of view unimpeded by the usual guns and power systems.

The upper barbette of the remote sighting system mounted in a Lancaster. Armed with twin 20 mm Hispano guns, the barbettes were linked to the tail sighting station by an electrical control system.

The BTH remote control system employed an Amplidyne (Ward Leonard) electrical system for gun and turret control which was to be linked with a 'convergence computer' designed by the Royal Aeronautical Establishment. When the gunner aligned his sight onto the target, electrical impulses would be transmitted to the computer, which would control the power system of the barbette and 20 mm guns to coincide with the alignment of the sight, allowing for deflection and the angular differences of the barbette and sight.

Meanwhile, in 1943 Vickers Armstrongs were asked to investigate the possibility of installing remotely controlled barbettes in the rear of engine nacelles, aimed by a sighting station in the tail. This system was developed by Vickers who used an electro-hydraulic linking controller initially, which was later dropped in favour of the Metropolitan Vickers all-electric 'Metadyne' system. However, testing by the RAE revealed that the distortions set up in the wing structure during flight seriously affected the accuracy of gun alignment of the nacelle barbettes, and the project was terminated.

By 1944 Boulton Paul had completed the fuselage barbettes which passed their functioning tests in July. The sighting and remote control systems, however, were not so advanced, for the requirements were so complex as to require completely new technology. The designers were gradually overcoming the problems, but were not very pleased when the Air Staff made a requirement for the inclusion of AGLT radar tracking in the system.

Development proceeded until the end of hostilities in 1945. The ground and air testing was delayed when operations ceased, but the system was finally

Vickers Warwick prototype with remote barbettes mounted in the engine nacelles. The tail sighting station was similar to that of the fuselage-mounted Boulton Paul/BTH system. (Vickers)

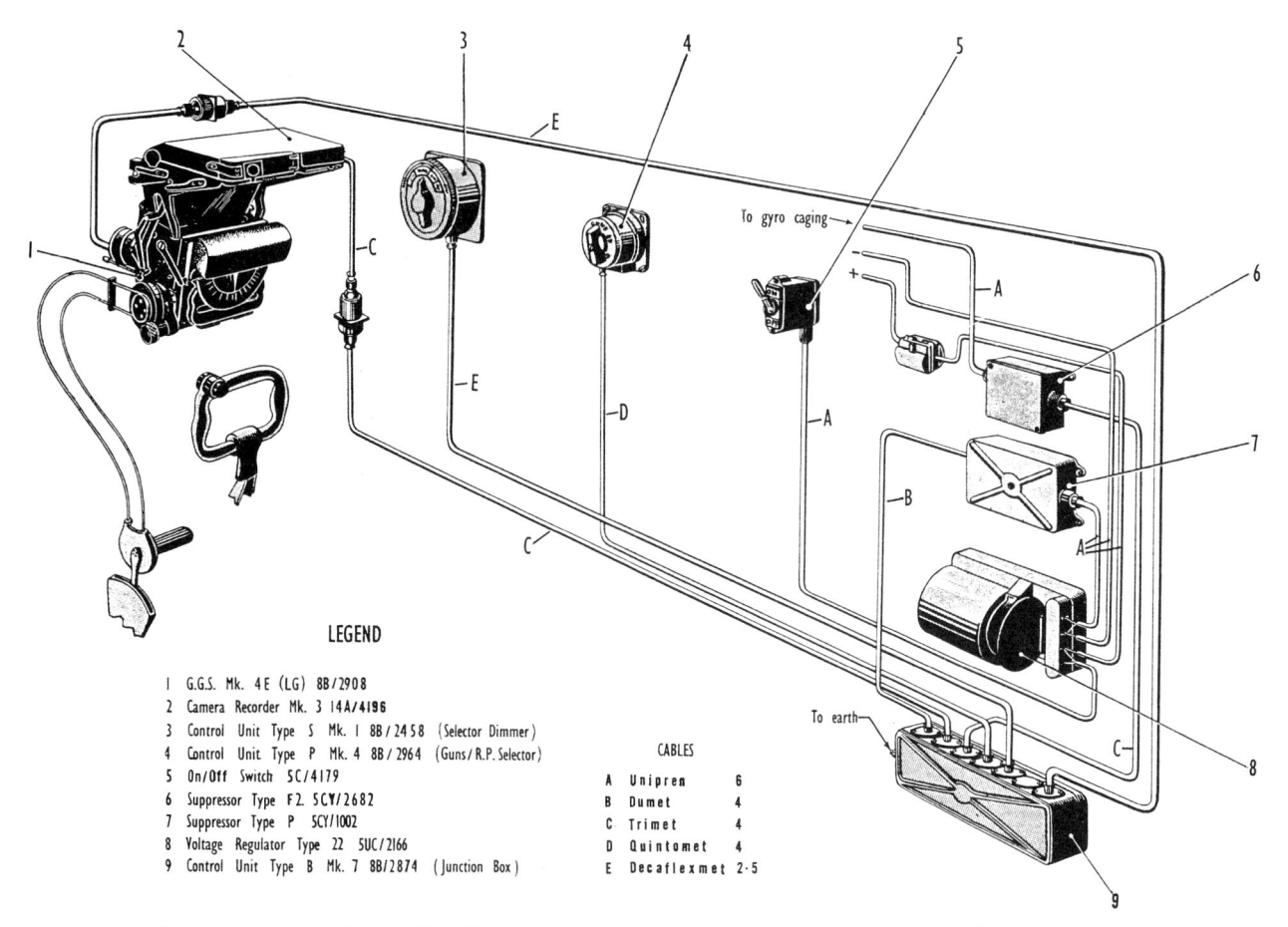

LEGEND

1. G.G.S. Mk. 4E (LG) 8B/2908
2. Camera Recorder Mk. 3 14A/4196
3. Control Unit Type S Mk. 1 8B/2458 (Selector Dimmer)
4. Control Unit Type P Mk. 4 8B/2964 (Guns/R.P. Selector)
5. On/Off Switch 5C/4179
6. Suppressor Type F2 5CY/2682
7. Suppressor Type P 5CY/1002
8. Voltage Regulator Type 22 5UC/2166
9. Control Unit Type B Mk. 7 8B/2874 (Junction Box)

CABLES

A Unipren 6
B Dumet 4
C Trimet 4
D Quintomet 4
E Decaflexmet 2·5

Above *The Mk 4E post-war fighter installation with sight recording camera fitted.*

Below *The GGS Mk 4E used on Meteor and Vampire aircraft.* (Ferranti)

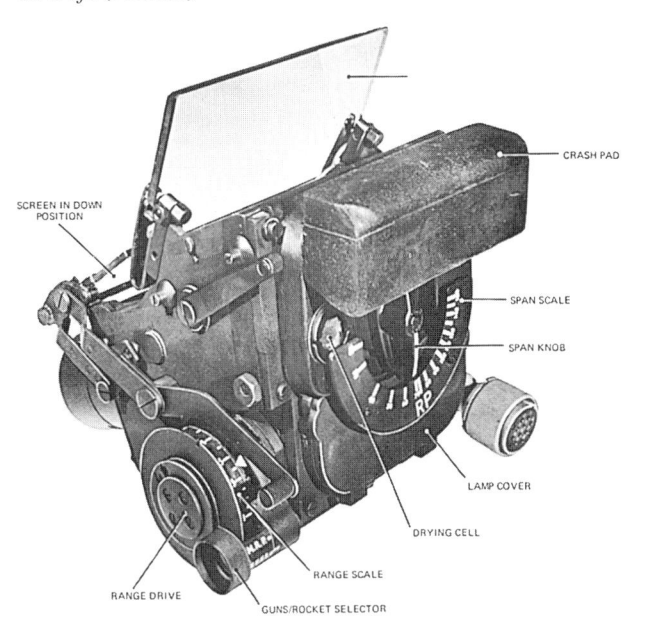

---

passed for Service use in late 1945. The system was however never adopted, for two reasons. Firstly, the development of radar sighting had reached a stage where the fire controller could carry out search and firing with the aid of radar from a central position in the aircraft; secondly, the jet bombers then being designed would not need guns for protection.

This project was the British equivalent to the American RCT system. That it was brought to a successful conclusion was a credit to all concerned, and if the war had continued there is no doubt that it would have been adopted by Bomber Command.

Post-war developments

As the European war drew to its close, production of gyro sights was switched to a Far Eastern version; instead of German types, relevant Japanese aircraft were shown on the span dial, and other small modifications were included. Existing contracts were terminated by mid-1945, but sights were soon needed for the new generation of jet fighters such as the Gloster Meteor and de Havilland Vampire. The GGS Mks 4B and 4C and Mk 5 were examples of the improved sights fitted to such aircraft. The Mk 4E was licensed to be manufactured in France by Sadir Carpentier of Paris, which was acquired by CSF, whose version equipped Dassault Mirages.

Radar sights

E. K. Cole Ltd (Ekco) was one of the main producers (others were GEC and Frazer-Nash) of airborne radar equipment during the war. The company decided to continue in this field after the end of hostilities, and in 1949 developed a radar ranging device for the Hawker Hunter and Supermarine Swift. It was similar to the AGLT, working through an output valve which drove servo mechan-

Left *A frame taken from a 16mm sight recording camera on a 4B. The target was a Meteor 8 in mock combat with another flown by F/O George Hastings from Stradishall in 1952. Hastings recalls that 'it was easy enough to get the wingspan of the target to correspond with the diamonds, but the strength of the 'g' one was pulling affected the position of the graticule. ... It took a great deal of skill to place the target accurately in the sight for as much as two seconds.'* (G. Hastings)

Below *Layout of the Mk 5. This was the first major redesign of the basic GGS, the ranging link being electrically controlled.* (Ferranti)

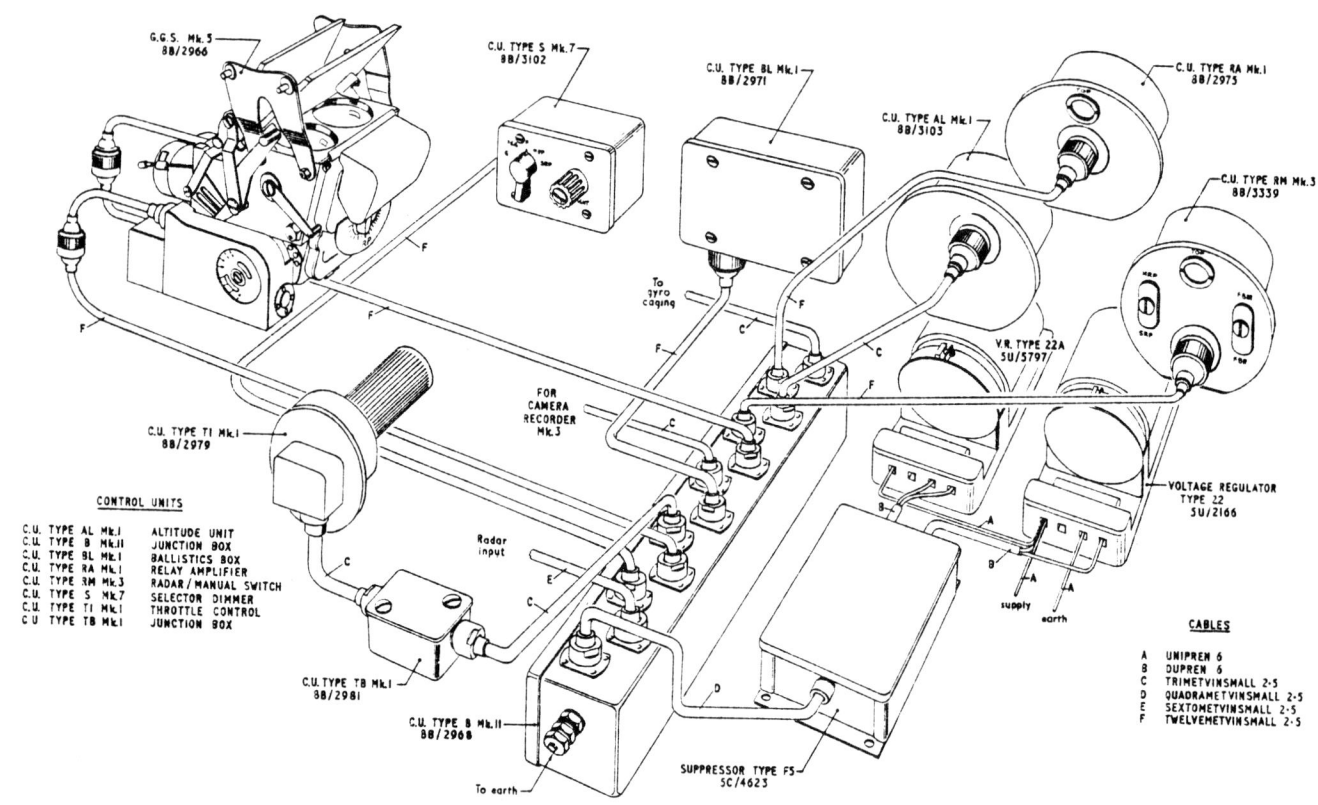

THE GUNSIGHTS

isms controlling the graticule circle of diamonds. This automatically gave the lead angle as well as the range. When the pilot set the target span and tracked it visually, the radar would keep the six diamonds round the target. The span setting was not an essential part of the process, as the range was fed directly into the predictor circuits, but it told the pilot that the radar was functioning correctly.

Other search radars were being developed of a more sophisticated nature than the simple ranging equipment. These were search and track systems, early versions of which were used by Mosquitos and Meteor NF11 aircraft. Later models were fitted to the Meteor NF14s and the Gloster Javelin all-weather fighter. However, the gyro sight continued to be the standard weapon-aiming system, and, just as in the AGLT, radar presentation was combined with the GGS screen in most cases. The arrangement in the Javelin was to display a collimated cathode-ray tube behind the combining glass of the sight so that the pilot could track the blip on the tube with the centre of the diamonds of the GGS. This was called the BPGS (Blind Positional Gyro Sighting) system, and utilized the GGS Mk 7 sight head.

Development of day fighters led to the GGS Mk 8, which was further developed to accommodate air-to-ground computation for the Hunter FGA Mk 9.

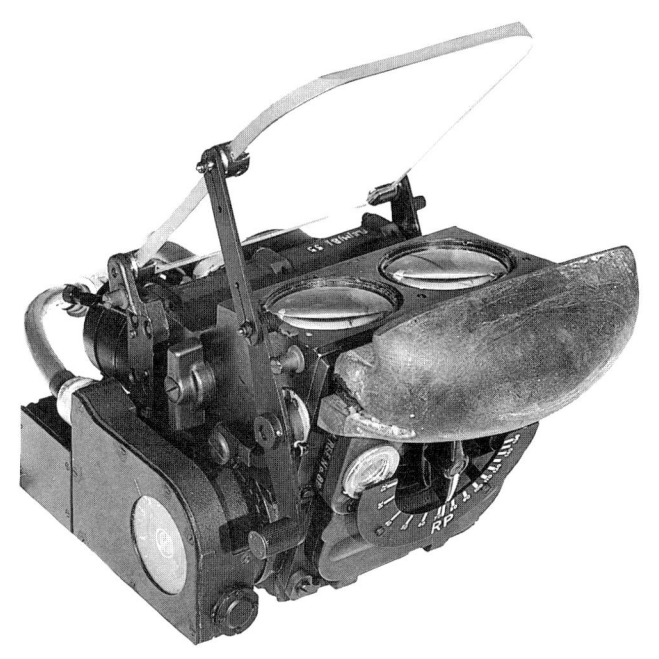

Above *The GGS Mk 7, used on the Gloster Javelin. The sight head had a search and track radar input.* (Ferranti)

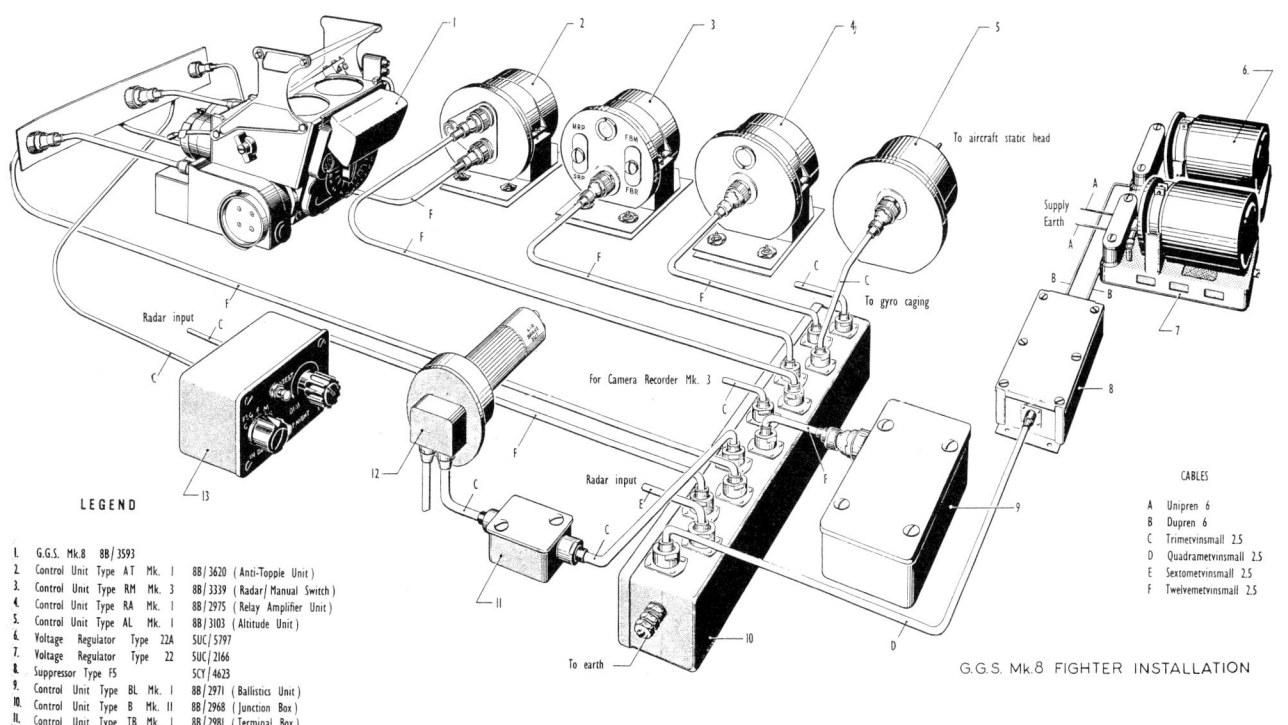

Below *Layout of the Mk 8 fighter installation, developed to accommodate air-to-ground computation for the Hunter FGA Mk 9.* (Ferranti)

LEGEND

1. G.G.S. Mk.8 — 8B/3593
2. Control Unit Type AT Mk. 1 — 8B/3620 (Anti-Topple Unit)
3. Control Unit Type RM Mk. 3 — 8B/3339 (Radar/Manual Switch)
4. Control Unit Type RA Mk. 1 — 8B/2975 (Relay Amplifier Unit)
5. Control Unit Type AL Mk. 1 — 8B/3103 (Altitude Unit)
6. Voltage Regulator Type 22A — 5UC/5797
7. Voltage Regulator Type 22 — 5UC/2166
8. Suppressor Type F5 — 5CY/4623
9. Control Unit Type BL Mk. 1 — 8B/2971 (Ballistics Unit)
10. Control Unit Type B Mk. II — 8B/2968 (Junction Box)
11. Control Unit Type TB Mk. 1 — 8B/2981 (Terminal Box)
12. Control Unit Type TI Mk. 2 — 8B/3231 (Throttle Unit)
13. Control Unit Type S Mk. 8 — 8B/3733 (Selector Dimmer Unit)

CABLES
A — Unipren 6
B — Dupren 6
C — Trimetvinsmall 2.5
D — Quadrametvinsmall 2.5
E — Sextometvinsmall 2.5
F — Twelvemetvinsmall 2.5

G.G.S. Mk.8 FIGHTER INSTALLATION

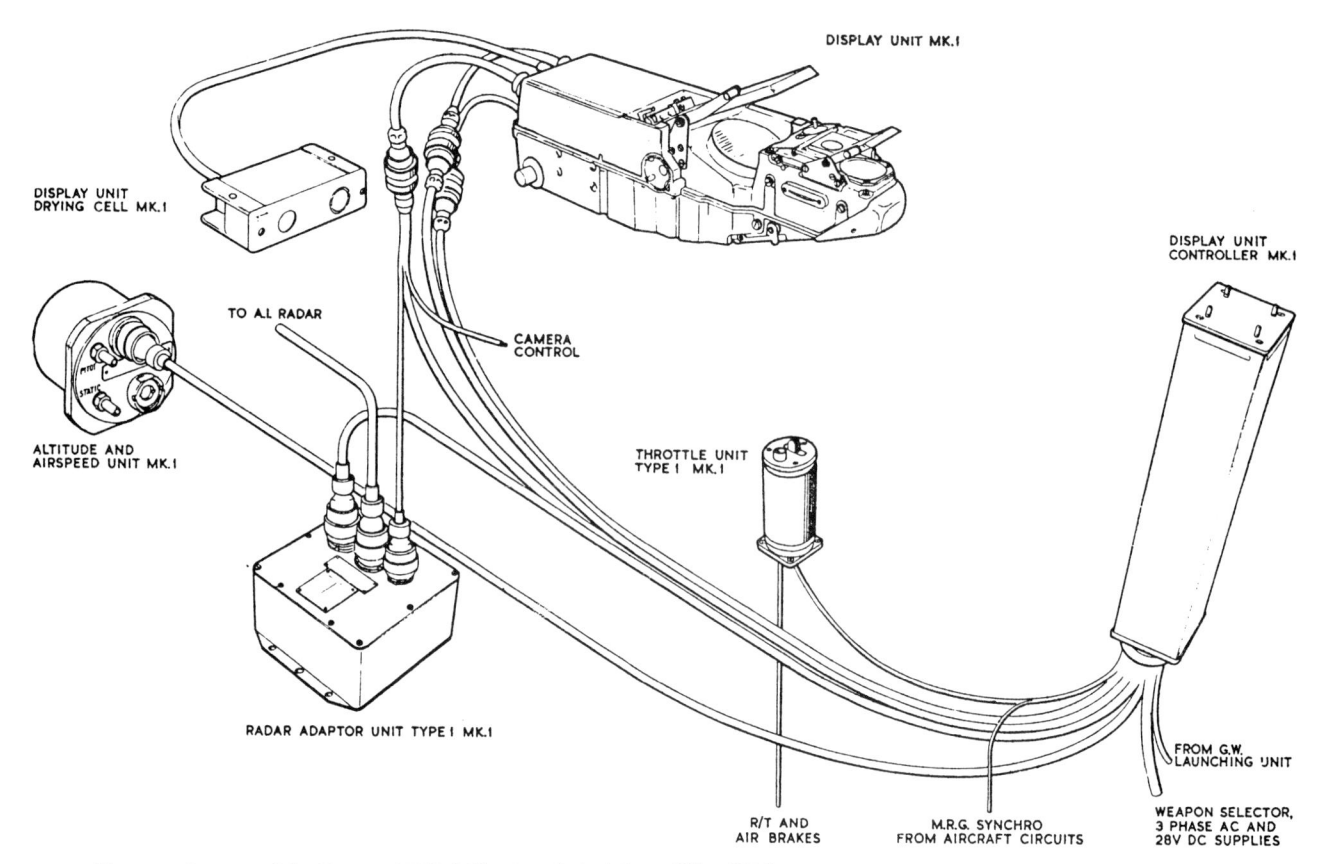

Above *Layout of the Ferranti PAS (Pilot Attack Sight) incorporating radar target indicating and blind attack facility.* (Ferranti)

Below *AIRPASS (Airborne Interception Radar PAS System) introduced the first monopulse radar for supersonic interceptors, using the PAS sight head.*

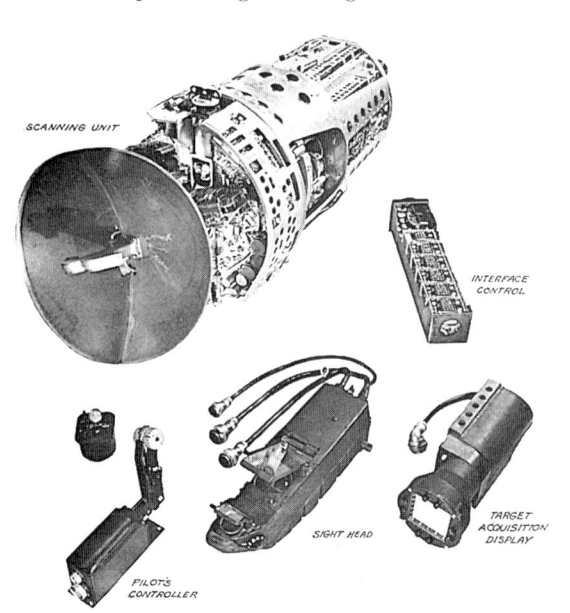

The PAS system

In 1957 Ferranti produced the Pilot Attack Sight. This was a visual (head-up) gyro sight for interceptors, designed for pursuit-type attacks with guns, rockets or guided missiles. It incorporated a radar target indicator presentation which permitted the approach to be made under blind conditions. Such an approach could lead (with a suitable weapon) to a blind attack, or be converted into a visual mode when the target was in view.

PAS Mk 1 could deal with speeds of up to Mach 2 at up to 70,000 ft, being designed for the ADEN gun firing at ranges of 200–1,200 yd (183–1,097 m). Once radar lock-on was established, target position and range were continually displayed, a winged indicator was tracked, and when the firing position was reached firing bracket indicators appeared, followed by a break-off signal. Should radar lock-on be lost, the closing-speed scale was replaced by a circle of diamonds, so that if visual contact was made the PAS became a conventional gyro sight.

AIRPASS

In 1954–8 Ferranti produced the first mono pulse radar for supersonic interceptors. AIRPASS (Airborne Interception Radar PAS System) combined this radar with the Pilot Attack Sight. Combining the cathode-ray display with an optical sight head in

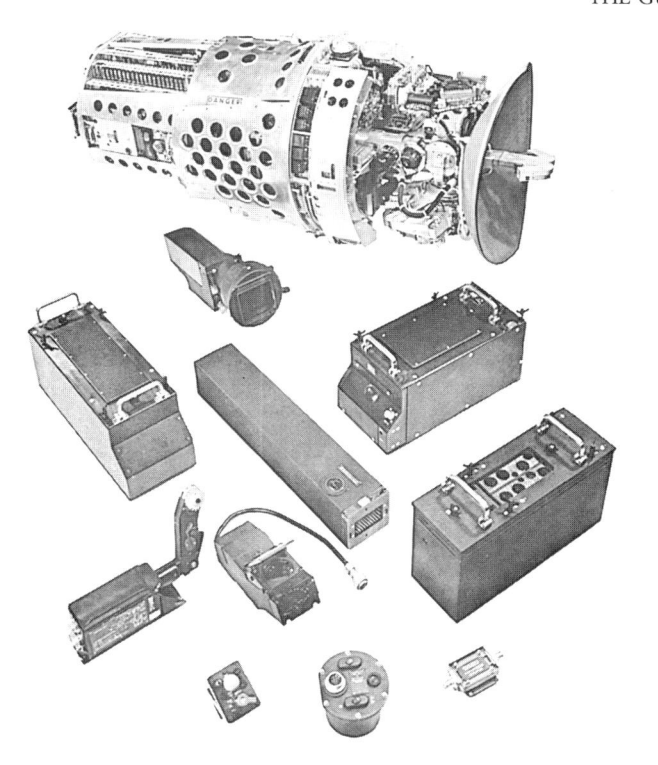

THE GUNSIGHTS

The LFS (Light Fighter Sight) designed for aircraft such as the Gnat and Strikemaster. It retained the basic gyro system of the PAS system.

AIRPASS Mk 2 components. This system enabled the pilot to carry out an attack without seeing the target.

the Javelin had not been very satisfactory, due to the difficulty of tracking the CRT aiming mark with the GGS graticule. In Airpass the scanner located the target and was locked on, elevation and azimuth differences between the aircraft and scanner axes being transformed into angular deflections of the small gyro in the sight head.

Head-down interception

In the early 1960s fighter design tended to fall into two classes. The first was the all-weather fighter with the latest blind-flying aids and search and track radar system, the second was the light fighter designed primarily for the ground attack role. Ferranti designed the Airpass 2 for the all-weather fighter. This provided a blind tracking lock-on facility in which the pilot could carry out the attack without seeing the target, and was known as the head-down interception. For the second, light fighter, category the company produced the Light Fighter Sight (LFS). This retained the basic gyro of the PAS system and was used on aircraft such as the Fiat G91 and the Folland Gnat. Later versions were fitted to the BAC Strikemaster and Aermacchi MB 326.

ISIS

From 1960 Ferranti developed the ISIS (Integrated Strike and Interception System) for air-to-ground operations. ISIS N of 1962 gave the pilot roll and pitch information, air data computers giving altitude and airspeed. This was first supplied to the Canadian Defence Forces for Northrop CF-5s. The ISIS N was followed by the ISIS B and F, which were tailored for specific aircraft and were supplied with various sight heads and computers. These systems achieved world-wide sales, and when the company perfected the laser designator ranging system, a new range of sights was produced. These were the

Cut-away of the ISIS N sight head. (Ferranti)

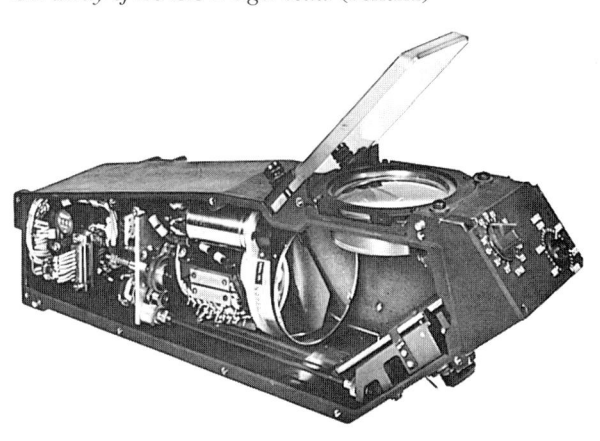

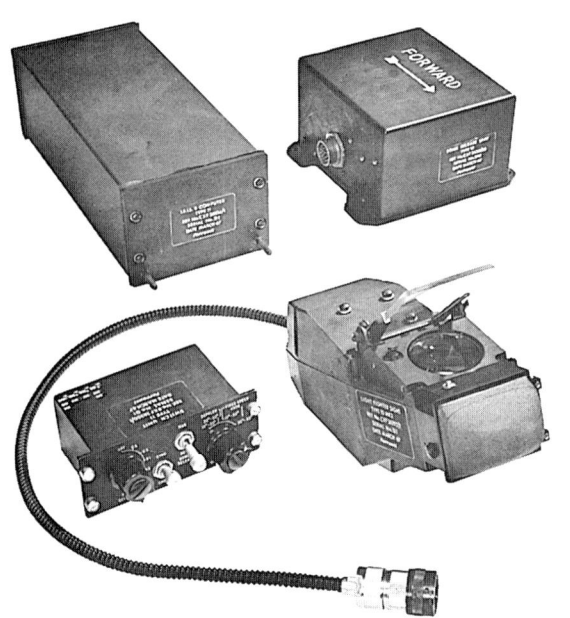

Above *The ISIS N (Integrated Strike and Interception System) installed in a CSF aircraft.* (Ferranti)

ISIS D and K which co-ordinated the laser ranging with two graticules, one indicating the aircraft axis and the other (steerable) controlling the laser, which steered the gyro: the pilot flew the aircraft to bring the two graticules together. The sight head then gave the pilot a computed impact point, allowing for the necessary depression angle. Proof of the accuracy of ISIS D-126R came in 1982 when Skyhawks of the Argentine Navy scored many direct hits on Royal Navy ships during the South Atlantic campaign. If the bombs had been correctly fused, many more ships and men would have been lost.

Left *The ISIS B installation.* (Ferranti)

THE GUNSIGHTS

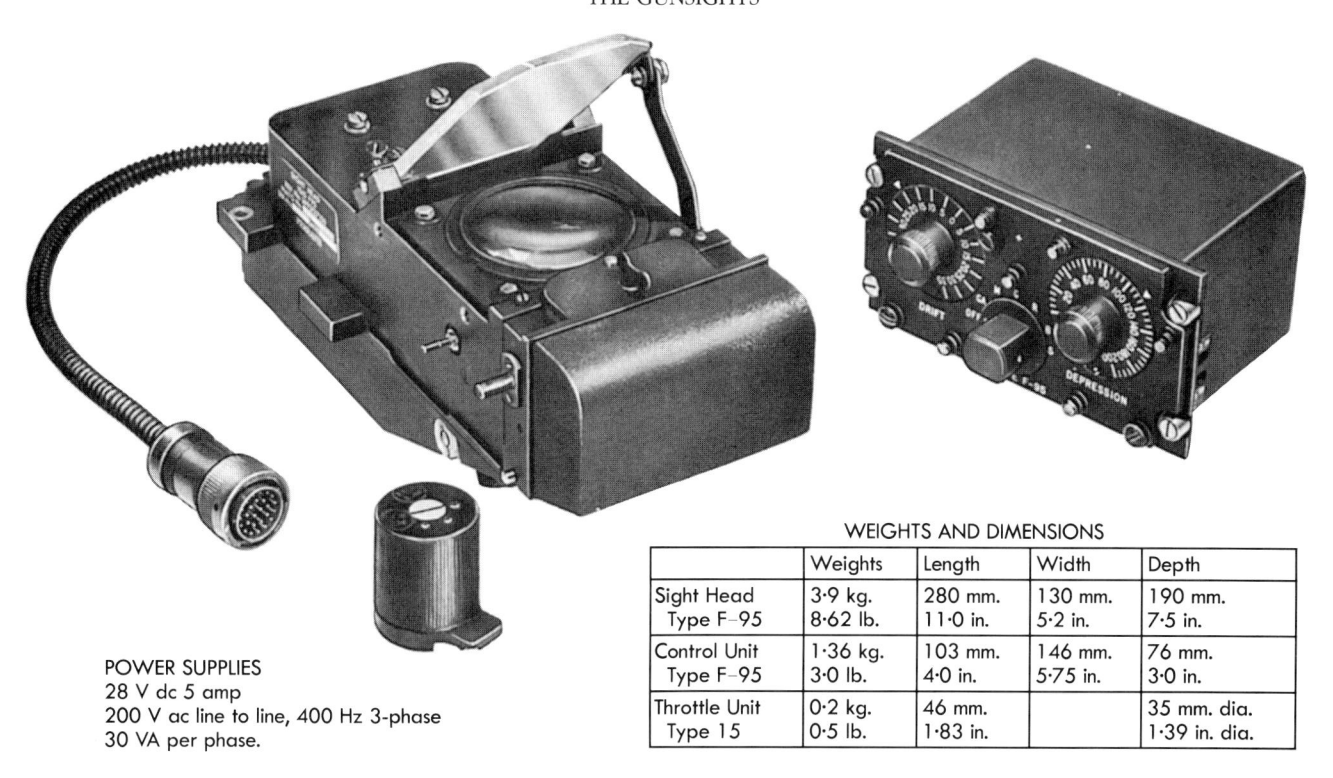

WEIGHTS AND DIMENSIONS

	Weights	Length	Width	Depth
Sight Head Type F-95	3·9 kg. / 8·62 lb.	280 mm. / 11·0 in.	130 mm. / 5·2 in.	190 mm. / 7·5 in.
Control Unit Type F-95	1·36 kg. / 3·0 lb.	103 mm. / 4·0 in.	146 mm. / 5·75 in.	76 mm. / 3·0 in.
Throttle Unit Type 15	0·2 kg. / 0·5 lb.	46 mm. / 1·83 in.		35 mm. dia. / 1·39 in. dia.

POWER SUPPLIES
28 V dc 5 amp
200 V ac line to line, 400 Hz 3-phase
30 VA per phase.

The ISIS F 195, similar to the ISIS B but giving a fully computed aiming point for guns, bombs or rockets. (Ferranti)

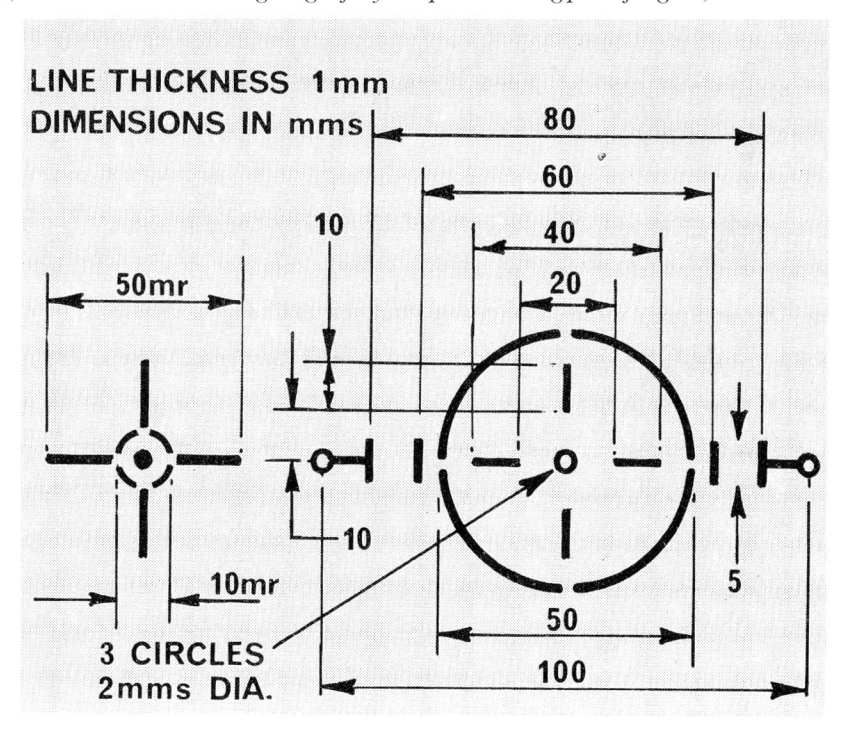

LINE THICKNESS 1mm
DIMENSIONS IN mms
80
60
40
20
10
50mr
10
10mr
5
50
100
3 CIRCLES
2mms DIA.

SIGHT RETICLES

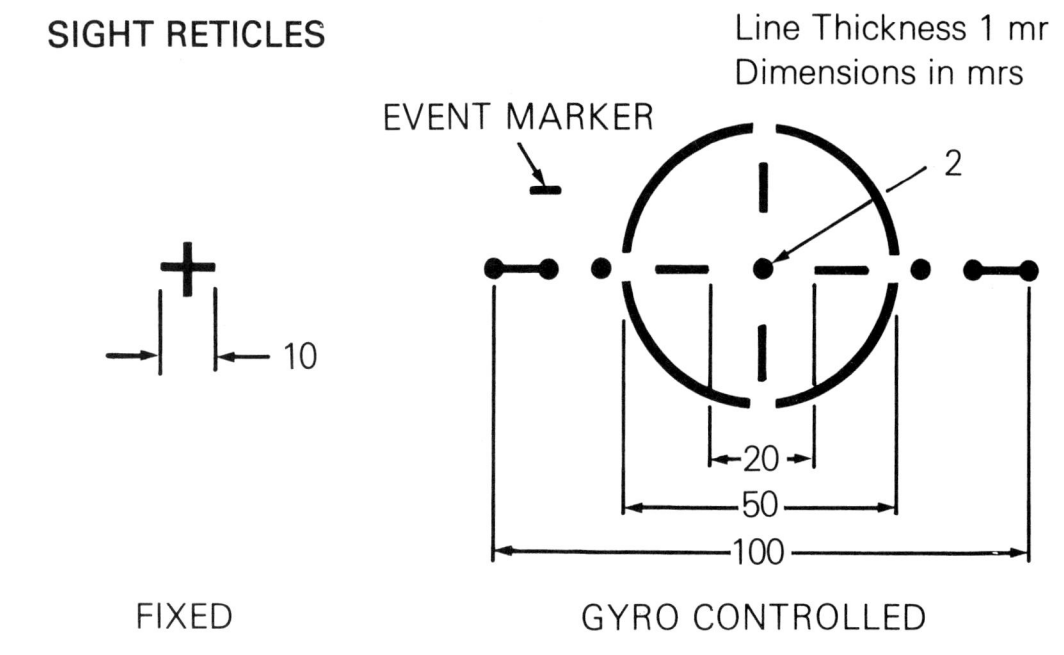

Line Thickness 1 mr
Dimensions in mrs

EVENT MARKER

2

10

−20−
−50−
−100−

FIXED

GYRO CONTROLLED

Above *Sighting graticule of the ISIS D 126R.*

Below *The ISIS D 150R system, which can provide lead-computed head-up aiming marks for most air-delivered weapons. Six weapon types could be selected with three pre-set range options. It was used by the Argentine Navy against the Royal Navy with devastating accuracy.* (Ferranti)

SIGHTHEAD D 09 RM

ELECTRONIC UNIT

THROTTLE UNIT

ISIS D 209RM, used on the Hunter. (Ferranti)

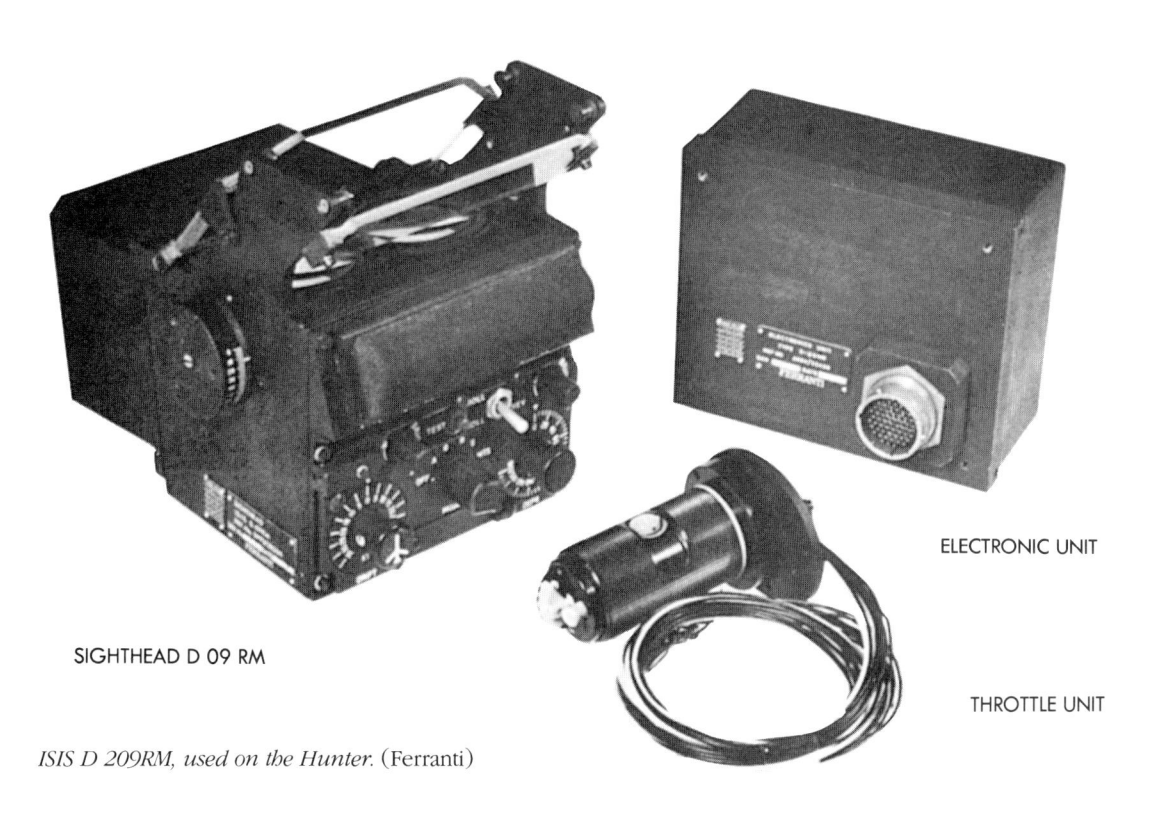

ISIS systems are still used by many of the world's air forces. The list below of users from 1968 to 1980 gives some idea of their widespread use:

Aircraft	System	Customer	First order
Canadair CF 5A	ISIS N	Canada	1968
Fiat G 91Y	ISIS B	Italy	1974
Northrop NF 5A	ISIS F-195R	Norway	1971
Aeromacchi 326G	ISIS F-126	Australia	1971
HAL HF 24	ISIS F-124	India	1971
SAAB 105	ISIS F-105	Austria	1971
MDC/LAS A-4S	ISIS D-101	Singapore	1972
LAS A-4C	ISIS D-126R	Argentina	1974
BAe Hawk	ISIS D-195R	United Kingdom	1974
HAL Ajeet	ISIS F-195R-3	India	1975
SAAB 105G	ISIS F-126-3	Austria	1976
Mostar Galeb	ISIS D-282-2	Yugoslavia	1978
BAe/LASS Hunter GA9	ISIS D-209RM	Singapore	1978
MIG 21 FL (HAL)	ISIS F-195R	India	1978
BAe Hawk	ISIS D-195R-3	Kenya	1978
BAe Hawk	ISIS D-195R-3	Indonesia	1979
HAL Kiran trainer	ISIS F-195R-3	India	1979
HAL Ajeet trainer	ISIS F-195R-3	India	1979
MDC A-4E	ISIS D-195	Indonesia	1980
MDC A-4C	ISIS D-101	Singapore	

Total orders equivalent to 1,400 systems (Ferranti Ltd, Edinburgh).

Head-up display

During the 1950s it was realized that future strike aircraft would have to attack at virtually ground level. To fill this role the Air Staff issued a specification for such an aircraft. It would have to be capable of pinpoint navigation at very low level, high-speed weapon aiming and delivery, and have a head-up display giving flight and operational information.

The first aircraft to fill this role was the Blackburn (Hawker Siddeley) Buccaneer, which was designed specifically for low-level operations to avoid radar detection, and was specially strengthened to withstand the stresses and buffeting of these sorties. It entered service with the Royal Navy in 1961, and was the first operational aircraft fitted with a HUD (head-up display).

The aircraft which the Air Staff hoped would serve as an effective deterrent in the coming decades was the TSR.2. It was to be the ultimate in low-level strike capability, all necessary flight and weapons information being displayed on a HUD developed by Marconi Avionics and Smiths Industries. An inertial velocity navigation system measured the flight parameters and fed them into a computer, the computer then processed this information into a form which could be displayed on a high-brightness CRT projected onto the HUD. To the chagrin of the Air Staff the TSR.2 project was cancelled in 1965 when the work was well under way. The Government had been advised that US Swing Wing GD IIIs would be more cost-efficient. Ironically, the contracts for the General Dynamics 111s were cancelled in 1968, and as a stop-gap measure Buccaneers were ordered for RAF strike squadrons.

NAVWASS

The expertise gained on TSR.2 led to the production of HUD systems for the American A-7D, A-4M, F-16 and many other aircraft. Meanwhile, the RAF Jaguar and Harrier were fitted with NAVWASS (Navigation and Weapon-Aiming Sub-System). This enabled the pilot to reach his target by a series of dog-leg courses, with barometric and radar height, compass headings, airspeed, vertical speed and angle of attack all presented on the reflector glass (which became known as the combining glass for obvious reasons). The weapon-aiming mode of the system is typical of many similar systems, being primarily concerned with the accurate release of bombs and other free-fall stores. The pilot was also given the point of aim for rockets, guns and guided missiles where appropriate, with allowances made for ballistics and height.

The NAVWASS of a Jaguar set to air-to-air mode.

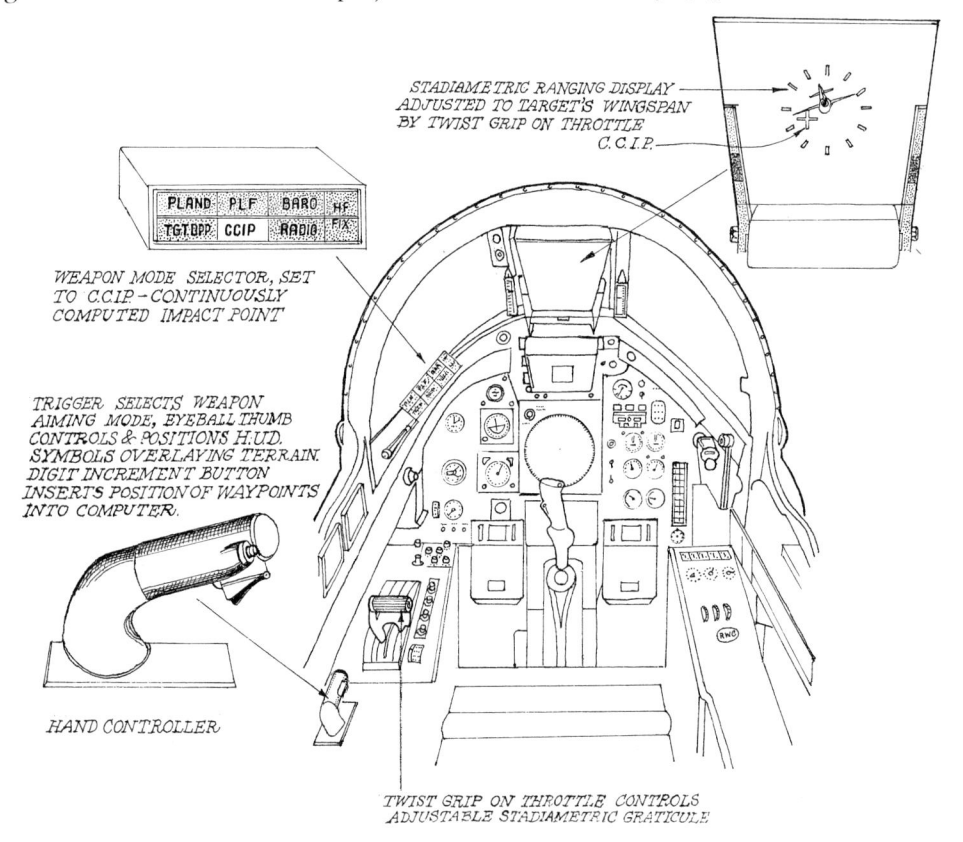

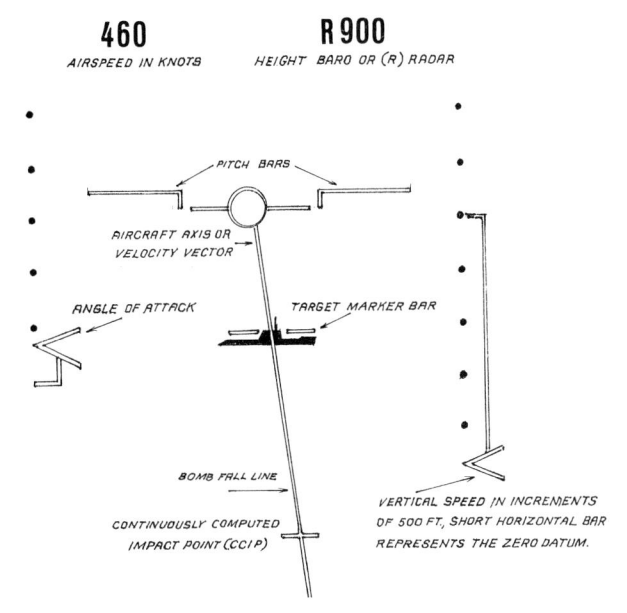

THE GUNSIGHTS

460 AIRSPEED IN KNOTS
R 900 HEIGHT BARO OR (R) RADAR

PITCH BARS
AIRCRAFT AXIS OR VELOCITY VECTOR
ANGLE OF ATTACK
TARGET MARKER BAR
BOMB FALL LINE
CONTINUOUSLY COMPUTED IMPACT POINT (CCIP)
VERTICAL SPEED IN INCREMENTS OF 500 FT, SHORT HORIZONTAL BAR REPRESENTS THE ZERO DATUM.

Above *In a target of opportunity attack the pilot lines up the target with the bomb fall line, and releases the bomb when the CCIP (Continuously Computed Impact Point) crosses the target.*

Below *GEC Avionics HUD in dogfight mode. The pilot can anticipate the target's movement, having a boresight cross, a missile diamond and a tracer line as well as closure rate and a 1-sec range indication. (GEC)*

On visually acquiring the ground target, the pilot selects 'guns' or 'rockets' on his weapon control panel. He then sets the HUD WS/DEP (wingspan/depression) knob to A/G (air to ground). The weapon mode selector is then set to CCIP (continuously computed impact point) and a 'pipper' appears in the centre of the aircraft symbol representing the CCIP for a projectile flight of 2 sec. The boresight cross on the HUD represents the aircraft's longitudinal axis. As the pilot tracks the target with the pipper, the depression reduces when the projectile time falls below 2 sec. The pipper then indicates the true impact point; using the stick trigger or commit button, the pilot fires the guns or rockets.

During air-to-air combat the same panel and aiming mode selections are made, with the target span set into the WS/DEP. On selection of CCIP a stadiametric ranging display appears on the HUD, all other symbols, with the exception of the angle of attack and the cross, being deleted. Ranging is carried out with a twist grip on the throttle control, as on the GGS, with which a circle of radial bars is made to frame the target. The computer then calculates gravity drop and deflection. Though the GGS and HDU are similar in stadiametric input, the accu-

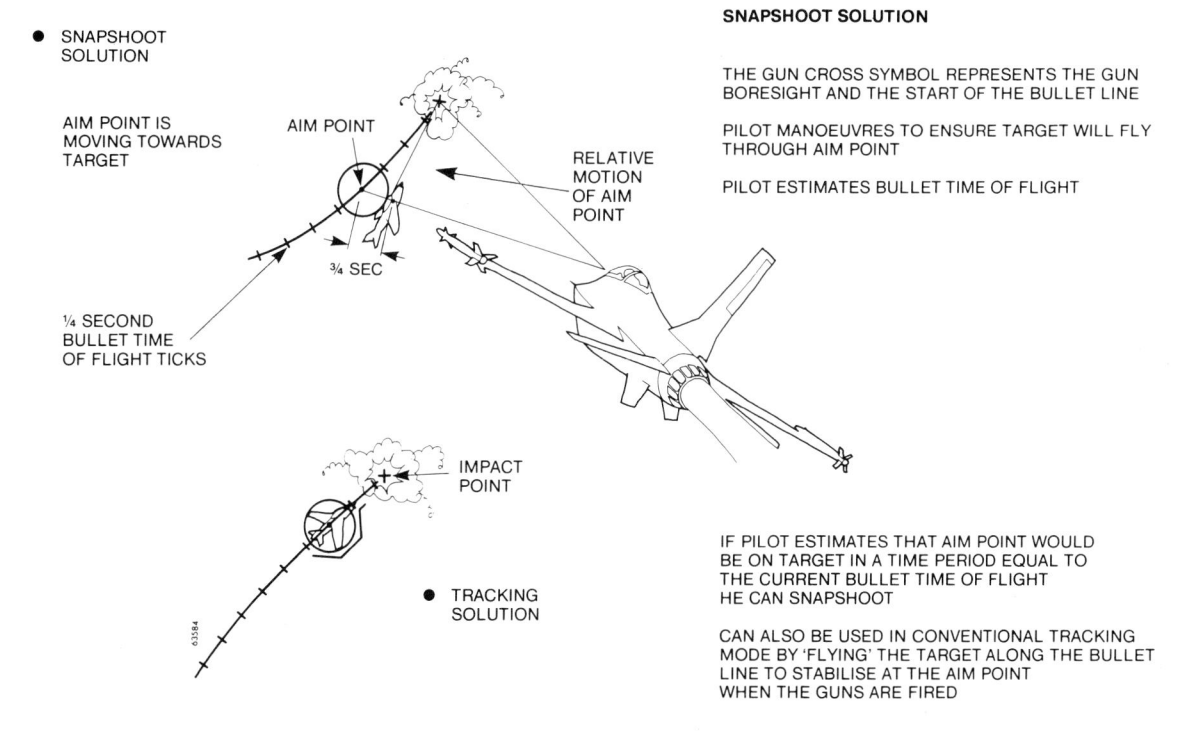

- SNAPSHOOT SOLUTION
AIM POINT IS MOVING TOWARDS TARGET
AIM POINT
¾ SEC
RELATIVE MOTION OF AIM POINT
¼ SECOND BULLET TIME OF FLIGHT TICKS
IMPACT POINT
- TRACKING SOLUTION

SNAPSHOOT SOLUTION

THE GUN CROSS SYMBOL REPRESENTS THE GUN BORESIGHT AND THE START OF THE BULLET LINE

PILOT MANOEUVRES TO ENSURE TARGET WILL FLY THROUGH AIM POINT

PILOT ESTIMATES BULLET TIME OF FLIGHT

IF PILOT ESTIMATES THAT AIM POINT WOULD BE ON TARGET IN A TIME PERIOD EQUAL TO THE CURRENT BULLET TIME OF FLIGHT HE CAN SNAPSHOOT

CAN ALSO BE USED IN CONVENTIONAL TRACKING MODE BY 'FLYING' THE TARGET ALONG THE BULLET LINE TO STABILISE AT THE AIM POINT WHEN THE GUNS ARE FIRED

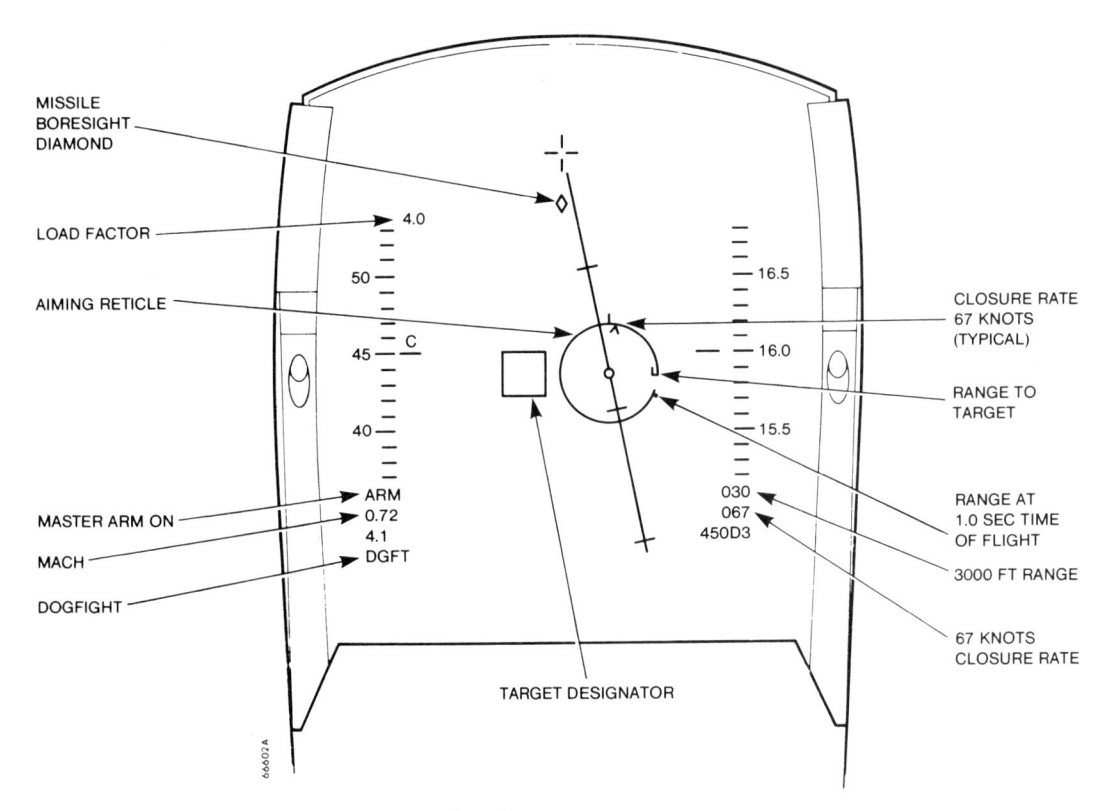

Above *GEC HUD display in dogfight mode.* (GEC)

Below *The elaborate optics of a GEC LANTIRN HUD in an A10 Warthog.* (GEC)

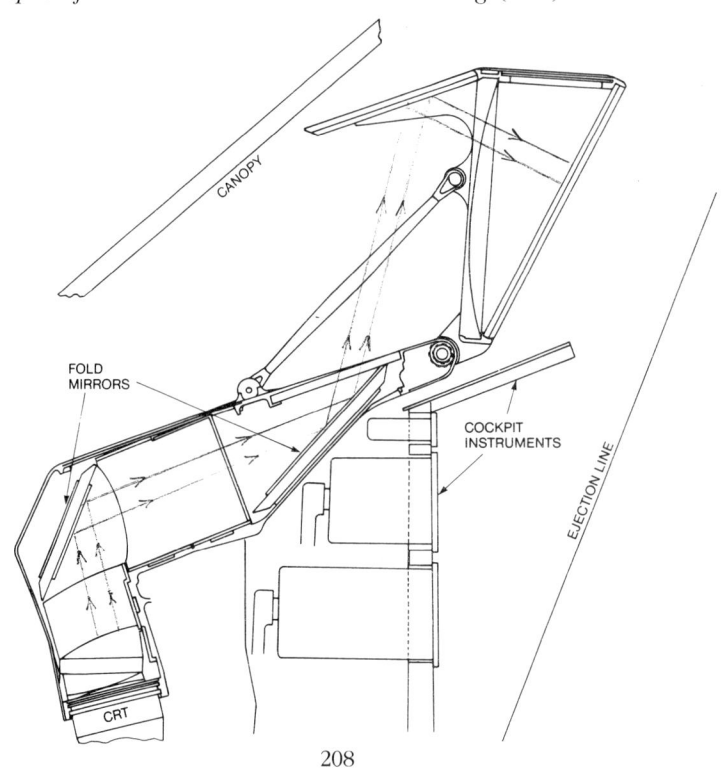

Pilot's view whilst tracking a target on a GEC PERIHUD of an F16 Falcon. (GEC)

racy of the electronic system is much superior. This is because the computer stores the firing envelope of ammunition or rockets, and it can also give an allowance for three-axis angular rates, bullet jump angle and gun offset from the aiming line.

NAVWASS has now been improved upon. When Marconi Avionics (now GEC Avionics) designed the HUD for the F-16, one requirement of the USAF was a snapshoot, or continuously computed impact line, showing where projectiles fired at various intervals in the previous few seconds would be now. This facility was requested after investigation into dogfight weapon-aiming demands. Using this system, the pilot could anticipate the enemy's movements during an engagement in which his own and the target's movement were always changing. He is therefore presented with a boresight cross, missile diamond, tracer line and aiming mark, as well as closure rate and 1-sec range indication.

The F-16-C and A-10 HUD systems produced by GEC Avionics are designed to suit the Martin Marietta LANTIRN (Low Altitude Navigation and Targeting Infra Red for Night) specification. Night operations require a maximum FOV (field of view), air-to-ground operations in particular needing a good viewing angle. The HUD reflective optics give a FOV of $30° \times 20°$, $10°$ wider than normal. The F-16-C HUD employs folding and holographic means to give a compact configuration. Another optical system used by GEC is the PERIHUD, whose combining glass contains beam-splitting surfaces which allow the pilot to 'look around' the screen.

Smiths Industries of Cricklewood have an impressive record for HUD systems. No less impressive is the company's record for the development and production of instruments for the RAF. As early as 1933 the company produced the type 704 Flight Recorder, which gave a cinematograph record of flight parameters – one of the first 'black box' systems. Smiths supply most of the HUDs for the RAF (the NAVWASS HUD was a company product) and the Harrier, Jaguar, Tornado and Euro Fighter 2000 all use forms of the Smiths HUDWAC (Head-Up Display Weapon Aiming Computer) system. For example, the type 1101 HUD for the Harrier GR3 and Jaguar GR-1 is very compact but uses a dual combining glass which gives an increased FOV in elevation. All the glass surfaces have an anti-reflective coating which reduces unwanted reflections from outside sources. The brightness of the

BRITISH AIRCRAFT ARMAMENT

THE GUNSIGHTS

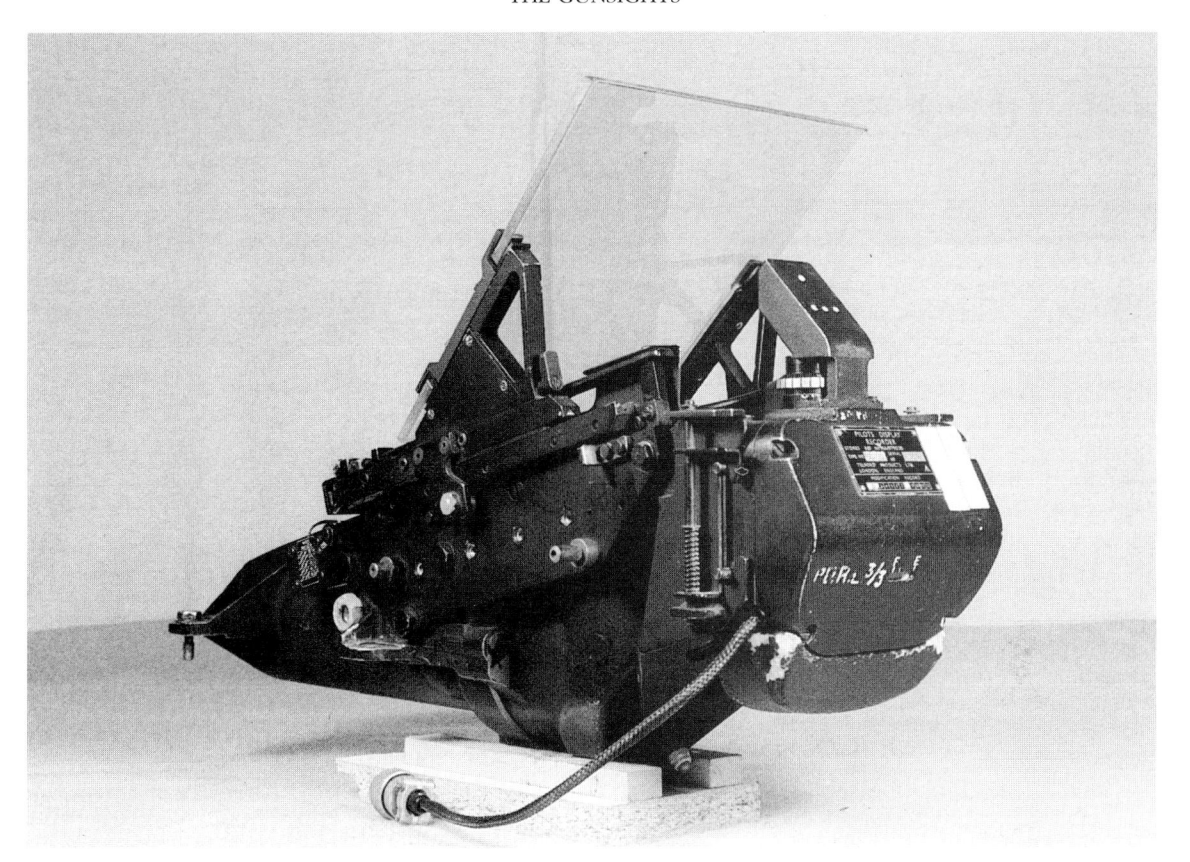

Above *The Smiths Type 1101 sight head with sight recording camera fixed in position.* (AGGAS)

Left *A Harrier cockpit showing the Smiths Type 1101 HUD, which was used with such success in the South Atlantic.* (Smiths Industries)

Right *Tornado front cockpit with the Smiths HUD giving the pilot weapon-aiming computing for every situation.* (Smiths Industries)

display is set by the pilot; thereafter, a solar cell maintains a constant contrast between the display and the background. A mounting bracket at the rear accepts either a recording camera or crash pad. The anode of the CRT takes 15,000 volts, being supplied by an extra high-tension unit of Smiths design.

There is a large potential market for such units to replace the electro-mechanical gyro-based systems. With an up-to-date weapon delivery and air-to-air gunnery facility, expensive aircraft replacement can be deferred. Retrofit conversions can be lucrative, and prospective customers are subjected to competitive offers from the major manufacturers. The UK enjoys a strong position in the aircraft weapon-aiming field.

BRITISH AIRCRAFT ARMAMENT

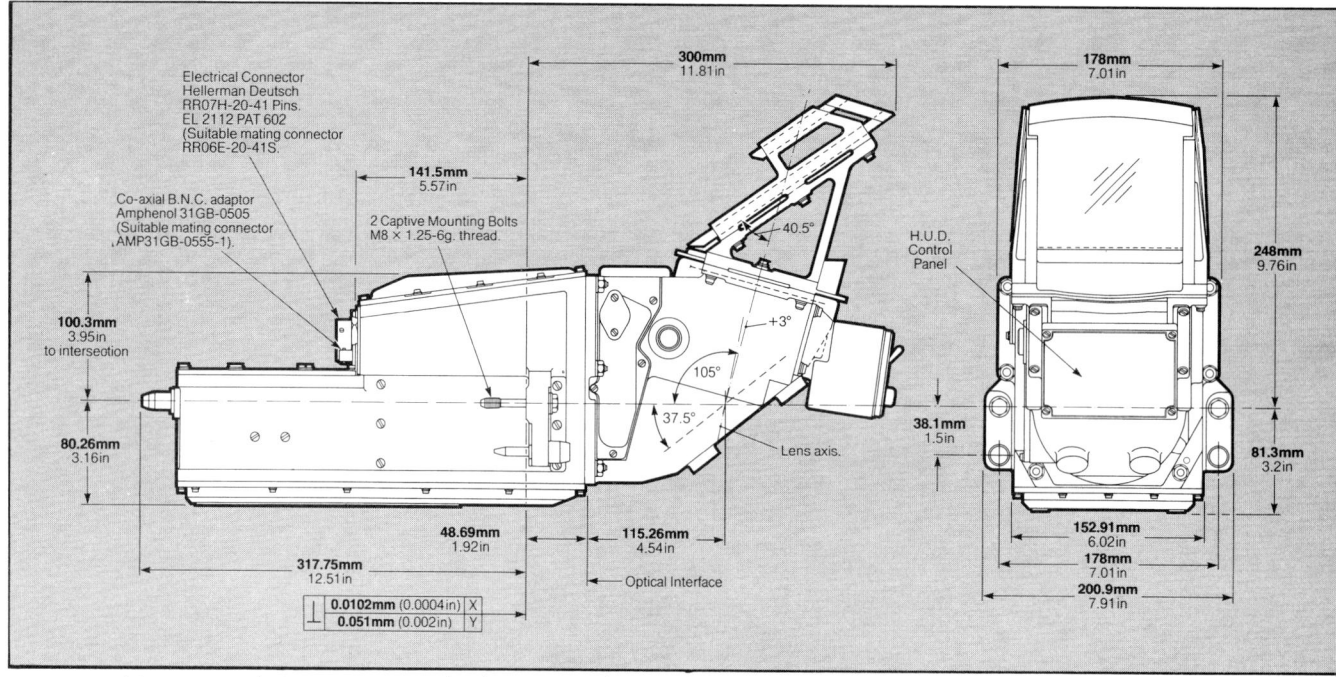

This page *The Type 1203 HUD for the BAE Hawk 200. This gives all air-to-air and air-to-ground weapon-aiming, having provision for a TV camera or pilot's display recorder.* (Smiths Industries)

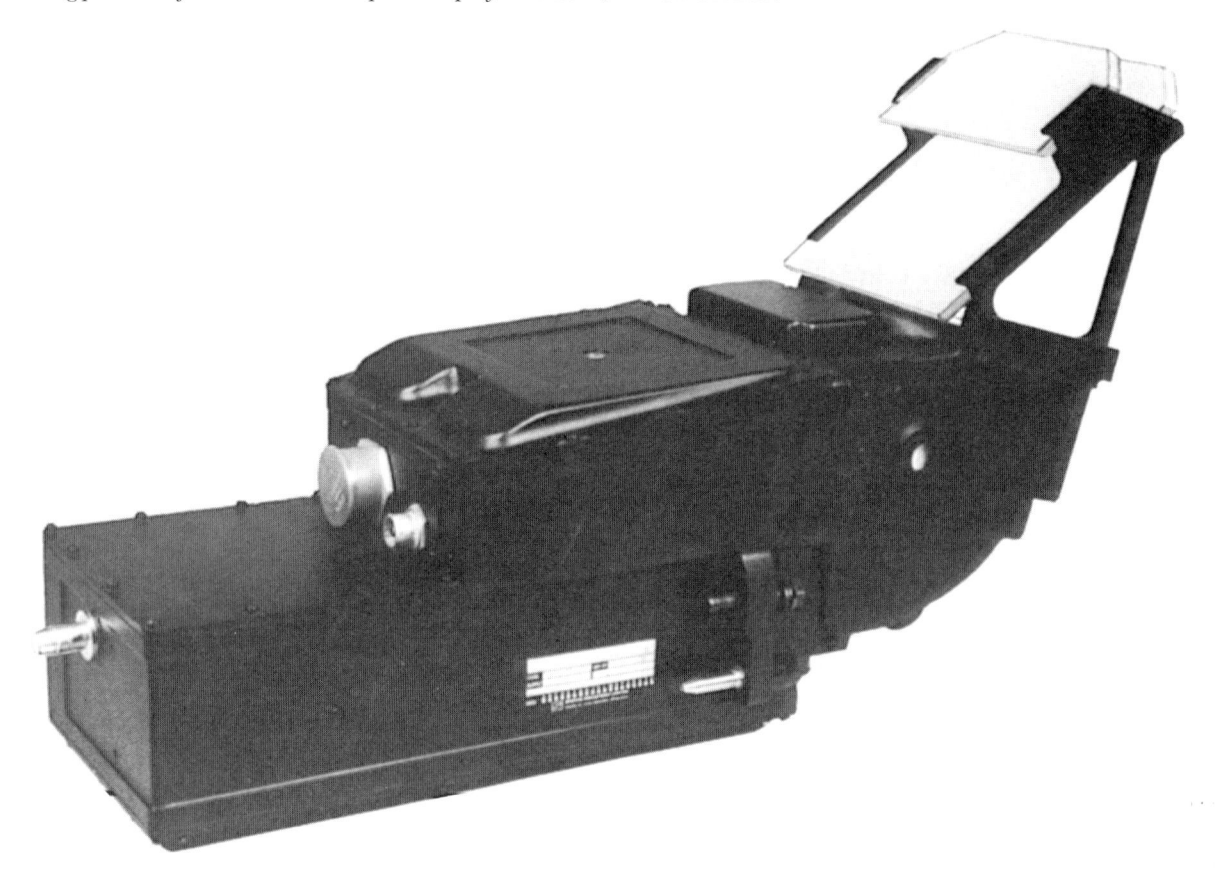

THE GUNSIGHTS

Above *The Smiths Type 1101 HUD, which is in widespread use world-wide.* (Smiths Industries)

Above *The Type 1202 was designed to US specifications for the McDonnell Douglas AV-8B. It features a dual combining glass, and the system is designed for easy replacement of the modular sections. Provision is made for a side-mounted recording camera.* (Smiths Industries)

Below *A moonless night view through SI low-light diffractive optics HUD in a Sea Harrier.* (Smiths Industries)

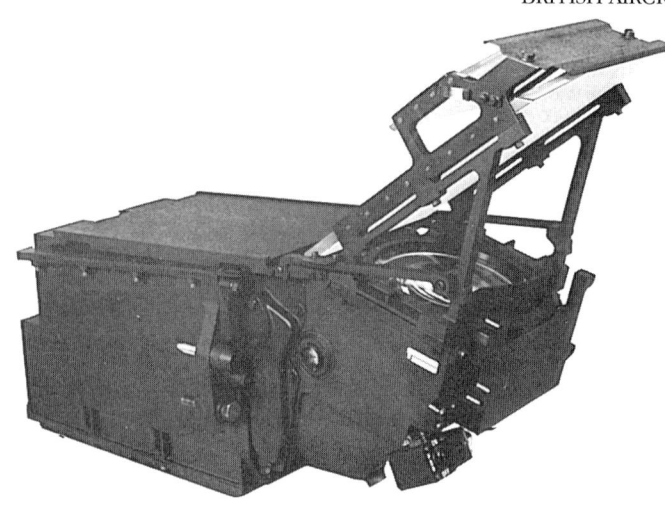

The Type 1301 is one of the most versatile HUDs currently available. In addition to the latest symbology, it gives a high 625-line day or night picture of the field of view. The lens unit consists of input and output groups giving high definition in all modes, with a quartz halogen lamp for a standby graticule. Provision is made for a variety of display recording cameras and a day/night filter assembly. (Smiths Industries)

Helmet-mounted sights

The ultimate sighting system is surely one in which the gunner has merely to look at the target to align the barrel of a turret gun or direct a missile at it. This has been achieved by the HMS (Helmet-Mounted Sight). Honeywell produced an HMS for US Navy F-4 fighters. This system, known as the Visual Acquisition Set, controls the radar antenna and seeker heads of Sidewinder missiles. The leading UK company in this field is Ferranti. The Ferranti HMS presents the operator with an aiming mark projected onto his helmet visor by a small cathode-ray tube. He can also be given cueing arrows directed from an external source. Ferranti also produce a helmet-mounted package incorporating a high-resolution CRT with optics giving a wide FOV for more complex computer and sensor imagery. With helicopters being used in increasingly offensive roles, the HMS is being used to control helicopter turrets such as the unit produced by Lucas, which is armed with a 0.5 Browning gun. Briefly, the Ferranti HMS operates as follows: A small (1 in (25 mm) cube) radiating element is fitted to the cockpit roof. This contains coils producing magnetic fields. On the operator's helmet is a small sensor containing coils which produce signals dependent on their orientation relative to the radiator. The sensor data is processed by the electronics unit, which determines the observer's line of sight angles; this information is then relayed to the powered controls of the turret, keeping the barrels slaved to the operator's visor. In later versions of this system the HMS controls a laser beam which is used to direct guns or missiles to the target.

Schematic layout of the Ferranti Helmet-Mounted Sight (HMS). (Ferranti)

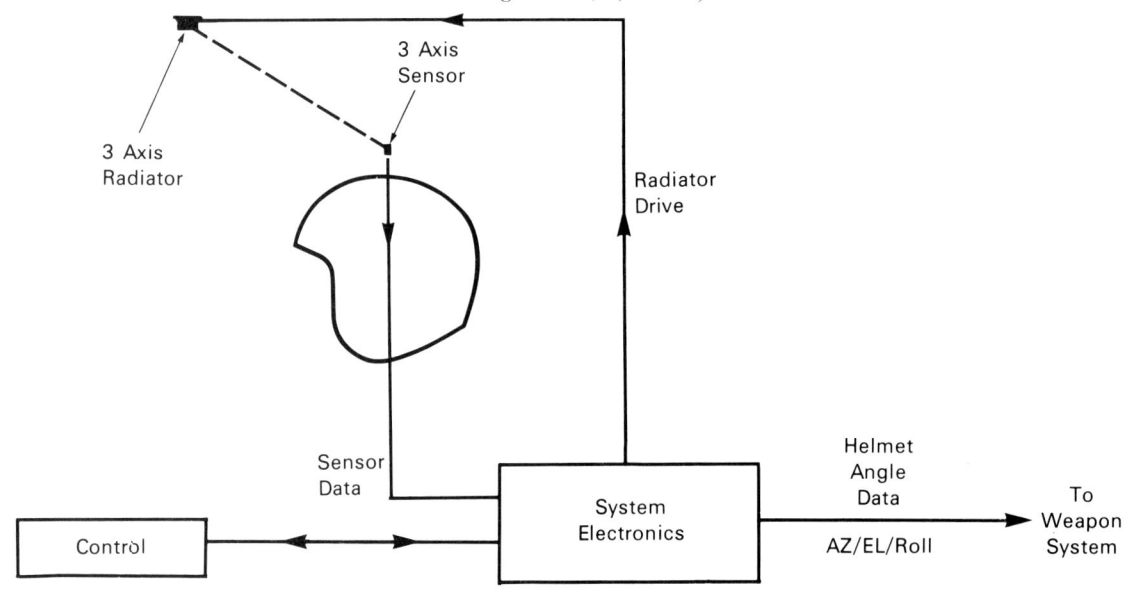

A Ferranti HMS with side-mounted CRT projector.

Laser ranging

Modern tactical strike aircraft are equipped with laser ranging and target-seeking equipment, which significantly reduces the pilot's workload and target-tracking time. It can even allow targets to be hit without being seen by the pilot. Hawker Typhoons in the Second World War were directed by radio to their targets by liaison officers operating with front-line troops. Much the same procedure is carried out today, but with the radio replaced by a laser beam. A FAC (Forward Air Controller) with front-line troops aims a laser designator at a selected target when the friendly aircraft approaches. The optical head of the airborne laser contains a scanner which picks up the reflected energy from the designator. This automatically places the target bar of the pilot's HUD over the target, and the pilot centralizes the target bar. As soon as the target-seeking lock is established, the letter 'T' appears on the HUD. An airborne ranging laser in the nose of the aircraft then starts to fire, and when the range lock occurs the letter 'R' also appears on the HUD. The pilot then simply alters course in azimuth to bring the bomb line through the target bar. The laser ranger functions automatically during the tracking phase and ceases only when automatic weapon release occurs; the pilot may not have seen the target he attacked. For normal strike operations the airborne laser ranger is directed onto the target by the pilot, and permits effective attack even after late target detection. This ranging system is made by Ferranti who also supply the type 105 Ranger used on American A10 aircraft.

In early 1990 the Ferranti Company sold its

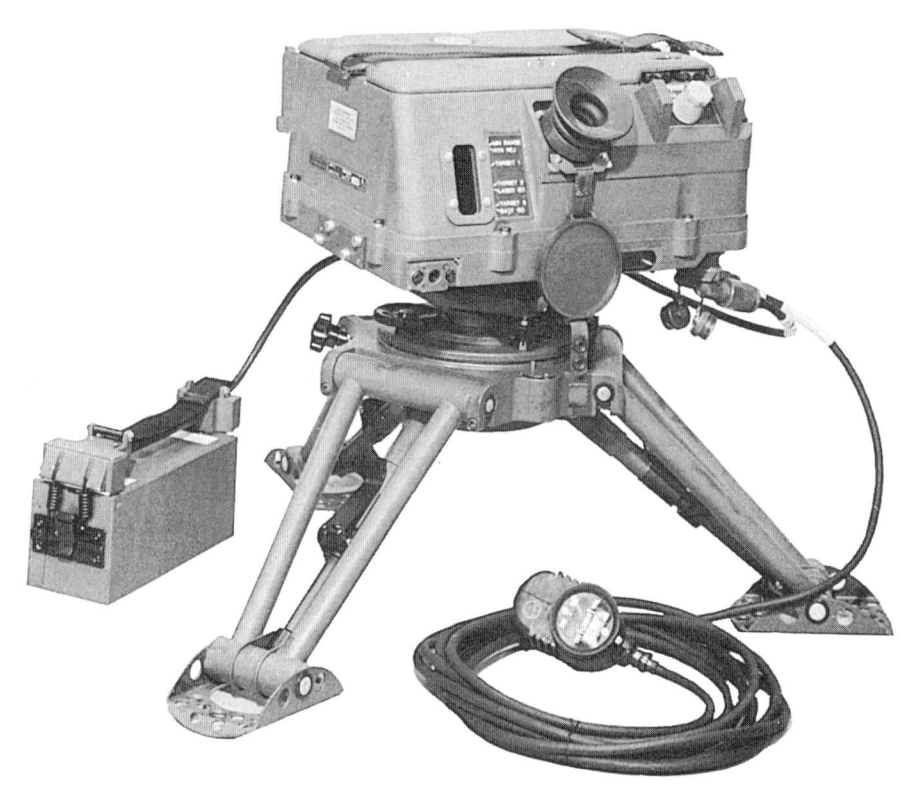

The Ferranti Ground Designator, used by forward observers to indicate targets for the laser tracker of strike aircraft. (Ferranti)

weapon-aiming and radar interests to the General Electric Co., which had already absorbed Marconi. This gave GEC a very strong manufacturing and design capacity for the future.

Whilst it can be said that Britain has led the world in innovative design in this field, many excellent weapon-aiming systems have been produced elsewhere. The United States has provided its air services with many highly effective sighting systems, and is now engaged in weapon-aiming schemes for use in outer space. The Soviet Union was renowned for the excellence of its aircraft weaponry. In the 1970s Russian fighter pilots were provided with advanced radars, passive infra-red search and track sensors, laser rangers and an HMS.

As guided missiles become ever more lethal, it is sometimes suggested that piloted aircraft will soon be surplus to requirements. One cannot but remember when the British Government announced that the English Electric Lightning would be the last manned fighter to see service in the RAF. Today there are more manned fighters in service world-wide than in any other peacetime period. Though guided missiles can do many of the tasks allotted to aircraft, anything designed by man can be countered if enough time and money is available. It is for this reason that no nation can afford to abandon aircraft gun development. A modern gun aimed with the latest gunsight is a very effective weapon.

All development turns a full circle. The rotary cannon is based on the Gatling gun of 1860. The RAF uses a free-mounted general purpose machine-gun to arm some of its helicopters. The blade sight of this gun is no different from those used by the military aviators of 1914.

Summary of the Development History of British Aircraft Gunsights

1900	Sir Howard Grubb invents the optical reflector sight for military use.
26 July 1912	First official air firing carried out at Farnborough by Geoffrey de Havilland from an FE2 with a Vickers Maxim gun.
1913	Commander Clark-Hall fits a naval tubular optical gunsight above the barrel of a Vickers Maxim, giving improved field of view.
1914	Basic infantry blade sights used on first aeroplanes sent to France with the British Expeditionary Force.
Late 1914	Vickers fit purpose-made parallel motion sight to Vickers gun on Experimental biplane No. 2.
1915	Frame or gate sight developed from naval quick-firing guns. Used by Allied and German aviators, giving first stadiametric ranging.
Late 1915	First ring and bead sights appear, elevated over gun for improved field of view.
Nov. 1915	First practical reflector sight patented by Henry Mallock and Vickers.
Jan. 1916	Dimension of ring element of ring and bead standardized to give various range and deflection allowances.
Early 1916	Prismatic tubular sight for experimental Vickers 1-in upward-firing gun.
Early 1916	RL tracer ammunition issued to selected squadrons, found to be effective only at very short range, withdrawn after complaints of misleading results. SPK tracer more effective but questionable value for accurate gunnery. Increased use as a deterrent against enemy attacks.
March 1916	Hutton night sight. Illuminated vee and pylon-mounted bead invented by Sgt. A. E. Hutton, 39 Squadron. Accepted for Service use.
April 1916	Norman vane sight adopted for use on free-mounted guns.
April 1916	Beliene tubular pilot's sight used briefly on DH2s.
April 1916	Scarff automatic compensating sight introduced, giving mechanical means of range and deflection allowance. Found to be too complicated.
June 1916	Le Prieur frame sight tested in RE8 aircraft at Martlesham. Found to have little advantage over simpler sights.
Sept. 1916	Aldis 1.8 optical sight accepted by RFC as standard fixed gunsight.
Oct. 1916	Hamilton sight, an adjustable frame-type sight with built-in allowance unit, used in Davis gun trials, but not proceeded with.
Early 1917	Neame illuminated night sight invented by Lt. H. B. Neame RFC. Used mainly by Home Defence Squadrons.
Mid-1917	Illuminated Norman Vane sight (pylon and ring illuminated).
June 1917	Secretan sight invented by Major K. Secretan RFC. Non-reflective large wooden ring and bead, no illumination.
Jan. 1918	Oigee reflector sight issued to selected German units for trials, first used operationally. Produced by Optische Anstal Oigee of Berlin.
Feb. 1919	Oigee evaluated by Britain and America.

1921	Messrs Barr & Stroud Ltd begin work on reflector-type sights to Grubb's patent.
1925	Double-purpose optical tube developed for long-range firing. Aldis-type sight with magnification facility.
Aug. 1926	Barr & Stroud GD1 pilot's sight – First aircraft sight by this company.
1926	Barr & Stroud type J free-gun reflector sight.
Mar. 1927	Barr & Stroud GD2 pilot's sight accepted for Service trials.
Aug. 1927	Small service development order for six Barr & Stroud GD2A pilot's sights with upright lens housing (reflector type).
1929	Barr & Stroud CJ1 free gunsight.
1931	Oigee announce the ENI reflector sight. Evaluated by US at McCook Field, leading to development of US 'N' series of reflector sights.
1934	American Army Air Force accept the N1 reflector sight for trials.
1934	Barr & Stroud GD5 pilot's reflector sight featuring parabolic mirror.
1934	Barr & Stroud GD12, a modified GD5, submitted for Hawker PV3 aircraft.
1934	German Retiflexer Revi 1 covertly produced for the re-emerging German Air Force.
1934	Norman Vane sight Mk III, featuring variable speed settings, accepted for use.
1935	Barr & Stroud GD6 free gunsight introduced with speed cranks and adjustable arm unit for deflection and range assistance. Sight head offset.
1935	Hawker introduce 36 in (914 mm) sight bar for Hart variants.
1936	O.P.L. reflector sight evaluated by French Air Force.
1936	Barr & Stroud GJ3 accepted for use in turrets and free guns.
1936	Baille-Lemaire reflector sight adopted by French for fixed guns.
1936	Barr & Stroud receive development order for Mk 1A reflector sight (sighting head offset for Lewis).
1936	Barr & Stroud GM1 evaluated at Martlesham.
Oct. 1936	RAE suggest switch to prismatic sights after reports of poor graticule acquisition on reflector sights. Submit two types for evaluation.
1936	GM2 accepted for squadron use, designated Reflector Sight Mk II in RAF.
1937	GJ3 designated Free Gun Reflector Sight Mk III. With protective hood.
1938	Messrs C. P. Goerz of Vienna accept contract to supply Air Ministry with GM2 sights to Barr & Stroud specifications. To be known as Mk III fixed gunsight.
Oct. 1938	Tests carried out on Gladiator aircraft to compare performance of reflector and prismatic sights. Messrs Ross produce four type A prismatic sights for these trials.
1939	Experimental gyro-controlled gunsight produced at RAE Farnborough.
Sept. 1939	Supply of GM2 Mk III from Goerz terminated on outbreak of war.
Oct. 1939	US Air Force adopt N-3 pilot's reflector sight.
1939	Mk IIIA introduced with revised hood and bakelite sight body.
1939	Periscopic tubular sight produced for Boulton Paul Type K under turret.
Oct. 1939	Experimental gyro sights fitted to Wellington and Spitfire for trials.
1939	Type B periscopic sight for FN 64 Lancaster under turret.
April 1940	Method devised to use tracer to detect movement of oncoming fighters to assist rear gunners.
March 1940	After protracted development, prismatic sights rejected for RAF use. (Redesign suggested with less interference to pilot's view, low priority.)
1940	Periscopic gunsight types A and B for FN rear defence under turrets.
1941	Circular reflector screen of pilot's Mk II replaced with rectangular glass screen, sun screen deleted, all Mk IIs converted with retrofit kits. Sight now Mk II*.
1941	Predictor sight designed for Vickers dorsal turret mounting 40 mm S gun.
June 1941	All work on prismatic sights terminated.
1941	Periscopic tubular sight designed at Farnborough for Blenheim FN 60 turret.

SUMMARY OF THE DEVELOPMENT HISTORY OF BRITISH AIRCRAFT GUNSIGHTS

1941	Predictor sight produced for Vickers Armstrongs 40 mm S gun turret (Nannini).
July 1941	Mk I gyro sight produced in Ferranti instrument factory.
Aug. 1941	First Mk I gyro sight, the Mk ID, issued to fighter squadrons for operational testing.
Sept. 1941	Mk II sight head with facility for depressed sight line adopted for rocket weapons, designated Mk IIL.
Feb. 1942	Mk IIIA* fitted with adjustable line of sight mechanism for air-to-ground rocket use; maker believed to be Ross Ltd.
April 1942	Over-the-nose sighting mirror system for Spitfire tested, not adopted.
May 1943	Air Ministry assess possibility of radar-assisted gunsighting in turrets.
June 1943	Farnborough produce redesigned gyro sights—Mk IIc for turrets, Mk IID for fighters. Ferranti given production contract with high priority for materials.
Nov. 1943	Ferranti commence production line at purpose-built factory in Edinburgh for Mk IIC & D gyro sights.
Dec. 1943	Windscreen projection system developed for air interception radar Mk 8, in which gunsight graticule and artificial horizon are displayed on the windscreen of night fighters, the first HUD.
Feb. 1944	Quantity production commences at Ferranti gyro factory, Edinburgh.
March 1944	American evaluation of gyro sight resulting in request for production rights in US.
Sept. 1944	Radar-assisted gunlaying device tested at RAF Defford in Lancaster *ND712* by TRE technicians.
Dec. 1944	Production AGLT units installed in rear turrets of 49, 16 and 635 Squadrons.
Feb. 1945	Production of US gyro sight begins, designated Mk 14.
Sept. 1945	Development of AGLT continued at reduced priority.
March 1949	Meteor and Vampire aircraft fitted with modified gyro sights, Mks 4B and 4C.
April 1949	Mk 4E gyro sight manufactured under licence in France by Sadire Carpentier of Paris.
Sept. 1949	E. K. Cole develop radar ranging sight for Hawker Hunter and Supermarine Swift.
March 1950	GGS Mk 7 adopted for limited use.
1951	Advanced radar ranging facility incorporated into GGS Mk 7 sight head.
1952	Mk 8 GGS introduced with limited air-to-ground computation.
1953	Ferranti introduce 'Airpass' system, first mono pulse radar for supersonic interception.
1955	Airpass 2 adopted in which pilot carries out blind attack without needing to observe target – known as head-down interception.
1957	Pilot Attack Sight (PAS) introduced – attack presentation linked with radar.
1958	Ferranti launch the Light Fighter Sight (LFS) basic gyro sight for use in lightweight ground attack aircraft.
1961	LFS type 5 used in Strikemaster.
1962	Ferranti ISIS N system using air data computers giving digital presentation of roll, pitch, altitude and airspeed first used on Northrop CF5.
1969	Ferranti ISIS B systems featuring permutation of computers and sighting heads.
1972	Isis D and K co-ordination of laser ranger with lead computing head.
1978	Navigation and Weapon-Aiming Sub-System (NAVWASS) giving flight parameters navigation and weapon-aiming direction. Developed by a consortium of Ferranti, GEC and Smiths for fighter and strike aircraft.
1985	GEC Avionics HUDWAC giving 'snap-shoot' and advanced weapon-aiming.
1986	Smiths Type 1202 developed for McDonald AV8B to US MIL specification.
1987	Ferranti helmet sighting system developed for 'hands off' weapon-aiming.
1987	Smiths Type 1203 HUD for BAe Hawk with provision for TV camera and recorder.
1991	Thermal imaging Airborne Laser Designator (TIALD) enables strikes to be held on boresight by auto tracker. Thermal imaging or TV during flight (GEC/Marconi Avionics).

Bibliography

Lt Col. G. M. Chinn, *The Machine-Gun* (US Bureau of Ordnance)

G. F. Wallace, *Guns of the Royal Air Force* (William Kimber, 1972)

James Tanner, RAFM, *British Aircraft Guns of WW2* (Arms & Armour Press, 1979)

Chris Chant, *Pictorial History of Air Warfare* (Octopus)

O. G. Thetford and E. J. Riding, *Aircraft of the Fighting Powers*, Vol. VII (Harborough Publishing Co., 1946)

Bill Gunston, *Aircraft Armament* (Salamander Books, 1987)

Dudley Pope, *Guns* (Spring Books, 1965)

Harry Woodman, *Early Aircraft Armament* (Arms & Armour Press, 1989)

H. F. King, *Armament of British Aircraft 1909–1939* (Putnam, 1971)

J. David Truby, *The Lewis Gun* (Paladin Press)

Col. Ross Whistler USAF Ret. US Aircraft Gunsights (restricted publication)

Index

A&AEE Farnborough Weapons Dept. 68
A/C Radar sighting 189
A1 Mk VIII radar 189
Aberporth firing range 88
AC-119 a/c Vietnam 35
AC-47 a/c Vietnam 35
Action of automatic guns diag. 110
Adams, E. S. R. Captain 53, 58, 65, 68, 83
Adams Willmott gun 47, 48
ADEN:
    gun 16
    25mm 98–100
    30mm Mks. 1–5, 87–96
    amm. details 95
    firing sequence 92–94
AGLT Mk III 195
Ainley, William 69, 88
Air Ministry Gun Section 63
Airborne Gunlaying, Turrets (AGLT) 191–195
Aircraft 6-pdr Class M (Molins) gun 85
Airpass system 200
Aldis Bros. 129
Aldis sight 129–132
Amplidyne power system 195
Armes Automatic Lewis 18
Armstrong Siddeley Aircraft Co. 77
Armstrong Whit. Siskin a/c 57
Armstrong Whitworth Co. 57
Arsiad synch system 27
Askania EZ/42 sight 187
Automatic Arms Company Ohio 16, 17
Avro 504 a/c 127

Bagshot, Bristol a/c 44
Balistite explosive 38
Banff RAF stn. 85
Barr & Stroud Free Gun (FG) sight:
    type J1 144
    GH6 145
    Mk III (RAF) series 147–151
    type GJ3 146
    GH6 145
    GJ1 (RAF Mk 1) 146
Barr & Stroud Pilots (P) sight:
    type GD1 136
    GM2 Mk III 158
    Mk II 156
    Mk II* 158, 159, 160
    Mk IIL 160
    type GD12 139
    type GD2b 137
    type GD5 138, 139, 155
    type GM1 155
    type GM2 156
    windscreen projector types 161
Barr & Stroud Ltd 133, 136, 144
Barwise, Sgt 128
Bawdsley Suffolk 189
BE 12 a/c 126
BE 2c a/c 19
BE 2 a/c 119

Beardmore, William 46
Beardmore-Farquhar gun 46, 47
Beaufighter a/c 72
Beaufort Bristol a/c 55
Becker cannon 63
Bellington, Major A.V. 27
Belt feed unit Bristol 67
Belt, disintegrating link 28
Benet-Mercier gun 18
Berthier, Andre 53
Birkigt, Marc 63
Birmingham Small Arms Co. 13, 18, 47
Blenheim Bristol a/c 55
Blind positional sighting 199
Bolas, Harrold 34
Bombay Bristol a/c 55
Boscombe Down 80, 83
Boulton Paul 195, 196
Bowen, Dr. E. G. 189
Boys 0.55 cal amm. 83
Bristol:
    B16 nose turret 71
    B17 turret 69
    Aviation Co. 106
    Fighter a/c 46, 57
British Arsiad synch system 27
British Manufacturing and Research Co. (BMARC) 65
British United Shoe Machinery Co. 79
Brooklands V.A. 71
Browning 0.50 M2 74–77
Browning, John 55, 74
Browning 0.303 cal Mk II* 56–61
BSA 0.50 gun 51, 52
Buckingham incendiary amm. 29, 111
Bullet types 0.303 cal 62

Camera gun:
    Thornton Pickard (Hythe) 142
    Williamson G22 143
    Williamson G42 144
    Williamson G45 153–155
Capper, Col. J. E. 117
Carl Ziess 141
Cauldron G6 a/c 33
CCIP Cont. Computed Impact Point 207
Chain gun 16
Challenger, Prof. George 27
Chandler, Capt. C. De Forest 17
Chauchat gun 46
Clark-Hall, Cdr. R. H. RNAS 33, 117, 118
Cocking & link clearance tools Browning Mk II* 60
College Park Maryland 27
Colt Browning 13, 56
Colt Browning:
    model 1918 57
    model 1921 0.50 cal 74
    model 40 & 40/2 58
Colt model 1895 gun 14, 15
Computing sights (US) 179
Constantinesco-Colley (CC) gear 28, 31, 107
Convergence Computer (RSS) 195

Coventry Ordnance Works:
 1½-Pdr gun 36, 42–46
 1-Pdr gun 36, 43, 44
Crayford factory, Vickers 13, 45, 49, 53
Cunningham, Dr. L. B. C. 165
Curtiss Hl flying boat 33
Curve of Pursuit 163

Darne gun (Fr) 52, 58
Davis, Commander Cleland USN 32
Davis gun 32–37
Davis gun carrier a/c 37
Dee, Dr. P. I. 192
DEFA 30mm gun (Fr) 88
Defford TRE 195
Degtyarev breech mech. 77
de Havilland, Geoffrey 42, 117
Deighton, Sgt. E. A. 22
Delicon 304RK 30mm gun 96
Desvignes incendiary amm. 29
DH3 a/c 35
Diehl & Dynamit Nobel 97
Donitz, Admiral 85
Dowding, Air Vice Marshal 53
DPG sight optical 133, 134
Driffield RAF stn. 60
Driscoll, Capt. J. USAAF 125
Drury, Wg. Cdr. (Dru) 72
Dunkerque RNAS stn. 34
Dunlop firing system 106
Dyot Battle Plane 35

E11 Browning gun cradle 74
Eastchurch RNAS stn. 13, 35, 46
Edmonton RCAF base 89
Edwards, George and William 26
Emerson APG-8 Radar sight 195
Enfield Royal Small Arms Factory 65, 88, 98
ENI sight (Oigee) 137

Fabrique Nationale Herstal (FN) 103
Fairey Battle a/c 54
Farnborough A&AEE 90, 117, 165
Farquhar Moubrey G. 46
FB5 gun bus a/c 13, 26
FE2 a/c 117
FE2d a/c 38, 39
FE4 a/c 43
FE5 a/c 43
Felixstowe RNAS stn. 33
Ferranti Ground designator 216
Ferranti Ltd. 16, 168, 171, 172, 176, 201, 214, 215
Ferranti/GEC takeover 215, 216
FH Recoilless gun 86
Fiat CR 42 a/c 22
Fighter Interception Unit 189
Film footage recorder (G45) 154
FN Mag 60/40 gun 103
Fokker, Anthony 119
Fokker E series a/c 26, 119
Folland, H. P. 43
Fonck, Rene 63
Ford, G. Capt. 172

Fort Halstead R&D est. 86
Forward Air Controller (FAC) 215
Foster rail over wing Lewis mounting 15, 20, 21
Foster, Sgt. R. G. 20
Frame sight (Gate) 119, 120, 121
Frazer Nash 106
Frazer Nash FN 25 turret 167

G1 Prismatic sight 151, 152
Garros, Roland 119
Gate sight (Frame) 119–121
Gatling gun 57
Gaunt, V. S. 45
GEC LANTIRN HUD 208
GEC Perihud 209
General Electric Co. (US) 101
General Ordnance Co. (US) 32, 33
GGS:
 Mk 5 197
 Mk 7 199
 Mk 8 199
 Mks 4B, 4C & 4E 197, 198
 Mks IIC & IID 170–177
GJ3 sight 146
Gloster Gladiator a/c 32
Gloster Meteor a/c 88
Gloster Javelin a/c 37
Gotha a/c 43, 128
Grahame, White 18, 56
Grain Isle of, RNAS stn. 33, 34
Graticule (Reticle) 129
Green, Dr. S. G. 76
Grubb, Sir Howard 135
Gun Section RAF 16
Gunsight Summary 217–219
Guynemer, Georges 63
Gyro Gunsight (GGS) Mk 1 165–169
Gyro Selector Dimmer 175
Gyro sight production (Ferranti) 176, 177

H2S Navigation 191
Hale, A. A. 172
Halstead Fort R&D Est. 86, 87
Hampton, Walter 79
Handley Page 0/400 a/c 35
Handley Page Hampden a/c 346
Harrier Bae a/c 94, 98
Harrier Cockpit 210
Hartmann, Eric 165
Hatfield de Havilland Base 84
Hawker, Maj. L. G. 28
Hazelton, George Lt. Cdr. 28
Head Down Interception (HDI) 201
Head stamps ammunition 63
Head Up Display (Hudwac) 209
Helmet Mounted Sight (HMS) 214
Herman Goering Werke (HG) 16
Heywood air compressor 107
Hillier, Flt. Lt. 73
Hispano Suiza 20mm gun 63–70
Hispano Suiza Company 16, 63, 65
Hives, Lord E. W. 70, 77
Hodgkin, Alan 189, 195

INDEX

Holland & Holland Ltd. 14, 15
Honeywell HMS 214
Horsley, Cpl. B. 124
Hotchkiss gun 58, 128
Hugget, Robert 86
Hughes Chain gun 104
Hunter Hawker a/c 89
Hurricane IID a/c 72
Hutton Night sight 126
Hutton, Sgt. Albert 127
Hythe Camera gun Mk III (Thornton Pickard) 143
Hythe Machine Gun School 35, 51

Identification of 0·303 cal ammunition 62, 63, 112
Inertial Velocity Nav. System 205
Integrated Strike & Intercept System (ISIS) 201
ISIS sights a/c:
   types 205
   type D 202
   type F 203
   type K 202
   type N 201
IWKA Mauser Werke 96

Japanese type 98 sight 186
Javelin, AW.a/c 86
Jones, Sir Melville 165, 172

K13 sight (US Gyro) 180
King, Rex D. H. 83
Kingsley Wood, Sir 54
Kynoch amm. 72

LANTIRN system 209
Laser Ranging (Ferranti) 215
Lee Enfield rifle 15
Lesnitzer, O. H. von 87
Lewis, Col. Isaac 16
Lewis Composite Twin gun 23
Lewis gun MkI 18
Lewis gun Mk II 19
Lewis gun Mk III 21
Lewis gun history 16-19
Light Fighter Sight (LFS) 201
Lightning English Electric a/c 90
Linder Dipl. Eng. 87
Link clearance tool Browning 46, 70
Long Recoil system 46, 70
Lovell, Sir Bernard 191
Low Diffractive optics Harier 213
Lowe, Flt. Lt. 89
Lulworth Range 72
Lysander Westland a/c 147

M39 US gun 87
Madsen gun 13, 58
Mair, Dr. 87
Mallock/Vickers reflector sight 135
Manton, Marcus 18
Martin, Captain 39
Martini Henry gun 15
Martlesham Heath A&AEE 45, 52
Mauser BK 27 96, 97

Mauser Werke Oberndorf 16, 96
Mauser IWKA Oberndorf 87
Mauser MG213c 97
Maxifort Solenoid firing unit 109
Maxim, Hiram 24
Maxim Vickers 15
McClean, Dr. Samuel 17
Melton Mowbray test facility 77
Metadyne System (Vickers) 196
Meteor Gloster 197-199
MG 108/15 gun 26
Milling, T. DeWitt 17
Mk 18 US Navy sight 174
Molins 6-pdr 32, 82-86
Molins belt feed 67
Molins, Desmond 84
Molins Machine Co. 65, 82
Morane Saulnier Co. 27
Mosquito XVIII a/c 32, 83
Muzzle attachment Browning 59

N1 sight (US) 140
N2A sight (US) 140
N2C sight (US) 141
N3 sight (US) 141
NATO 4173 ammunition 98
Neame night-sight 128
Newland, Sgt. A. 22
Night sighting 125
Norman Vane sight 123
Norman, Lt. G. H. 123

Oerlikon 304 RK 30mm 96
Oerlikon Company 63
Oigee sight 135, 138
Optische Antal Oigee 135
Orfordness RNAS stn. 43
Over the nose sighting (Spitfire) 159
Own Speed Factor 122

Palmer gun firing system 106
Parnall, Hendy Heck 147
Pendine Sands test ranges 80, 81
Periscopic sight type B 178
Periscopic sight type A 177
Pershing, (US) General John J. 74
Perth Blackburn flying boat 46
Picton RCAF base 53
Pilot Attack Sight (PAS) 200
Plessey belt making m/c 61
Pomeroy explosive amm. 112
Pontiac Div. GM 87
Poole Ordnance Factory 68
Porte Baby flying boat 33
Predannack RAF stn. 85
Prideaux, William 28
Prismatic sight type G1 151, 152
Prismatic sights 151-153
Purdy and Co. 128
Pynches, F/Lt. 46

Radar ranging A10 215
Rau, Felix 82

RCL gun 86, 87
Rea, F/Lt. 46
Reflector sight development 134, 140
Remote Sighting System 195
Reticle (Graticule) 129
Retiflexer REV1 1 sight 140
Revi 12D & 12F sights 185
Revi 16B, 186
Rheinmetal Werke 86
Ring and Bead sight 121, 122, 142, 164
Ring and Bead US systems 180
Rolls-Royce 0.50 in cal recoil operated gun 81–82
Rolls-Royce 40mm gun 77
Rolls-Royce 89
Rolls-Royce 0.50 in cal gas operated gun 81
Rowbotham, W. A. 79
Royal Flying Corps (RFC) 118
Royal Naval Air Service 21
RTS explosive ammunition 111

SACO 0.50 in cal gun 76
Salford Electrical Co. 158
Salisbury Hall (DH) museum 82
Salmond, General 57
Saunier, Raymond 119
Savage Arms Co. Utica USA 18, 22
Scarff Gallows Mounting 133
Scarff, W/O RNAS 132
Scarff Compensating sights 132
Scarff-Dibrovsky sync gear 27
Scimitar Supermarine a/c 89
SE5 a/c 130
Sea Fury Hawker a/c 68
Secretan night sight 129
Seddon, Cdr. RNAS 33
Serby, J. E. 83
Short 184 Seaplane 34
Short Cromarty seaplane 34
Short S81 a/c 23
Sight rec. camera Mk II Gyro 173, 174
Siskin AW a/c 31, 32
Small Arms Committee War Off. 118
Smiths:
    1202 AV-8B 213
    1301 HUD 214
    Industries Cricklewood 209
    type 1203 HUD Hawk 212
    HUD Tornado 211
    Type 1101 sighthead 211, 213
Snapshoot system 207
Sopwith 1½ Strutter 27
Sopwith seaplane no. 127 a/c 23
Sopwith-Kauper sync gear 27
Sorley, Air Marshal Sir Ralph 83
Spitfire over the nose sighting 163
Srb-a-Stys sight 156, 157
Stadiometric Ranging 148
Steel, Sgt. W. 124
Stellingwerf, Lt. 18
Summary of British a/c guns 105–106
Swift Supermarine a/c 89

Tellier flying boat 43
Tetse Mosquito a/c 83

Thompson, Major H. S. V. 16, 36, 53, 57, 64
Thompson compensating sight 187, 188
Thornton Pickard Co 144
Tizzard, Henry 165
Tornado ADS 96
Tornado a/c cockpit HUD 211
Tracer ammunition 161–162
Tracer ranging (Bomber Comm) 161–162
Typhoon Hawker a/c 69

US:
    Computer sights 179
    Navy Mk 8 181
    Navy Mk 9 181
    RCT Washington B29 System 182–185
    reflector sights list 181, 182
    retiflector sight N8A 183

Vampire a/c control handle 109
Vane Sight Norman 20
Veeder rounds counter 29, 30
Viale, Mario Spirito 77, 80
Vickers:
    Crayford 1.59in cal (rocket) gun 38, 39
    FB1 a/c 118
    1½-pdr. gun 24
    1-inch gun 41
    1-pdr. gun 40, 41
    Albert and son Ltd. 24, 28
    Armstrongs Ltd. 70
    Balloon gun 29
    Class C 0.5 in. cal gun 48, 49
    Class F 49, 50
    Class K VGO gun 52, 53, 54
    Class S 40mm gun 70–74
    FB5 a/c 26
    gun action 24
    Maxim .45 in cal gun 13, 25
    Mk II gun 30
    Mk III gun 31
    Mks 1&1* gun 24–28
    type 161 a/c 45
    Valentia a/c 161
    Vimi a/c 43
    exp. fighting biplane EFB 26
Vickers/Mallock reflector sight 135
Village Inn Radar 192–195
Voisin a/c 42
Voison a/c 42
Vulcan GAU.4 gun GE 101, 102

Wackett, W/Cdr. L. J. 165
Wallace, G. F. 68, 80, 83
Wellesley Vickers 22
West, Jacob 138
Whirlwind Westland a/c 66
Winchester Arms Co. 74
Windscreen project (Al) 189
Woolwich Arsenal 18, 33, 77, 83
Woolwich Tracer Amm. 123
Wright Biplane 17

Z type IR detector 194
Zeppelin 13, 15, 41, 43, 117, 126